AF820365

# Methods in Molecular Biology

*Series Editor*
John M. Walker
School of Life and Medical Sciences
University of Hertfordshire
Hatfield, Hertfordshire, AL10 9AB, UK

For further volumes:
http://www.springer.com/series/7651

# Non-Natural Nucleic Acids

## Methods and Protocols

Edited by

**Nathaniel Shank**

*Department of Chemistry and Biochemistry, Georgia Southern University, Savannah, GA, USA*

Humana Press

*Editor*
Nathaniel Shank
Department of Chemistry and Biochemistry
Georgia Southern University
Savannah, GA, USA

ISSN 1064-3745 ISSN 1940-6029 (electronic)
Methods in Molecular Biology
ISBN 978-1-4939-9215-7 ISBN 978-1-4939-9216-4 (eBook)
https://doi.org/10.1007/978-1-4939-9216-4

© Springer Science+Business Media, LLC, part of Springer Nature 2019
This work is subject to copyright. All rights are reserved by the Publisher, whether the whole or part of the material is concerned, specifically the rights of translation, reprinting, reuse of illustrations, recitation, broadcasting, reproduction on microfilms or in any other physical way, and transmission or information storage and retrieval, electronic adaptation, computer software, or by similar or dissimilar methodology now known or hereafter developed.
The use of general descriptive names, registered names, trademarks, service marks, etc. in this publication does not imply, even in the absence of a specific statement, that such names are exempt from the relevant protective laws and regulations and therefore free for general use.
The publisher, the authors, and the editors are safe to assume that the advice and information in this book are believed to be true and accurate at the date of publication. Neither the publisher nor the authors or the editors give a warranty, express or implied, with respect to the material contained herein or for any errors or omissions that may have been made. The publisher remains neutral with regard to jurisdictional claims in published maps and institutional affiliations.

This Humana Press imprint is published by the registered company Springer Science+Business Media, LLC, part of Springer Nature.
The registered company address is: 233 Spring Street, New York, NY 10013, U.S.A.

# Preface

DNA has become a unique protagonist in the realm of bioorganic chemistry because of its predicable and defined structure, succinct programmability, and known interactions and chemistries. This has led to non-natural DNA mimics, either in part or in whole, becoming integral parts of nanoscaffolds, detection devices, and critical probes in a host of biological applications. This volume aims to highlight the diverse and exciting umbrella of non-natural nucleic acids. As nature has served to inspire, it is hoped that this volume will serve as a means to guide researchers toward a viable method when exploring their own inquiries and to serve as a springboard for new endeavors.

***Savannah, GA, USA*** ***Nathaniel Shank***

# Contents

## Contributors


NA AN • *Department of Chemistry, University of Utah, Salt Lake City, UT, USA*

DANIEL H. APPELLA • *National Institute of Diabetes and Digestive and Kidney Diseases, National Institutes of Health, Bethesda, MD, USA*

ANDREY V. ARALOV • *Shemyakin-Ovchinnikov Institute of Bioorganic Chemistry, Russian Academy of Sciences, Moscow, Russia*

DEV P. ARYA • *Laboratory of Medicinal Chemistry, Department of Chemistry, Clemson University, Clemson, SC, USA*

HIROYUKI ASANUMA • *Department of Biomolecular Engineering, Graduate School of Engineering, Nagoya University, Nagoya, Japan*

ANTHONY BERDIS • *Department of Chemistry, Cleveland State University, Cleveland, OH, USA*

TANAYA BOSE • *CSIR-National Chemical Laboratory, Pune, Maharashtra, India*

CYNTHIA J. BURROWS • *Department of Chemistry, University of Utah, Salt Lake City, UT, USA*

HANA CAHOVÁ • *Institute of Organic Chemistry and Biochemistry, Czech Academy of Sciences, Prague, Czech Republic*

JOHN CHAPUT • *Department of Pharmaceutical Sciences, University of California, Irvine, Irvine, CA, USA; Department of Chemistry, University of California, Irvine, Irvine, CA, USA; Department of Molecular Biology and Biochemistry, University of California, Irvine, Irvine, CA, USA*

MARYLINE CHEMAMA • *Department of Chemistry, University of California, Riverside, Riverside, CA, USA*

JUNG-SUK CHOI • *Department of Chemistry, Cleveland State University, Cleveland, OH, USA*

VLADIMIR G. DEDKOV • *Central Research Institute of Epidemiology, Moscow, Russia*

STELLA DIAFA • *Department of Chemistry and Biochemistry, University of Bern, Bern, Switzerland*

ANDREW V. DIX • *National Institute of Diabetes and Digestive and Kidney Diseases, National Institutes of Health, Bethesda, MD, USA*

ETHAN A. ENGLUND • *Clinical Research Management, Inc, Gaithersburg, MD, USA*

DAMIEN EVÉQUOZ • *Department of Chemistry and Biochemistry, University of Bern, Bern, Switzerland*

AARON M. FLEMING • *Department of Chemistry, University of Utah, Salt Lake City, UT, USA*

HAILEY L. GAHLON • *Department of Health Sciences and Technology, ETH Zurich, Zurich, Switzerland*

KENT S. GATES • *Department of Chemistry, University of Missouri, Columbia, MO, USA; Department of Biochemistry, University of Missouri, Columbia, MO, USA*

KARA M. GEORGE ROSENKER • *National Institute of Diabetes and Digestive and Kidney Diseases, National Institutes of Health, Bethesda, MD, USA*

XU GUO • *Department of Chemistry, University of Missouri, Columbia, MO, USA*

YOSHIYUKI HARI • *Faculty of Pharmaceutical Sciences, Tokushima Bunri University, Yamashiro-cho, Tokushima, Japan*

MICHAL HOCEK • *Institute of Organic Chemistry and Biochemistry, Czech Academy of Sciences, Prague, Czech Republic*
MARCEL HOLLENSTEIN • *Department of Structural Biology and Chemistry, Laboratory for Bioorganic Chemistry of Nucleic Acids, CNRS UMR 3523, Institut Pasteur, Paris, France*
KURT HOUSH • *Department of Chemistry, University of Missouri, Columbia, MO, USA*
HIROMU KASHIDA • *Department of Biomolecular Engineering, Graduate School of Engineering, Nagoya University, Nagoya, Japan; PRESTO, Japan Science and Technology Agency, Kawaguchi, Saitama, Japan*
VAIJAYANTI A. KUMAR • *CSIR-National Chemical Laboratory, Organic Chemistry Division, Pune, Maharashtra, India*
JAYANTA KUNDU • *Department of Organic Chemistry, Indian Association for the Cultivation of Science, Jadavpur, Kolkata, India*
MICHAEL P. LEDBETTER • *Department of Chemistry, The Scripps Research Institute, La Jolla, CA, USA*
CHRISTIAN J. LEUMANN • *Department of Chemistry and Biochemistry, University of Bern, Bern, Switzerland*
YI LIAO • *Department of Chemistry, University of Utah, Salt Lake City, UT, USA*
KRISHNAGOPAL MAITI • *Laboratory of Medicinal Chemistry, Department of Chemistry, Clemson University, Clemson, SC, USA*
DENIS A. MALYSHEV • *Department of Chemistry, The Scripps Research Institute, La Jolla, CA, USA*
GÜNTER MAYER • *Chemical Biology, Life and Medical Sciences Institute (LIMES), University of Bonn, Bonn, Germany*
HUI MEI • *Department of Chemistry, Stanford University, Stanford, CA, USA*
PETRA MÉNOVÁ • *Institute of Organic Chemistry and Biochemistry, Czech Academy of Sciences, Prague, Czech Republic*
KEIJI MURAYAMA • *Department of Biomolecular Engineering, Graduate School of Engineering, Nagoya University, Nagoya, Japan*
BAPPADITYA NANDI • *Department of Organic Chemistry, Indian Association for the Cultivation of Science, Jadavpur, Kolkata, India*
MARYAM IMANI NEJAD • *Department of Chemistry, University of Missouri, Columbia, MO, USA*
CHRISTOPHER NEL • *Department of Chemistry, University of Missouri, Columbia, MO, USA*
SATOSHI OBIKA • *Graduate School of Pharmaceutical Sciences, Osaka University, Osaka, Japan*
TAKASHI OSAWA • *Faculty of Pharmaceutical Sciences, Tokushima Bunri University, Yamashiro-cho, Tokushima, Japan*
IGOR G. PANYUTIN • *Department of Radiology and Imaging Sciences, Clinical Center, National Institutes of Health, Bethesda, MD, USA*
SANKHA PATTANAYAK • *Department of Organic Chemistry, Indian Association for the Cultivation of Science, Jadavpur, Kolkata, India*
SIBASISH PAUL • *Department of Organic Chemistry, Indian Association for the Cultivation of Science, Jadavpur, Kolkata, India*
FRANZISKA PFEIFFER • *Chemical Biology, Life and Medical Sciences Institute (LIMES), University of Bonn, Bonn, Germany*
NATHAN E. PRICE • *Department of Chemistry, University of California Riverside, Riverside, CA*

ELIZABETH E. RASTEDE • *National Institute of Diabetes and Digestive and Kidney Diseases, National Institutes of Health, Bethesda, MD, USA*
FLOYD E. ROMESBERG • *Department of Chemistry, The Scripps Research Institute, La Jolla, CA, USA*
NICOLE C. ROSECRANS • *Department of Chemistry, University of Utah, Salt Lake City, UT, USA*
MALTE ROSENTHAL • *Chemical Biology, Life and Medical Sciences Institute (LIMES), University of Bonn, Bonn, Germany*
NATHANIEL SHANK • *Department of Chemistry and Biochemistry, Georgia Southern University, Savannah, GA, USA*
GERMAN A. SHIPULIN • *Central Research Institute of Epidemiology, Moscow, Russia; Federal State Budgetary Institution "Center for Strategic Planning and Management of Biomedical Health Risks" of the Ministry of Health, Moscow, Russia*
SURAJIT SINHA • *Department of Organic Chemistry, Indian Association for the Cultivation of Science, Jadavpur, Kolkata, India*
RENATUS W. SINKELDAM • *Office of Technology Management, Washington University in St. Louis, St. Louis, MO, USA*
SHANA J. STURLA • *Department of Health Sciences and Technology, ETH Zurich, Zurich, Switzerland*
CHRISTOPHER SWITZER • *Department of Chemistry, University of California, Riverside, Riverside, CA, USA*
YITZHAK TOR • *Department of Chemistry and Biochemistry, University of California, San Diego, La Jolla, CA, USA*
ANNA M. VARIZHUK • *Research and Clinical Center for Physical Chemical Medicine, Moscow, Russia; Engelhardt Institute of Molecular Biology, Russian Academy of Sciences, Moscow, Russia*
MILAN VRÁBEL • *Institute of Organic Chemistry and Biochemistry, Czech Academy of Sciences, Prague, Czech Republic*
YINSHENG WANG • *Department of Chemistry, University of California Riverside, Riverside, CA*
DERRICK WATKINS • *Laboratory of Medicinal Chemistry, Department of Chemistry, Clemson University, Clemson, SC, USA*
ZHIYU YANG • *Department of Chemistry, University of Missouri, Columbia, MO, USA*
TIMOFEI S. ZATSEPIN • *Skolkovo Institute of Science and Technology, Skolkovo, Moscow, Russia; Department of Chemistry, Lomonosov Moscow State University, Moscow, Russia*

# Chapter 1

# Synthesis and Enzymatic Characterization of Sugar-Modified Nucleoside Triphosphate Analogs

**Stella Diafa, Damien Evéquoz, Christian J. Leumann, and Marcel Hollenstein**

## Abstract

Chemical modification of nucleic acids can be achieved by the enzymatic polymerization of modified nucleoside triphosphates (dN*TPs). This approach obviates some of the requirements and drawbacks imposed by the more traditional solid-phase synthesis of oligonucleotides. Here, we describe the protocol that is necessary to synthesize dN*TPs and evaluate their substrate acceptance by polymerases for their subsequent use in various applications including selection experiments to identify aptamers. The protocol is exemplified for a sugar-constrained nucleoside analog, 7′,5′-bc-TTP.

**Key words** Nucleoside triphosphate, Primer extension reactions, DNA polymerases, XNA, Solid-phase synthesis, Bicyclo-DNA

## 1 Introduction

Initially, interests in the chemical modification of nucleic acids were spurred by the need for more resistant and potent therapeutic and antiviral agents [1–3]. Since then, chemical functionalization of nucleic acids has advanced as a key step in a number of applications including the development of potent and serum-resistant aptamers and DNAzymes [4, 5], questioning the origin of life [6] and the creation of novel biomaterials [7]. Traditionally, chemical modifications were incorporated into DNA or RNA by solid-phase synthesis of oligonucleotides using activated phosphoramidite building blocks. While larger amounts of oligonucleotides, decorated with a number of functional groups, are accessible, this method presents some limitations: (1) only rather short sequences ($\leq$100 nucleotides) can be synthesized, (2) some sensitive functional groups are not compatible with the rather harsh conditions that are required particularly for the deprotection and oxidation steps, and (3) there is no direct replication of the oligonucleotides

Nathaniel Shank (ed.), *Non-Natural Nucleic Acids: Methods and Protocols*, Methods in Molecular Biology, vol. 1973, https://doi.org/10.1007/978-1-4939-9216-4_1, © Springer Science+Business Media, LLC, part of Springer Nature 2019

or a facile creation of multiple modified sequences at once which thus precludes its use in evolution experiments [8]. As a strongly developing alternative, nucleic acids can be functionalized through the polymerase-mediated incorporation of modified nucleoside triphosphates (dN*TPs) [9, 10]. Application of this methodology has permitted to introduce an impressive variety of functional groups ranging from small amino acid-like residues [11–13] or electrochemical tags [14, 15] to entire oligonucleotides [16, 17] and even polymerases [18]. The only prerequisite for the enzymatic synthesis of modified DNA is that the dN*TPs are recognized as a substrate by the polymerase.

In this report, we describe the entire protocol—from synthesis to biochemical characterization—required for the evaluation of new dN*TPs in order to endow nucleic acids with additional modifications and properties. The protocol is exemplified with the modified triphosphate 7′,5′-bc-TTP [19, 20], which is a member of the general bicyclo-DNA family of conformationally constrained nucleosides [21], but can be broadly applied to any modified analog.

## 2 Materials

Use ultrapure water of 18 MΩ-cm resistivity for the preparation of all buffers as well as for the reactions. All anhydrous organic solvents used for reactions were obtained by filtration through activated aluminum oxide or by storage over activated molecular sieves (4 Å). All elution buffers should be stored at 4 °C.

### 2.1 Synthesis of a Bicyclo-DNA Triphosphate

1. Two round bottom flasks (5 mL volume), one round bottom flask (25 mL volume), two magnetic stir bars (1 cm in length), two septa, two balloons, syringes, needles, two Falcon™ tubes of 15 mL volume.
2. A Schlenk line connected to a trap and to a vacuum pump.
3. A rotary evaporator also connected to a vacuum pump.
4. A 5′-hydroxy-nucleoside.
5. Pyridine.
6. Bis(tri-*n*-butylammonium) pyrophosphate.
7. 1,4-Dioxane.
8. 2-Chloro-1,3,2-benzodioxaphosphorin-4-one.
9. Dimethylformamide (DMF).
10. *n*-tri-butylamine (*n*-$Bu_3N$).
11. Iodine pearls.

12. Ammonium hydroxide solution ($NH_4OH$): ~28–30% $NH_3$ in $H_2O$.
13. $Na_2S_2O_3$: 10% solution in water.
14. $NaClO_4$: 2% solution in acetone.
15. A centrifuge.

#### 2.2 Reversed-Phase High-Performance Liquid Chromatography (RP-HPLC)

1. A 100 mL round bottom flask.
2. A freeze dryer connected to a vacuum pump.
3. Stock solution: 1 M triethylammonium bicarbonate (TEAB), pH = 7.5 (*see* **Note 1**).
4. Elution buffer A: 50 mM TEAB. Store the bottle at 4 °C.
5. Elution buffer B: 50 mM TEAB in ACN/$H_2O$ 1:1.
6. A core-shell, C18 semi-preparative RP-HPLC column (5 μm, C18, 100 Å).
7. An HPLC running system.

#### 2.3 Anion-Exchange High-Performance Liquid Chromatography (AE-HPLC)

1. Elution buffer B (for the purification of triphosphates): 2 M ammonium acetate (AA) in $H_2O$.
2. Elution buffer A (for the purification of triphosphates): 20 mM AA in $H_2O$.
3. Elution buffer A (for the purification of oligonucleotides): 25 mM Trizma® base (tris(hydroxymethyl)aminomethane), pH = 8. Filter the buffers through a glass microfiber filter.
4. Elution buffer B (for the purification of oligonucleotides): 25 mM Trizma® base, 1.25 M NaCl, pH = 8. Filter the buffers through a glass microfiber filter.
5. A semi-preparative column (4 μm, 9.0 × 250 mm) containing quaternary amine functionalized beads and equipped with a guard column (4.0 × 50 mm).
6. Sep-Pak® C18 Cartridge for desalting.

#### 2.4 Denaturing Polyacrylamide Gel Electrophoresis (PAGE)

1. Vertical Gel Electrophoresis running system: two gel plates, two spacers, one comb, a gel chamber, a power supply with two electrodes that can be connected to the gel chamber.
2. Gel dryer.
3. Phosphorimager.
4. 20% acrylamide solution. Store the solution at 4 °C.
5. Diluent solution, 1 L: In a 1 L bottle equipped with a lid and a stir magnet, add 420 g urea, 590 mL $H_2O$, and 100 mL of 10× TBE buffer. Stir at rt. for 1 h until the urea has been dissolved. Store the bottle at 4 °C.

6. Running buffer, 10× TBE. Transfer the buffer to a 5 L container and store it at rt.
7. APS (ammonium persulfate): 10% solution in $H_2O$ (*see* **Note 8**).
8. TMED (tetramethylethylenediamine).

### 2.5 $^{32}$P-Labeling of DNA Primers

Appropriate care should be taken when working with radioactive material. The labeling should be performed in certified labs designated for this purpose.

1. $^{32}$P-γ-ATP (270 μCi).
2. T4-PNK enzyme (T4 polynucleotide kinase).
3. Buffer A: 70 mM Tris–HCl, 10 mM $MgCl_2$, 5 mM DTT, pH = 7.6.
4. Spin column (0.5 mL loading volume).
5. Thermoshaker.

### 2.6 Primer Extension Reactions

1. Annealed oligonucleotides: In a 1.5 mL Eppendorf tube, add 10 pmol of unlabeled primer, 10 pmol of template, and 1 pmol of the radiolabeled primer. Mix gently, heat the oligonucleotides at 95 °C for 2 min, and let them slowly cool down to rt. This solution will be diluted in the reaction mixture to give a final concentration of 500 nM (*see* **Note 10**).
2. Loading buffer: 70% formamide, 50 mM EDTA, 0.1% bromophenol, 0.1% xylene cyanol.
3. dNTPs, 1 mM solutions.
4. Therminator™ DNA polymerase.
5. 10× reaction buffer: 200 mM Tris–HCl, 100 mM $(NH_4)_2SO_4$, 100 mM KCl, 20 mM $MgSO_4$, 1% Triton® X-100, pH = 8.8.
6. $MnCl_2$, 100 mM solution.

### 2.7 Solid-Phase Synthesis of an Oligonucleotide Containing Bicyclo-DNA Residues

1. DNA synthesizer, columns for the solid support, natural DNA phosphoramidites (dT, $dC^{4bz}$, $dG^{2DMF}$, $dA^{6Bz}$), modified bc-DNA phosphoramidite,{Evequoz, 2017 #9} solid support (dmf-dG-Q-CPG 500), bottles and vials for the synthesizer, 1.5 mL Eppendorf tubes, grease. All the glassware should be dried overnight in an oven and carefully purged with argon prior to use. All the solutions should be stored under an argon atmosphere.
2. Molecular sieves: wash approximately 150 g of molecular sieves (beads of ~2 mm diameter, 4 Å pore diameter) with 500 mL of hexane, filtrate off the solvent, and activate the molecular sieves by heating to 200 °C overnight under vacuum (<1 mbar). Store the molecular sieves under argon.

3. Dried acetonitrile: fill a 1 L bottle with anhydrous acetonitrile. Add approximately 50 g of the previously activated molecular sieves, and store the solution for a least one night before any further use.
4. Dried dichloroethane: fill a 500 mL bottle with dichloroethane. Add approximately 25 g of the previously activated molecular sieves, and store the solution for a least one night before any further use.
5. Detritylation solution: 3% dichloroacetic acid in dichloroethane.
6. Capping solution A: 0.5 M solution of 4-(dimethylamino) pyridine in dry acetonitrile.
7. Capping solution B: acetonitrile, 2,4,6-trimethylpyridine and acetic anhydride in a 5:3:2 ratio.
8. Oxidation solution: 0.01 M solution of iodine in 2,4,6-trimethylpyridine/acetonitrile/water.
9. DNA phosphoramidite solutions: 0.1 M phosphoramidite in acetonitrile, in 10 mL vials.
10. bc-DNA phosphoramidite solution: 0.1 M phosphoramidite in dichloroethane—in a 10 mL vial, add the bc-DNA phosphoramidite, and transfer with a syringe purged with argon the previously prepared dried dichloroethane solution. Add four beads of activated molecular sieves.
11. Coupling agent solution: 0.25 M 5-(ethylthio)-1*H*-tetrazole (ETT) in acetonitrile—in a 50 mL vial, add ETT, and transfer with a syringe purged with argon the previously prepared dried acetonitrile solution. Add approximately 2.5 g of molecular sieves per 100 mL.

## 3 Methods

### 3.1 Synthesis

The reaction is based on the protocol of Ludwig-Eckstein [22]. All the glassware should be dried overnight in an oven prior to usage. Use dry solvents for all steps of the reaction (*see* Subheadings 2.1 and 2.7). Perform the reactions under inert atmosphere of argon.

1. In a 5 mL round bottom flask, co-evaporate on a rotary evaporator 1 equivalent of the 5′-hydroxy-nucleoside (*see* **Note 2**) two times with 0.3 mL of anhydrous pyridine; then add a magnetic stir bar, and let it dry overnight on a Schlenk line.
2. In a 5 mL round bottom flask containing a magnetic stir bar, add 1.3 equivalent of bis-(tri-*n*-butylammonium) pyrophosphate (*see* **Note 3**), and let it dry overnight on a Schlenk line.

3. Dissolve the 5′-hydroxy-nucleoside in 0.2 mL pyridine. Add 0.4 mL of 1,4-dioxane and 1.1 equivalent of 2-chloro-1,3,2-benzodioxaphosphorin-4-one (*see* **Note 4**), and stir the yellow suspension for 45 min at rt.
4. Dissolve the pyrophosphate in 170 μL DMF and 58 μL of *n*-tributylamine ($n$-$Bu_3N$), and add the new solution to the suspension containing the nucleoside. Stir at rt for 1 h.
5. Add 1.6 equivalent of $I_2$ as a solution in pyridine/$H_2O$ (98:2) and stir for 30 min at rt.
6. Add a 10% $Na_2S_2O_3$ solution (0.5 mL) to quench the excess of $I_2$.
7. Evaporate the solvents in the flask until dryness using a rotary evaporator connected to a vacuum pump (*see* **Note 5**).
8. Dissolve the residue in 5 mL $H_2O$ and transfer it to a 25 mL round bottom flask. Stir at rt for 30 min.
9. Add 10 mL $NH_4OH$ (28–30%) and stir for 2 h at rt.
10. Evaporate the solvents to dryness under vacuum.
11. Add 2 mL $H_2O$ into the flask to dissolve the residue, and split the content between two Falcon™ tubes. Fill the tubes with a 2% $NaClO_4$ solution in acetone, mix, and centrifuge for 30 min at 16,000 × *g*. Discard the organic layer, redissolve the yellow precipitate in the $NaClO_4$ solution, and repeat the precipitation. Remove the organic phase, dissolve the residues in $H_2O$, and combine them in one tube with a total volume of 4 mL. Store in the freezer until the end and during the purification by RP-HPLC or AE-HPLC.

### 3.2 Purification by RP-HPLC

1. Bring buffers A and B to rt, and degas them for 20 min by stirring them under vacuum prior to their usage (*see* Subheading 2.2).
2. Carry out the purification of the triphosphate in aliquots of 1 mL on a C18 semi-prep reverse phase column equipped with a guard column.
3. Inject 200 μL of the crude solution per round, and use the following gradient conditions for the purification: 0–8 min, 0% B; 8–48 min, 0–15% B; 48–50 min, 15–100% B (*see* **Note 6**) (Fig. 1).
4. Collect the fractions corresponding to the triphosphate peak ($R_t$ = 43 min) in round bottom flasks, and store them in the freezer until the end of the purification.
5. Freeze-dry all the collected aliquots, and co-evaporate with $H_2O$ two times to sublime the triethylammonium bicarbonate salt.
6. Confirm the presence of the triphosphate in the sample by MALDI-MS (*see* **Note 7**).

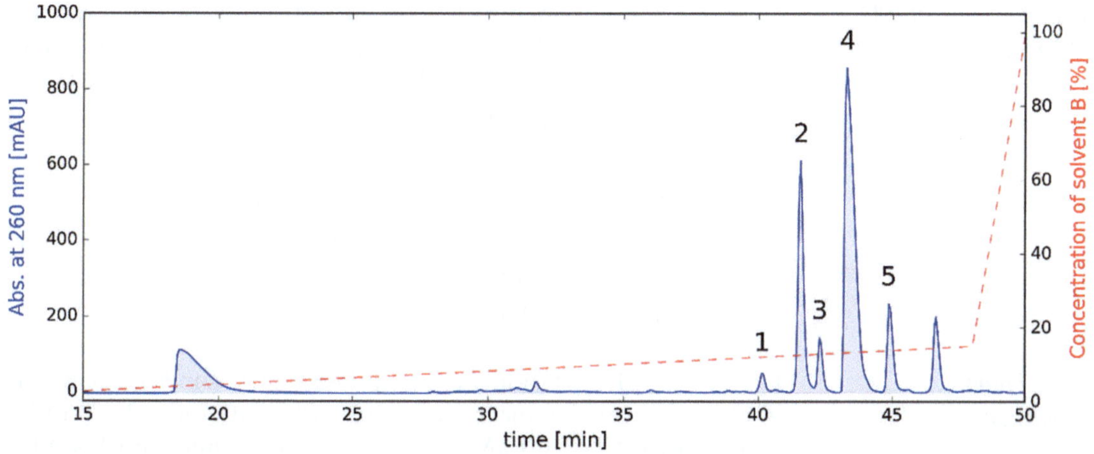


**Fig. 1** Representative chromatogram of a 7′,5′-bc-TTP purification by RP-HPLC. The peaks correspond to (**1**) 7′,5′-bc-T-monophosphate, (**2**) 7′,5′-bc-T-phosphonate, (**3**) 7′,5′-bc-T-diphosphate, (**4**) 7′,5′-bc-T-triphosphate, and (**5**) 7′,5′-bc-tetraphosphate

### *3.3 Purification by AE-HPLC*

1. Repeat the same steps as described in Subheading 3.2 but using a linear gradient reaching 30% of eluent B in 30 min (*see* **Note 6** and Subheading 2.3).

### *3.4 10% PAGE Analysis*

1. Mix equal volumes of the 20% acrylamide solution and the diluent in a 250 mL beaker (*see* Subheading 2.4).
2. Add 1% of the volume 10% APS solution to initiate the radical reaction, and continue stirring the mixture.
3. Add 0.1% of the volume TMED to begin the polymerization.
4. Take up the solution with a 60 mL syringe, cast a gel, and let the gel polymerize for 4 h at rt.
5. Fix the gel plates on the gel chamber, and load 1× TBE buffer by diluting the 10× running buffer.
6. Preheat the gel for 30 min at 20 W to remove unpolymerized acrylamide (*see* **Note 9**).
7. Using a syringe, wash the wells with the running buffer to remove any bubbles or polyacrylamide particles trapped in them.
8. Load the samples.
9. Run the gel at 40 W for 2.5 h.
10. Take down the gel, remove the gel plates carefully, and transfer the gel on a Whatman® blotting paper. Dry the gel for 1 h under vacuum using a gel dryer.
11. Visualize the gel on a Phosphorimager.

### 3.5 Labeling of DNA Primers

1. In a 1.5 mL Eppendorf tube, add 30 pmol of primer, 3 μL buffer A, 3 μL $^{32}$P-γ-ATP, and 1 μL PNK, and fill up with $H_2O$ up to 30 μL total volume. Mix at rt (*see* Subheading 2.5).
2. Incubate the Eppendorf tube at 37 °C for 40 min using a thermoshaker.
3. Heat the sample to 95 °C for 5 min to quench the enzyme.
4. Purify the labeled primer using a spin column and following the instructions of the manufacturer.
5. Store the label in the freezer.

### 3.6 Primer Extensions

1. In a 1.5 mL Eppendorf tube containing the annealed oligonucleotides (final concentration of 500 nM), add 1× buffer, 1 mM $MnCl_2$, 100 μM dNTPs, 2 U of enzyme, and $H_2O$ for a total reaction volume of 20 μL, and mix at rt (*see* **Note 11** and Subheading 2.6).
2. Incubate the reactions using the thermoshaker at 55 °C with continuous stirring for 1 h.
3. Stop the reactions by adding equal volume of loading buffer.
4. Denature the DNA strands by heating the reaction tubes at 95 °C for 5 min and rapidly cooling them down to 0 °C with an ice bath. Centrifuge the reactions for 2 s before loading them on the gel.
5. Load 8 μL of each reaction on a 10% polyacrylamide gel well.
6. Visualize the radiolabeled extended primers with phosphorimaging (Fig. 2).

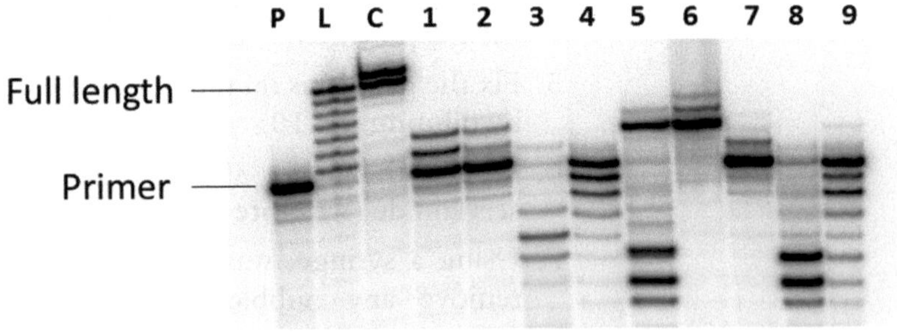


**Fig. 2** Representative 20% PAGE image as visualized by phosphorimaging showing several primer extension reactions performed with different polymerases (lanes 1–9). P contains only the primer, L is a ladder formed to assist the determination of the number of incorporations, and C is a control reaction performed with the wild-type dTTP to exclude possible error factors attributed to any of the additives

### 3.7 Solid-Phase Synthesis of an Oligonucleotide Containing Bicyclo-DNA Residues

Capping, oxidation, and detritylation should be performed with standard conditions and following the protocols recommended by the manufacturer of the gene assembler (*see* Subheading 2.7). Natural DNA phosphoramidites are coupled using a 1.5 min step. The modified bc-DNA requires an extended coupling time of 12 min. To ensure optimal yields, the acetonitrile, phosphoramidite, and coupling agent solutions should be maintained under strict anhydrous conditions, and the system should be kept under an inert atmosphere of argon.

1. Flush the DNA synthesizer for 10 min with argon and then fix the reagent bottles. Add a fine layer of grease to the top of the vials and fix them. Make sure the DNA synthesizer is under a slight overpressure of argon.
2. Wash the system by running acetonitrile. Prime all the reagent lines and then wash the system again with acetonitrile.
3. Run a test sequence: place a column filled with 1.3 μmol of solid support into the reactor chamber, and run a sequence having each natural nucleobase in a consecutive manner (e.g., 5′-AAT TCC GGG-3′). Monitor the coupling efficiency by recording the trityl absorbance at 436 nm during the detritylation step. The coupling efficiency should be over 98% (*see* **Note 12**).
4. Run the solid-phase synthesis: enter the desired sequence (e.g., 5′-GGX ACA XGT CAT AGC TGT TTC CTG-3′). Make sure that the correct protocols are selected for each nucleobase and that the final detritylation will be performed. Place a column filled with 1.3 μmol of solid support into the reactor chamber. Run the synthesis and monitor the coupling efficiency as before.
5. Dry the solid support: at the end of the synthesis, take out the column from the reactor chamber, and place it into a 1.5 mL Eppendorf tube. Centrifuge for 2 min at 1,500 × *g*.
6. Deprotect and cleave the oligonucleotide from solid support: open the column and transfer the solid support into a 1.5 mL Eppendorf tube equipped with a screw cap. Add 1.5 mL of a 33% ammonia solution (*see* **Note 13**). Incubate the Eppendorf tube at 55 °C shaking at 300 rpm for 16 h.
7. Cool down the Eppendorf tube to 0 °C, centrifuge it for 10 min at 1,500 × *g*, and collect the supernatant. To further wash the solid support, add 0.5 mL $H_2O$ to the beads, shake the Eppendorf tube, centrifuge it for 10 min at 1,500 × *g*, and collect the supernatant. Repeat this step a second time. Freeze-dry the combined supernatant, resuspend the crude oligonucleotide in 1 mL $H_2O$, and filter the resulting solution with a 13 mm syringe filter (0.45 μm pore size). The crude oligonucleotide can be purified by ion-exchange HPLC or by 20% PAGE.

### 3.8 Purification of Oligonucleotides by Ion-Exchange HPLC

1. Bring buffers A and B to rt, and degas them for 20 min by stirring them under vacuum prior to their usage (*see* Subheading 2.3).
2. Equilibrate the semi-preparative DNAPac 200 column with a solvent mixture of 20% B.
3. Inject 300 μL of the crude oligonucleotide solution per round, and use the following gradient conditions for the purification: 0–7 min, 20% B; 7–35 min, 20–55% B; 35–40 min, 100% B.
4. Collect the top of the peak corresponding to the full-length oligonucleotide in a 15 mL Falcon™ tube. Equilibrate the column with the starting condition before performing another round.
5. Desalt the combined fractions with a Sep-Pak® C18 Cartridge following the manufacturer's instructions. Freeze-dry the collected fractions, and resuspend the purified oligonucleotide in 1 mL $H_2O$.
6. Determine the concentration by measuring the absorbance at 260 nm with a NanoDrop™ spectrophotometer, using the extinction coefficient of the corresponding natural DNA oligonucleotide.
7. Confirm the mass by $ESI^-$ mass spectrometry or LC-MS as shown in Fig 3.

## 4 Notes

1. The stock solution can be stored in the fridge for a few months. Eventually the $CO_2$ will escape the flask and lead to an increase of the pH, which will cause difficulties to the purification of the triphosphate. We recommend measuring the pH value of the solution every time prior to the preparation of the elution buffers.
2. We find that the reaction works best in the range of 20–60 mg of nucleoside (i.e., 60–190 μmol).
3. The nucleophilic bis-(tri-*n*-butylammonium) pyrophosphate is a key reagent in the synthesis. Since it is a rather hygroscopic solid, it is advised to thoroughly dry the reagent for at least 16 h under vacuum prior to use. Alternatively, a 0.5 M solution in DMF can be prepared which can be dried overnight with activated 4 Å molecular sieves. Lastly, if low triphosphate yields are regularly obtained, the reagent should be prepared freshly prior to use as described in the literature [23, 24].
4. The reaction is significantly influenced by the nature of this reagent. We find that it works best when it is in the form of colorless crystals than in the form of a powder. Additionally,

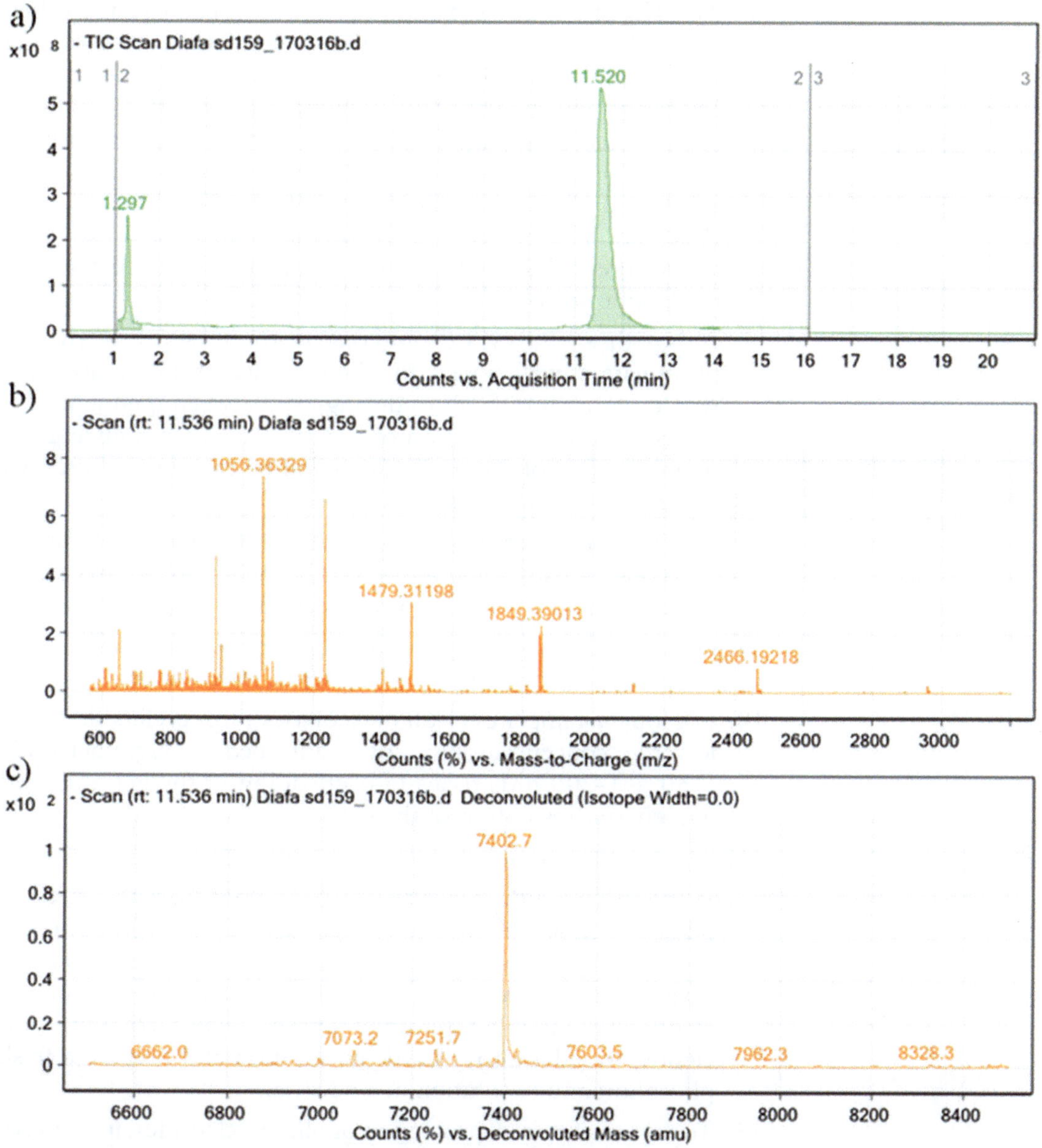


**Fig. 3** Representative result of an oligonucleotide characterization by LC-MS: (**a**) Chromatogram of the sample: the peak with a $R_t$ of 11.52 min corresponds to the oligonucleotide; (**b**) Mass spectrum: the higher peaks should correspond to *m*/*z* of the most abundant oligonucleotide isotope; (**c**) Deconvoluted mass spectrum: the higher peak should correspond to the molecular weight of the oligonucleotide

upon oxidation, a white coat can be formed around the crystals, preventing them from reacting. In such case, it is advised that the surface of the crystal be scratched prior to the reaction.

5. Ensure that the temperature of the water bath does not exceed the 30 °C to avoid thermal degradation of the triphosphate moiety.

6. The $R_t$s of the di- and triphosphate are close to each other. Depending on the resolution of the column, the corresponding peaks might be overlapping. In this case, the separation can be improved by reducing the volume of the crude solution injected and by adjusting the elution conditions. Alternatively, purification can be performed by AE-HPLC where a better separation of the diphosphate from the triphosphate is usually observed. The use of the volatile ammonium acetate buffer system precludes a tedious desalting protocol of the resulting nucleotide analogs.
7. Although an ESI-MS could confirm the presence of the triphosphate as well, a peak for the presence of the diphosphate would render it difficult to distinguish between fragmentation pattern and contamination of the sample. By minimizing the fragmentation of the molecule, it will be easier to ensure that successful separation has taken place.
8. We recommend preparing a fresh APS solution prior to use or to prepare small aliquots that can be stored in the freezer to avoid problems with the polymerization of the gels.
9. Depending on the temperature and the power source available, the power can be adjusted to higher or lower values.
10. For longer templates (>60 nucleobases), it is preferable to use lower concentrations of annealed oligonucleotide (150 nM–200 nM) to avoid overloading of the gel wells and get a better resolution of the gel.
11. The enzyme should be added last as it initiates the reaction.
12. If the DNA synthesizer has not been used over an extended period of time, the first couplings can be affected due to the remaining presence of moisture. If the initial test sequence does not lead to satisfactory yields, run a second test sequence. If one natural DNA phosphoramidite is not coupling efficiently, we advise to discard the solution and prepare a fresh phosphoramidite solution in a new vial.
13. To ensure proper deprotection of oligonucleotides, it is important to use a fresh solution of concentrated ammonia. For general use, we advise to prepare 1.5 mL aliquots from a new concentrated ammonia solution.

## Acknowledgments

This work was supported by Institut Pasteur start-up funds to MH and the Swiss National Science Foundation (grant n°: 200020-146646).

## References


1. Taskova M, Mantsiou A, Astakhova K (2017) Synthetic nucleic acid analogues in gene therapy: an update for peptide-oligonucleotide conjugates. Chembiochem 18:1671–1682
2. Khvorova A, Watts JK (2017) The chemical evolution of oligonucleotide therapies of clinical utility. Nat Biotechnol 35:238–248
3. Goyenvalle A, Griffith G, Babbs A et al (2015) Functional correction in mouse models of muscular dystrophy using exon-skipping tricyclo-DNA oligomers. Nat Med 21:270–275
4. Diafa S, Hollenstein M (2015) Generation of aptamers with an expanded chemical repertoire. Molecules 20:16643–16671
5. Herdewijn P, Marliere P (2009) Toward safe genetically modified organisms through the chemical diversification of nucleic acids. Chem Biodivers 6:791–808
6. Jia TZ, Fahrenbach AC, Kamat NP, Adamala KP, Szostak JW (2016) Oligoarginine peptides slow strand annealing and assist non-enzymatic RNA replication. Nat Chem 8:915–921
7. Chen T, Romesberg FE (2017) Enzymatic synthesis, amplification, and application of dna with a functionalized backbone. Angew Chem Int Ed 56(45):14046–14051
8. Loakes D, Holliger P (2009) Polymerase engineering: towards the encoded synthesis of unnatural biopolymers. Chem Commun 31:4619–4631
9. Hocek M (2014) Synthesis of base-modified 2′-deoxyribonucleoside triphosphates and their use in enzymatic synthesis of modified DNA for applications in bioanalysis and chemical biology. J Org Chem 79:9914–9921
10. Hollenstein M (2012) Nucleoside triphosphates–building blocks for the modification of nucleic acids. Molecules 17:13569–13591
11. Jäger S, Rasched G, Kornreich-Leshem H, Engeser M, Thum O, Famulok M (2005) A versatile toolbox for variable DNA functionalization at high density. J Am Chem Soc 127:15071–15082
12. Hollenstein M, Hipolito CJ, Lam CH, Perrin DM (2009) A self-cleaving DNA enzyme modified with amines, guanidines and imidazoles operates independently of divalent metal cations (M2+). Nucleic Acids Res 37:1638–1649
13. Sakthivel K, Barbas CF (1998) Expanding the potential of DNA for binding and catalysis: highly functionalized dUTP derivatives that are substrates for thermostable DNA polymerases. Angew Chem Int Ed 37:2872–2875
14. Hocek M, Fojta M (2011) Nucleobase modification as redox DNA labelling for electrochemical detection. Chem Soc Rev 40:5802–5814
15. Balintová J, Špaček J, Pohl R et al (2015) Azidophenyl as a click-transformable redox label of DNA suitable for electrochemical detection of DNA–protein interactions. Chem Sci 6:575–587
16. Baccaro A, Steck A-L, Marx A (2012) Barcoded nucleotides. Angew Chem Int Ed 51:254–257
17. Verga D, Welter M, Steck AL, Marx A (2015) DNA polymerase-catalyzed incorporation of nucleotides modified with a G-quadruplex-derived DNAzyme. Chem Commun 51:7379–7381
18. Welter M, Verga D, Marx A (2016) Sequence-specific incorporation of enzyme-nucleotide chimera by DNA polymerases. Angew Chem Int Ed 55:10131–10135
19. Evequoz D, Leumann CJ (2017) Probing the backbone topology of DNA: synthesis and properties of 7 ’,5 ’-Bicyclo-DNA. Chem Eur J 23:7953–7968
20. Diafa S, Evéquoz D, Leumann CJ, Hollenstein M (2017) Enzymatic synthesis of 7′,5’-Bicyclo-DNA oligonucleotides. Chem Asian J 12:1347–1352
21. Leumann CJ (2002) DNA analogues: from supramolecular principles to biological properties. Bioorg Med Chem 10:841–854
22. Ludwig J, Eckstein F (1989) Rapid and efficient synthesis of nucleoside 5′-0-(1-thiotriphosphates), 5′-triphosphates and 2′,3′-cyclophosphorothioates using 2-chloro-4H-1,3,2-benzodioxaphosphorin-4-one. J Org Chem 54:631–635
23. Sarac I, Meier C (2015) Efficient automated solid-phase synthesis of DNA and RNA 5-triphosphates. Chem Eur J 21:16421–16426
24. Sarac I, Meier C (2016) Solid-phase synthesis of DNA and RNA 5’-O-triphosphates using cycloSal chemistry. Curr Protoc Nucleic Acid Chem 64:4.67.61–64.67.13

# Chapter 2

# Synthesis of Site-Specific Crown Ether Adducts to DNA Abasic Sites: 8-Oxo-7,8-Dihydro-2′-Deoxyguanosine and 2′-Deoxycytidine

**Na An, Aaron M. Fleming, Nicole C. Rosecrans, Yi Liao, and Cynthia J. Burrows**

## Abstract

Formation of adducts to DNA is of great benefit to DNA sequencing and damage detection technology and to enzymology. Here we describe the synthesis and characterization procedures of 18-crown-6 adducts formed to abasic (AP) sites, 8-oxo-7,8-dihydro-2′-deoxyguanosine (OG), and 2′-deoxycytidine (C) residues in DNA oligodeoxynucleotides. These crown ether adducts were used as site-specific modifications to facilitate nanopore technology. The methods described can be readily expanded to attach other suitable primary amines of interest.

**Key words** 18-crown-6, Site-specific chemistry, DNA adducts, Abasic sites, 2′-Deoxycytidine, 8-Oxo-7,8-dihydro-2′-deoxyguanosine, DNA sequencing

## 1 Introduction

Post-synthetic site-specific modification of oligodeoxynucleotides (ODNs) has been well investigated due to its important role in studying the mechanism of DNA repair enzymes [1–5], the study of various carcinogens [6–8], the development of DNA sequencing, and modification detection technologies [9–12], and furthermore a number of ODN modifications are useful in surface and analytical chemistry [13, 14]. Herein, we describe step-by-step synthetic protocols for forming 18-crown-6 (18c6) adducts to DNA abasic (AP) sites, 8-oxo-7,8-dihydro-2′-deoxyguanosine (OG), and 2′-deoxycytidine (C) via site-specific chemistry (Fig. 1). We used the commercially available primary amine derivative of 18c6, 2-aminomethyl-18-crown-6, to demonstrate these reactions; however, these protocols can also be readily used to attach other suitable primary amines of interest with similar yields.

Nathaniel Shank (ed.), *Non-Natural Nucleic Acids: Methods and Protocols*, Methods in Molecular Biology, vol. 1973, https://doi.org/10.1007/978-1-4939-9216-4_2, © Springer Science+Business Media, LLC, part of Springer Nature 2019

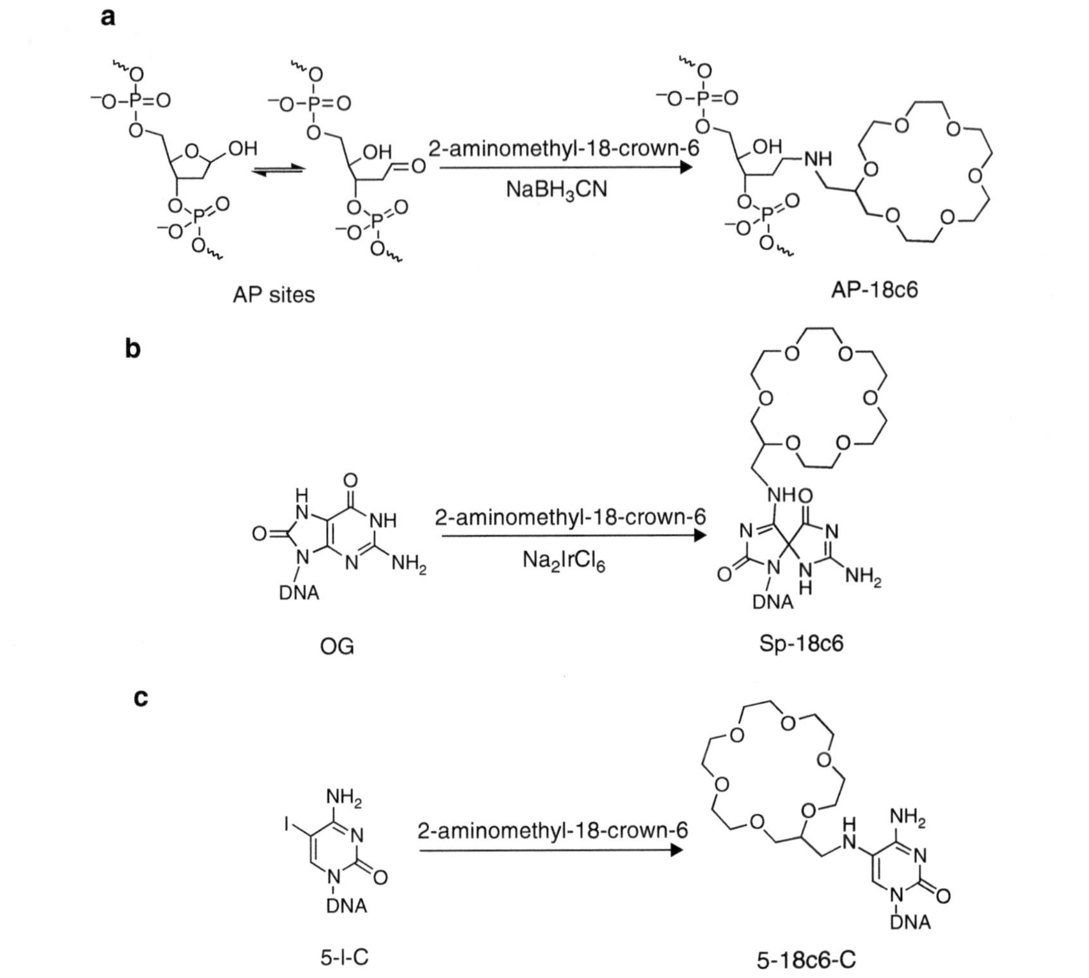


**Fig. 1** Synthetic schemes for making 18c6 adducts to AP sites (**a**), OG (**b**) and 5-I-C (**c**) via site-specific chemistry

Thus, these methods provide high potential to use these three nucleotides as "convertible nucleosides" in DNA chemistry [15].

These adducts were designed to better understand the interactions between DNA and the ion channel-forming bacterial protein α-hemolysin (α-HL) as an approach to advance nanopore technology in our laboratory and others [16–18]. Single-stranded DNA (ssDNA) can be electrically drawn through α-HL, generating blockages to the open channel current that are characteristic for the identity of the nucleotides [19]. To investigate closely the electrical signature of a specific nucleotide, streptavidin can be used to complex biotinylated ssDNA, which is too large in dimension to enter the ion channel [19]. As a result, the ssDNA chain is immobilized inside α-HL, positioning the modified site at the sensing zone (Fig. 2a) [20, 21]. The blockage current was recorded

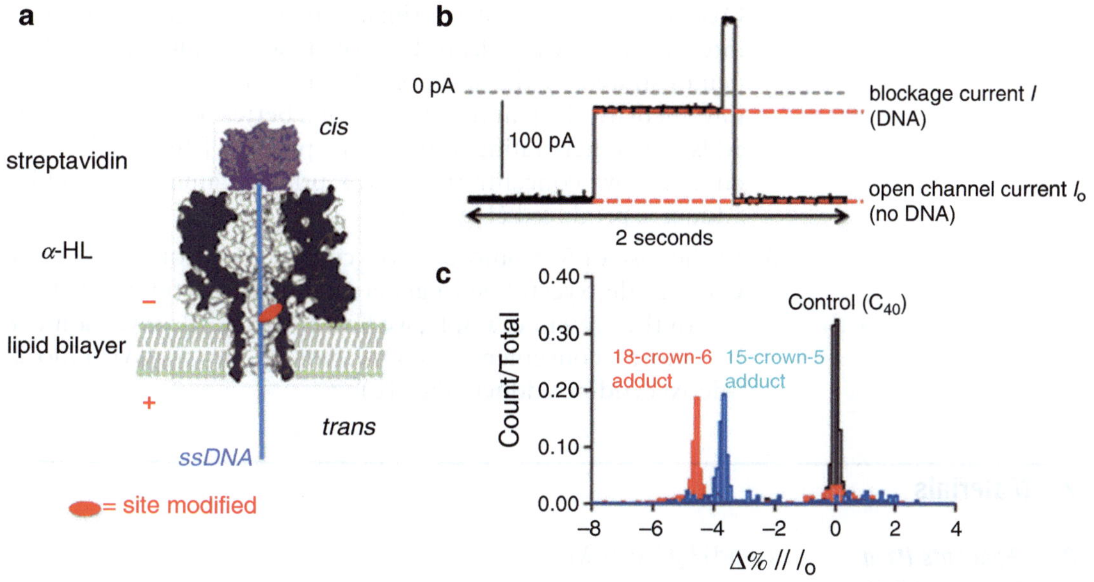


**Fig. 2** Illustrations of the nanopore immobilization experiment. (**a**) Experimental setup. (**b**) Typical current-time trace of an immobilization event. (**c**) Percentage residual current histograms of 18-crown-6 and 15-crown-5 adducts of AP sites, compared to a control DNA $C_{40}$ (%$I/I_o$ set to 0)

for 2 s (Fig. 2b), and then the voltage was briefly reversed to release the DNA allowing another strand to be captured. The blockage current so obtained was then compared to that of an unmodified control strand ($C_{40}$, Fig. 2c). By labeling AP sites and OG with these adducts, we were able to generate modulated electrical signatures (Fig. 2c) [11, 22, 23]. The 18c6 moiety was unique in its ability to interact with the electrolyte, providing clear changes in electrical current when the crown ether was present in the ion channel (Fig. 2c). A variety of primary amines were studied, from simple alkyl amines, polyamines, amino acids, short peptides, and aminoglycosides [22–24]. While these amines all readily formed adducts to DNA under the reaction conditions described here, only the crown ethers were able to slow the entry of the nucleotide into the constriction zone of the wild-type α-HL ion channel, yielding significant current modulations and providing readout of the presence of the modified nucleotide. This report therefore focuses on adduct formation with the 18c6 pendant amine.

Three methods to introduce the 18c6 adduct are illustrated:

1. Reductive amination, which is appropriate for covalent attachment of a primary amine to an aldehyde group, such as the AP sites generated by the treatment of a 2′deoxyuridine (U) containing ODN with uracil-DNA glycosylase (UDG) (Fig. 1a).

2. Oxidative addition of a primary amine to the electrophilic intermediate created from base oxidation of OG. This is best demonstrated by the addition of amines to the OG oxidation intermediate [25]; in the absence of a better nucleophile, $H_2O$ adds at C5, generating a spirocyclic product [26]. In the presence of low concentrations of a primary amine, the amine adduct is produced (Fig. 1b).
3. In the case of 5-bromo-2′-deoxycytidine, ammonia can add to C5 of 2′-deoxycytidine to generate 5-amino-cytidine [27].

   In the present example, we find 5-iodo-2′-deoxycytidine to be more conveniently converted to a 5-alkylamino-2-′-deoxycytidine adduct (Fig. 1c).

## 2 Materials

### 2.1 Reagents Used for 18-crown-6 DNA Adduct Synthesis

1. $ddH_2O$ (18 MΩ).
2. Sodium phosphate buffer solution (PBS).
3. Hydrochloric acid (HCl).
4. Phosphoric acid ($H_3PO_4$).
5. Sodium hydroxide (NaOH).
6. Ethanol.
7. Tris(hydroxymethyl)aminomethane (Tris).
8. Ethylenediaminetetraacetic acid disodium salt ($Na_2EDTA$).
9. Sodium acetate (NaOAc).
10. Ammonium acetate ($NH_4OAc$).
11. Dithiothreitol (DTT).
12. Acetonitrile ($CH_3CN$).
13. Ammonium hydroxide ($NH_4OH$).
14. Sodium hexachloroiridate ($Na_2IrCl_6$).
15. Sodium cyanoborohydride ($NaBH_3CN$).
16. 2-aminomethyl-18-crown-6 (18c6).
17. Sodium chloride (NaCl).
18. Potassium iodide (KI).
19. 3-(*N*-Morpholino)propanesulfonic acid (MOPS).
20. Potassium peroxymonosulfate as oxone ($KHSO_5$).
21. Uracil-DNA glycosylase (UDG) supplied by New England Biolabs (NEB).

### 2.2 DNA Preparation and Purification Procedures

1. DNA synthesis: The 3′-biotinylated oligodeoxynucleotides (ODN) used here were synthesized from commercially available phosphoramidites by the DNA-Peptide Core Facility at the University of Utah. Their sequences are 5′-CCCCC

CCCCC CCCCC CCCCC CCCCC CXCCC CCCCC CCCCC-Btn, where X = U, OG, or 5-iodo-2′-deoxycytidine (5-I-C) and Btn = biotin-triethylene glycol (TEG) linker. After synthesis, each ODN was cleaved from the synthetic column and deprotected according to the manufacturer's protocol.

2. DNA purification: The ODNs were purified with a semi-preparative anion-exchange HPLC column (A = 10% $CH_3CN$/90% dd$H_2O$, B = 20-mM PBS (pH 7.0), 1-M NaCl in 10% $CH_3CN$/90% dd$H_2O$, flow rate = 3 mL/min). A linear gradient of 25–100% B over 30 min was used while monitoring the absorbance at 260 nm. Desalination was performed with a dialysis cassette possessing 7000 molecular weight cutoff (MWCO) membrane against dd$H_2O$ at 4 °C for 2 d and changing the dd$H_2O$ three times daily, followed by vacuum concentration to reduce the volume.
3. DNA characterization: The concentrations of the ODNs were measured in dd$H_2O$ by UV-vis spectroscopy at 260 nm under room temperature (24 °C). The corresponding extinction coefficients for the strands were determined from their primary sequence with the assumption that the modified bases had the same extinction coefficients as their parent base (i.e., OG and G have the same $\varepsilon_{260nm}$). The identities of the ODNs were determined by negative ion electron spray mass spectrometry ($ESI^-$-MS) on a Micromass Quattro II mass spectrometer equipped with Zspray API source in the mass spectrometry laboratory at the Department of Chemistry, University of Utah. All ODNs were stored at −20 °C.

### 2.3 Preparation of Reactive Reagents and Buffers

Three reactions described in this article require different buffer conditions and reagents elaborated below:

#### 2.3.1 18c6 Adduct to DNA AP Sites

1. UDG buffer: 20-mM PBS (pH 8.0), 1-mM DTT, and 1-mM EDTA.
2. Reductive amination reaction buffer: 150-mM MOPS (pH 6.5).
3. Stock solutions of reagents: 2-M $NaBH_3CN$ solution and 2-M NaOH.

#### 2.3.2 18c6 Adduct to OG

1. Reaction buffer: 75-mM PBS (pH 8.0).
2. Stock solutions of reagents: 200-mM $Na_2IrCl_6$ and 50-mM $Na_2EDTA$ (pH 8.0).

#### 2.3.3 18c6 Adduct to 5-I-C

1. 1 M 2-aminomethyl-18-crown-6.

## 3 Methods

### 3.1 Synthesis of AP-18c6 Adduct

1. Thermally equilibrate a 100-μL solution of UDG buffer with 10-μM U-containing ODN and 1 unit UDG at 37 °C for 30 min to generate AP sites, and then dialyze the strands against $ddH_2O$ at 4 °C for 12 h (*see* **Notes 1** and **2**).
2. Lyophilize the sample to dryness, and resuspend the resulting AP-containing ODN in 100-μL MOPS buffer; then introduce by pipette 20 μL of 1-M 2-aminomethyl-18-crown-6 into the reaction, immediately followed by the addition of 5 μL of 2-M stock $NaBH_3CN$ solution (*see* **Note 3**). Keep the reaction at 37 °C for 24 h.
3. Add 10 μL of 2 M NaOH solution to cleave the unreacted ODNs to generate strand breaks. Keep the mixture at 90 °C for 10 min.
4. Dialyze the reaction mixture against $ddH_2O$ in a 7000 MWCO membrane dialysis cassette at 4 °C for 12 h before HPLC analysis (*see* **Notes 4–6**).

### 3.2 Synthesis of Sp-18c6 Adduct

1. Incubate a 100-μL solution of 75-mM PBS (pH 8) containing 10-μM OG-containing ODN and 2-mM aminomethyl-18-crown-6 at 45 °C for 30 min to achieve thermo-equilibrium (*see* **Note** 7).
2. Add 200-mM stock $Na_2IrCl_6$ solution in three 7.5-μL aliquots into the reaction mixture, and then incubate at 45 °C for another 30 min (*see* **Note 8**).
3. Add stock $Na_2EDTA$ to a final concentration of 1 mM to quench the reaction.
4. Dialyze the reaction mixture against $ddH_2O$ at 4 °C for 12 h before HPLC analysis (*see* **Notes 4**, **5**, and **9**).

### 3.3 Synthesis of 5-I-C and Conversion to an Amine Adduct in a Nucleoside Model

The iodination of 2′-deoxycytidine was achieved by following published protocols with the following modification [28, 29].

1. Briefly, a reaction with 1-mM 2′-deoxycytidine nucleoside, 8-mM KI, and 8-mM $KHSO_5$ was incubated in PBS (pH 6.0) at 45 °C for 24 h.

   The identity and site of iodination on the C heterocycle was established by $ESI^+$-MS and $^1$H-NMR on the nucleoside model reaction. $ESI^+$-MS $[M+H^+]$ calcd mass = 354.1, found = 354.0. $^1$H NMR ($D_2O$) δ 7.22 (s, 1H), 6.20 (dd, 1H), 4.45 (ddd, 1H), 4.08 (ddd, 1H), 3.84 (td, 2H), 2.45 (td, 1H), 2.33 (td, 1H). UV-vis: $\lambda_{max}$ = 294.0 nm.
2. The amination of 5-I-C was achieved by incubating the 5-I-C nucleoside with 1 M 2-aminomethyl-18-crown-6 in $ddH_2O$ at 55 °C for 24 h.

3. The identity and site of amination on 5-I-C was determined by $ESI^+$-MS and $^1$H-NMR from the nucleoside model reaction conducted when 5-I-C was allowed to react with methylamine. $ESI^+$-MS [$M+H^+$] calcd mass = 257.3, found = 257.1. $^1$H NMR ($D_2O$) δ 7.22 (s, 1H), 6.20 (dd, 1H), 4.45 (ddd, 1H), 4.08 (ddd, 1H), 3.84 (td, 2H), 2.56 (s, 3H), 2.45 (td, 1H), 2.33 (td, 1H). UV-vis: $\lambda_{max}$ = 305.0 nm.

### 3.4 Synthesis of 5-I-C and Conversion to an Amine Adduct in an ODN

In order to make site-specific 5-18c6-C adducts, 5-I-C was incorporated into the ODN through solid-phase synthesis; alternatively, site-specific iodination can be conducted by adding an n-1 complement to the strand such that the desired C residue is present as a bulged nucleotide [17]. The attachment of 18c6 is described below, which was adapted from published protocols [27]:

1. Incubate a 10-μM ODN containing 5-I-C and 1-M 2-aminomethyl-18-crown-6 in $ddH_2O$ at 55 °C for 24 h.
2. Dialyze the reaction mixture against $ddH_2O$ in a 7000 MWCO dialysis cassette at 4 °C for 12 h before HPLC analysis.

### 3.5 Purification and Characterization of Adducted ODNs

An analytical anion-exchange column can be used to analyze these reactions with the following mobile-phase conditions: A = 10% $CH_3CN$/90% $ddH_2O$ and B = 20-mM PBS (pH 8.0), 1-M NaCl in 10% $CH_3CN$/90% $ddH_2O$. A linear gradient of 25–100% B over 30 min (flow rate = 1 mL/min) can be used while monitoring the absorbance at 260 nm.

1. The chromatograms of these reactions should be compared to those of their corresponding starting materials, and new peaks (products) can be identified and collected (Fig. 3).
2. Dialyze the collected product solutions against $ddH_2O$ in a 7000 MWCO dialysis cassette at 4 °C for 24 h before vacuum concentration to reduce the volume.

### 3.6 Preparation of the ODNs for $ESI^-$-MS Analysis

Ethanol precipitation was used to prepare the ODNs for $ESI^-$-MS, which is briefly described below:

1. Add 3-M $NH_4OAc$ (1/10 volume of the ODN, pH 5.2) to achieve a final concentration of 0.3 M, and mix well.
2. Add 2–3 volumes of 100% ethanol and mix well.
3. Keep the tube at −20 °C for 30 min.
4. Spin in a centrifuge at 12,000 × *g* at 4 °C for 20 min, and carefully decant supernatant.
5. Add 1-mL 70% ethanol (−20 °C), mix, and then spin briefly. Decant supernatant. Repeat twice.
6. Vacuum dry pellet and resuspend in 3-mM $NH_4OAc$.

   All the reaction products were characterized with $ESI^-$-MS.

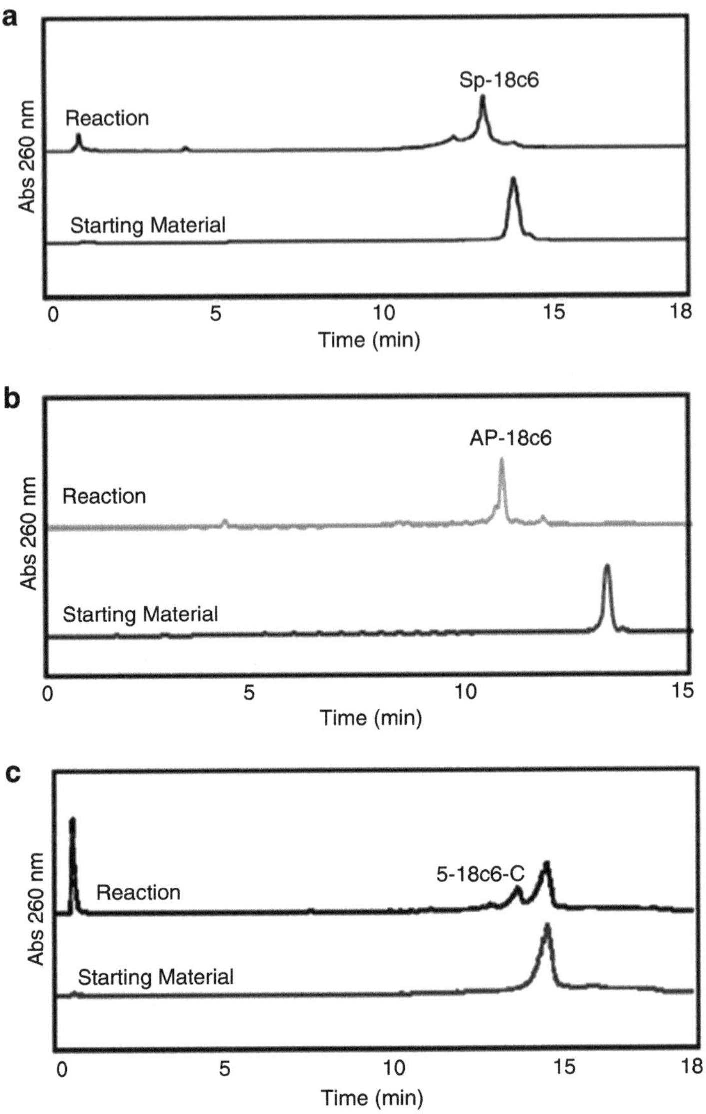


**Fig. 3** HPLC chromatograms of these three reactions, AP-18c6 (**a**), Sp-18c6 (**b**), and 5-18c6-C (**c**), and their comparisons to starting materials

## 4 Notes

1. During the synthesis of AP-18c6 adduct, the pH (6.5) we chose to use is based on a series of pH-dependent studies. DNA AP sites are vulnerable to elimination under basic conditions, generating strand breaks. On the other hand, higher pH facilitates the amination process. Therefore, a pH ranging from 5.5 to 6.5 produced the highest product yield.

2. It is highly recommended to use PBS instead of Tris buffer in the UDG treatment step. Being a primary amine, Tris can aminate AP sites, thus generating an undesirable side product. Similarly, we also recommend not using Tris buffer for OG-containing ODNs as well.
3. In order to guarantee a good yield in the reductive amination reaction, we recommend that one could add a second aliquot of $NaBH_3CN$ (100 mM, 10 μmol) 12 h after the first addition.
4. Similar reactions have been conducted with heterosequence ODN strands previously, and we observed similar yields with ODNs up to 50 nucleotides in length.
5. The yield decreases with the length of the ODN, and the ability to purify the adducts by HPLC also diminishes.
6. Similar product yields were achieved when the 2-aminomethyl-18-crown-6 in the AP and OG reactions was substituted with $N^{\alpha}$-OAc-*O*-methyllysine, spermine, spermidine, *O*-methyl-arginine, glucosamine, Gly-Pro-Arg-Pro carboxamide, and 2-aminomethyl-15-crown-5.
7. Do not use Tris buffer because it is a primary amine and will compete for adduct formation during the oxidation reaction to label OG with 2-aminomethyl-18-crown-6.
8. The biotin linker at the 3′ terminus of the ODNs is used to immobilize the molecules inside the α-HL ion channel by forming a complex with streptavidin during the nanopore measurements, and it is not relevant to the chemical reactions. However, it needs noting that during the synthesis of Sp-18c6, biotin can be oxidized with $Na_2IrCl_6$ yielding an M+16 product that does not change its ability to bind streptavidin [30].
9. The Sp-18c6 adduct is recommended to be studied immediately, due to its instability [25, 31–33].

## Acknowledgment

This work was supported by grants from the National Institutes of Health, HG005095 and GM093099.

### References

1. Ono T, Wang S, Koo C, Engstrom L, David SS, Kool ET (2012) Direct fluorescence monitoring of DNA base excision repair. Angew Chem Int Ed 51:1689–1692
2. McKibbin PL, Kobori A, Taniguchi Y, Kool ET, David SS (2012) Surprising repair activities of nonpolar analogs of 8-oxoG expose features of recognition and catalysis by base excision repair glycosylases. J Am Chem Soc 134:1653–1661
3. Erlanson DA, Chen L, Verdine GL (1993) DNA methylation through a locally unpaired intermediate. J Am Chem Soc 115:12583–12584
4. Xu YZ, Zheng Q, Swann PF (1992) Synthesis of DNA containing modified bases by post-

synthetic substitution. Synthesis of oligomers containing 4-substituted thymine: O4-alkylthymine, 5-methylcytosine, N4-dimethylamino-5-methylcytosine, and 4-thiothymine. J Org Chem 57:3839–3845

5. McKibbin PL, Fleming AM, Towheed MA, Van Houten B, Burrows CJ, David SS (2013) Repair of hydantoin lesions and their amine adducts in DNA by base and nucleotide excision repair. J Am Chem Soc 135:13851–13861
6. Gates KS (2009) An overview of chemical processes that damage cellular DNA: spontaneous hydrolysis, alkylation, and reactions with radicals. Chem Res Toxicol 22:1747–1760
7. Pirogov N, Shafirovich V, Kolbanovskiy A, Solntsev K, Courtney SA, Amin S, Geacintov NE (1998) Role of hydrophobic effects in the reaction of a polynuclear aromatic diol epoxide with oligodeoxynucleotides in aqueous solutions. Chem Res Toxicol 11:381–388
8. Omumi A, Beach DG, Baker M, Gabryelski W, Manderville RA (2011) Postsynthetic guanine arylation of DNA by Suzuki-Miyaura cross-coupling. J Am Chem Soc 133:42–50
9. Korlach J, Turner SW (2012) Going beyond five bases in DNA sequencing. Curr Opin Struct Biol 22:251–261
10. Kumar S, Tao C, Chien M, Hellner B, Balijepalli A, Robertson JWF, Li Z, Russo JJ, Reiner JE, Kasianowicz JJ et al (2012) PEG-labeled nucleotides and nanopore detection for single molecule DNA sequencing by synthesis. Sci Rep 2:684
11. An N, Fleming AM, White HS, Burrows CJ (2015) Nanopore detection of 8-oxoguanine in the human telomere repeat sequence. ACS Nano 9:4296–4307
12. Riedl J, Ding Y, Fleming AM, Burrows CJ (2015) Identification of DNA lesions using a third base pair for amplification and nanopore sequencing. Nat Commun 6:8807
13. Lee HJ, Wark AW, Corn RM (2006) Creating advanced multifunctional biosensors with surface enzymatic transformations. Langmuir 22:5241–5250
14. Cederquist KB, Keating CD (2009) Curvature effects in DNA: au nanoparticle conjugates. ACS Nano 3:256–260
15. Allerson CR, Chen SL, Verdine GL (1997) A chemical method for site-specific modification of RNA: the convertible nucleoside approach. J Am Chem Soc 119:7423–7433
16. An N, White HS, Burrows CJ (2012) Modulation of the current signatures of DNA abasic site adducts in the a-hemolysin ion channel. Chem Commun 48:11410–11412
17. Mitchell N, Howorka S (2008) Chemical tags facilitate the sensing of individual DNA strands with nanopores. Angew Chem Int Ed Engl 47:5565–5568
18. Zeng T, Liu L, Li T, Li Y, Gao J, Zhao Y, Wu H-C (2015) Detection of 5-methylcytosine and 5-hydroxymethylcytosine in DNA via host-guest interactions inside alpha-hemolysin nanopores. Chem Sci 6:5628–5634
19. Purnell R, Mehta K, Schmidt J (2008) Nucleotide identification and orientation discrimination of DNA homopolymers immobilized in a protein nanopore. Nano Lett 8:3029–3034
20. Stoddart D, Heron A, Mikhailova E, Maglia G, Bayley H (2009) Single-nucleotide discrimination in immobilized DNA oligonucleotides with a biological nanopore. Proc Natl Acad Sci U S A 106:7702–7707
21. Henrickson SE, Misakian M, Robertson B, Kasianowicz JJ (2000) Driven DNA transport into an asymmetric nanometer-scale pore. Phys Res Lett 85:3057–3060
22. An N, Fleming AM, White HS, Burrows CJ (2012) Crown ether-electrolyte interactions permit nanopore detection of individual DNA abasic sites in single molecules. Proc Natl Acad Sci U S A 109:11504–11509
23. Schibel AEP, An N, Jin Q, Fleming AM, Burrows CJ, White HS (2010) Nanopore detection of 8-oxo-7,8-dihydro-2′-deoxyguanosine in immobilized single-stranded DNA via adduct formation to the DNA damage site. J Am Chem Soc 132:17992–17995
24. An N, White HS, Burrows CJ (2012) Modulation of the current signatures of DNA abasic site adducts in the alpha-hemolysin ion channel. Chem Commun 48:11410–11412
25. Hosford ME, Muller JG, Burrows CJ (2004) Spermine participates in oxidative damage of guanosine and 8-oxoguanosine leading to deoxyribosylurea formation. J Am Chem Soc 126:9540–9541
26. Luo W, Muller JG, Rachlin EM, Burrows CJ (2000) Characterization of spiroiminodihydantoin as a product of one-electron oxidation of 8-oxo-7,8-dihydroguanosine. Org Lett 2:613–616
27. Goldman D, Kalman TI (1983) Formation of 5- and 6-aminocytosine nucleosides and nucleotides form the corresponding 5-bromocytosine derivatives: synthesis and reaction mechanism. Nucleosides Nucleotides 2:175–187
28. Ross SA, Burrows CJ (1997) Bromination of pyrimidines using bromide and monoperoxysulfate: a competition study between cytidine,

uridine, and thymidine. Tetrahedron Lett 38:2805–2808

29. Ross SA, Burrows CJ (1996) Cytosine-specific chemical probing of DNA using bromide and monoperoxysulfate. Nucleic Acids Res 24:5062–5063
30. Upadhya K, Khattak IK, Mullah B (2005) Oxidation of biotin during oligonucleotide synthesis. Nucleosides Nucleotides Nucleic Acids 24:919–922
31. Xue L, Greenberg MM (2007) Facile quantification of lesions derived from 2′-deoxyguanosine in DNA. J Am Chem Soc 129:7010–7011
32. Zhang B, Guo L, Greenberg MM (2012) Quantification of 8-oxodGuo lesions in double-stranded DNA using a photoelectrochemical DNA sensor. Anal Chem 84:6048–6053
33. Fleming AM, Armentrout EI, Zhu J, Muller JG, Burrows CJ (2015) Spirodi(iminohydantoin) products from oxidation of 2-′-deoxyguanosine in the presence of $NH_4Cl$ in nucleoside and oligodeoxynucleotide contexts. J Org Chem 80:711–721

# Chapter 3

# Synthesis of a Fluorescent Cytidine TNA Triphosphate Analogue


## Hui Mei and John Chaput



## Abstract

Threose nucleic acid (TNA) is refractory to nuclease digestion and capable of undergoing Darwinian evolution, which together make it a promising system for diagnostic and therapeutic applications that require high biological stability. Expanding the sequence/chemical diversity of TNA would enable the development of functional TNA molecules with enhanced physicochemical properties. Recently, we have reported the synthesis and polymerase activity of a fluorescent cytidine TNA triphosphate analogue (1,3-diaza-2-oxo-phenothiazine, $tC^fTP$) that maintains Watson-Crick base pairing with guanine. It not only improves TNA synthesis by enabling an engineered TNA polymerase to read through sequential G-repeats in a DNA template but it also introduces new physicochemical properties, such as increased hydrophobic character and fluorescence. Here, we describe the synthesis protocol for $tC^fTP$, which includes three silica gel purifications, two precipitations, and one HPLC purification. Starting from protected threofuranosyl sugar, desired TNA nucleoside was obtained in a Vorbrüggen glycosylation reaction. After deprotection, nucleoside was further converted to 3′-monophosphate, activated by 2-methylimidazole, and subsequently treated with pyrophosphate to afford the desired 3′-triphosphate ($tC^fTP$).



**Keywords** Xeno-nucleic acids (XNA), Threose nucleic acid (TNA), Nucleoside triphosphate, 1,3-diaza-2-oxo-phenothiazine, Fluorescence


## 1 Introduction

Threose nucleic acid (TNA) is an artificial genetic polymer that has a backbone structure composed of repeating units of α-L-threose sugars that are vicinally linked by 2′,3′-phosphodiester bonds (Fig. 1a) [1]. Despite a backbone repeat unit that is one atom (or bond) shorter than that of DNA or RNA, TNA is capable of forming stable antiparallel Watson-Crick duplex structures with itself and with complementary strands of DNA and RNA [2, 3]. Similar to DNA/RNA, TNA is also able to undergo Darwinian evolution in vitro to produce affinity reagents with strong ligand-binding activity [4]. This property, coupled with a backbone structure that is refractory to nuclease digestion [5], makes TNA an excellent candidate for therapeutic and diagnostic applications [6].


Nathaniel Shank (ed.), *Non-Natural Nucleic Acids: Methods and Protocols*, Methods in Molecular Biology, vol. 1973, https://doi.org/10.1007/978-1-4939-9216-4_3,


© Springer Science+Business Media, LLC, part of Springer Nature 2019

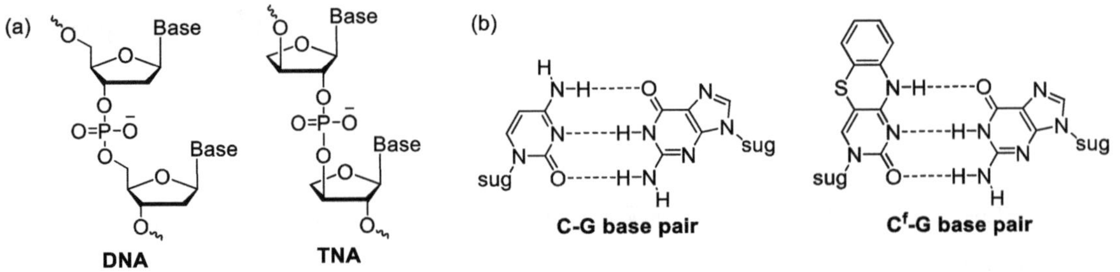


**Fig. 1** Molecular structures, (**a**) constitutional structure for the linearized backbone of DNA and TNA, (**b**) Watson-Crick base pairs for C:G and $C^f$:G. $C^f$ is the cytosine analogue, 1,3-diaza-2-oxo-phenothiazine

Recently, we have developed an engineered polymerase from the thermophilic species *Thermococcus kodakarensis*, termed Kod-RI, which can copy DNA templates into TNA in just 3 h [7]. However, despite the enhanced activity of Kod-RI over previous TNA polymerases, the synthesis of completely unbiased TNA libraries remains limited by chain termination events that occur when the polymerase encounters sequential G-nucleotides in the DNA template [8]. As a possible solution to this problem, we sought to develop analogues of TNA cytidine triphosphate (tCTP) that would stabilize the tC:dG base pair in the enzyme active site. We found that the cytidine triphosphate analogue 1,3-diaza-2-oxo-phenothiazine ($tC^fTP$), a fluorescent tricyclic aromatic ring system that maintains Watson-Crick base pairing with guanine (Fig. 1b) [9, 10], can efficiently read through sequential G-nucleotides in a polymerase-mediated TNA synthesis reaction [11]. TNA templates synthesized with $tC^fTP$ replicate with 98.4% overall fidelity, indicating that in vitro selection experiments could be performed using fluorescent 1,3-diaza-2-oxo-phenothiazine as a modified base [11]. Together, these results provide a platform for synthesizing unbiased TNA libraries with enhanced hydrophobic and fluorescent properties.

In this chapter, we provide the protocol for chemical synthesis of 1,3-diaza-2-oxo-phenothiazine TNA nucleoside 3′-triphosphate ($tC^fTP$). To this end, protected α-L-threofuranosyl-1,3-diaza-2-oxo-phenothiazine nucleoside was prepared from the universal glycosyl donor and 1,3-diaza-2-oxo-phenothiazine in a Vorbrüggen glycosylation reaction [12]. After deprotection, nucleoside was subsequently converted to 3′-monophosphate, which was then converted to 3′-phosphoro-(2-methyl)-imidazolide [13]. Subsequent displacement of the imidazole residue with the pyrophosphate resulted in $tC^fTP$. After HPLC purification, lyophilization, and sodium precipitation, the desired triphosphate was obtained as a sodium salt.

## 2 Materials

### 2.1 General Methods

Nonaqueous reactions were performed using oven-dried glassware under an atmosphere of argon or nitrogen. All fine chemicals are purchased from commercial suppliers and used without further purification. Anhydrous solvents were purchased or generated in-house using standard methods. Reactions were monitored by thin layer chromatography using UV-activated TLC plates with silica gel 60 F254 and aluminum backing. Flash column chromatography was performed using SiliCycle 40–60 mesh silica gel.

### 2.2 Reagents for tC$^f$TP Synthesis

1. Acetonitrile.
2. Bis(trimethylsilyl)acetamide (BSA).
3. Trimethylsilyl trifluoromethanesulfonate (TMSOTf).
4. Dichloromethane ($CH_2Cl_2$).
5. Sodium bicarbonate ($NaHCO_3$).
6. Sodium sulfate ($Na_2SO_4$).
7. Ethyl acetate (EtOAc).
8. Hexane.
9. Tetrahydrofuran (THF).
10. Tetrabutylammonium fluoride (TBAF).
11. Methanol (MeOH).
12. 4-(Dimethylamino)pyridine (DMAP).
13. N,N-Diisopropylethylamine (DIPEA).
14. 2-Cyanoethyl N,N-diisopropylchlorophosphoramidite.
15. Acetone.
16. 3-Hydroxylpropionitrile.
17. Tetrazole.
18. Hydrogen peroxide ($H_2O_2$).
19. Ammonium hydroxide.
20. 2-Methylimidazole.
21. Triphenylphosphine.
22. 2,2′-Dipyridyl disulfide.
23. Dimethyl sulfoxide (DMSO).
24. N,N-Dimethylformamide (DMF).
25. Diethyl ether.
26. Triethylamine.
27. Sodium perchlorate ($NaClO_4$).
28. Tributylammonium pyrophosphate.

29. Tributylamine.
30. 0.1 M triethylammonium acetate buffer, pH 7.0 (TEAA).
31. Acetonitrile HPLC grade.

### 2.3 Consumables and Instruments

1. 10, 50, 100, and 250 mL round-bottom flasks, oven-dried.
2. Gas balloon.
3. Syringe and needle.
4. Magnetic stir plate and stir bar.
5. Filter paper.
6. Büchner funnel.
7. 100 mL separatory funnel.
8. UV-activated TLC plates with silica gel 60 F254 and aluminum backing (Sigma-Aldrich).
9. 40–60 mesh silica gel (SiliCycle).
10. Glass chromatography columns.
11. UV lamp, 254 nm.
12. Rotary evaporator equipped with a vacuum pump.
13. Sealed tube and stopper.
14. Lyophilizer.
15. Centrifuge 5702 (Eppendorf).
16. Ultimate 3000 semi-preparative HPLC (Thermo Fisher Scientific).
17. NanoDrop 2000c (Thermo Fisher Scientific).

## 3 Procedures

Synthesis of $tC^{f}TP$ (Fig. 2).

### 3.1 3-(2′-O-Benzoyl-3′-O-tert-butyldiphenylsilyl-α-L-threofuranosyl)-1,3-diaza-2-oxophenothiazine (3)

1. Threose sugar donor 1-*O*-acetyl-2-*O*-benzoyl-3-*O*-*tert*-butyldiphenylsilyl-L-threofuranose **1** [12] and 1,3-diaza-2-oxophenothiazine **2** [14] were prepared according to the reported procedures.
2. Add 0.66 g nucleobase (**2**) (3.03 mmol) in an oven-dried 100 mL round-bottom flask.
3. Place a magnetic stir bar into the reaction flask.
4. Introduce an argon atmosphere into the flask, close with a rubber stopper, and maintain it throughout the reaction.
5. Add 40 mL anhydrous acetonitrile through syringe.
6. Add 1.85 mL (7.57 mmol) bis(trimethylsilyl)acetamide (BSA) to the flask, and stir the mixture at 75 °C for 2 h in an oil bath.

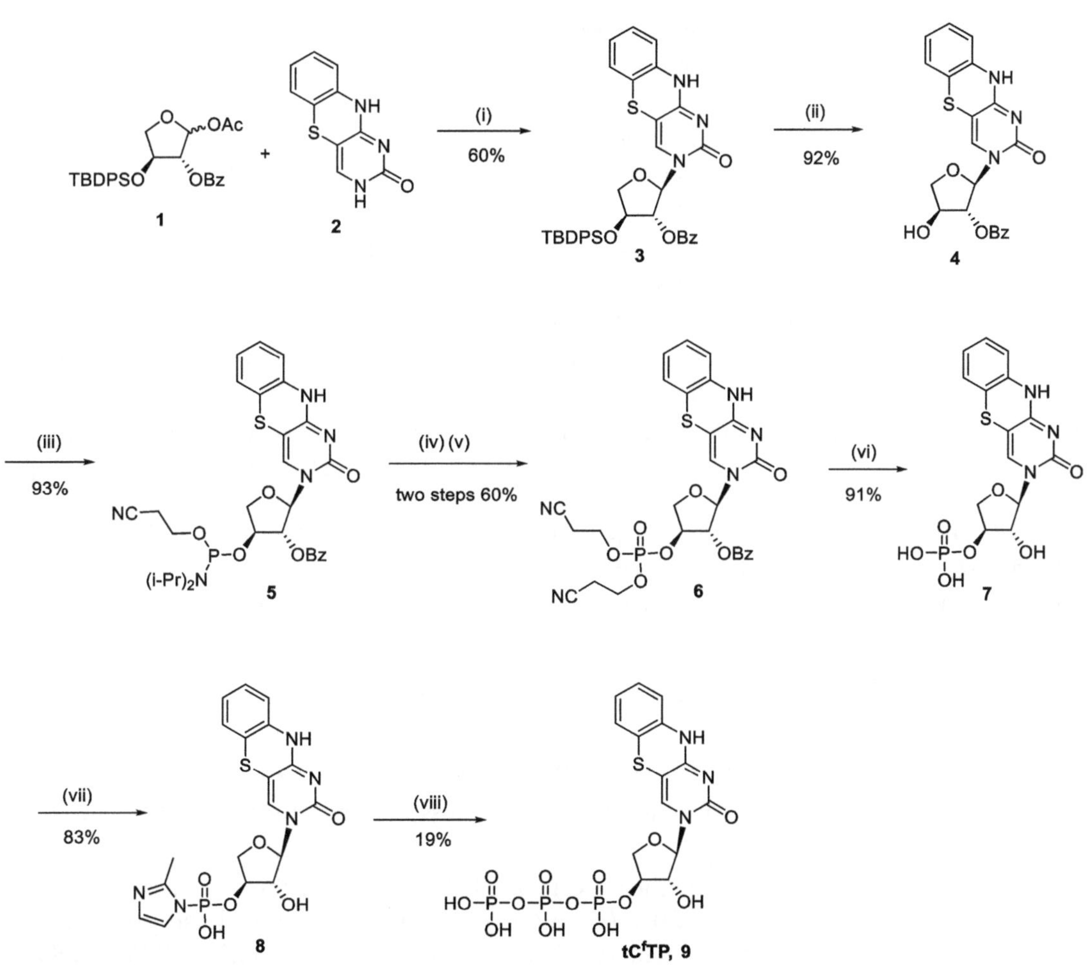


**Fig. 2** Synthesis of tC$^f$TP. Reagents and conditions. (i) N,O-bis(trimethylsilyl)-acetamide (BSA), trimethylsilyl-triflate (TMSOTf), $CH_3CN$, 65 °C, 5 h; (ii) 1 M TBAF in THF, 0 °C, 1 h; (iii) $NH_3$-MeOH, rt. 3.5 h (iii) $NC(CH_2)_2OP(Cl)N(i\text{-}Pr)_2$, $(i\text{-}Pr)_2EtN$, DMAP, rt, 40 min; (iv) 3-hydroxypropio-nitrile, tetrazole, $CH_3CN$, rt, 3 h; (v) $H_2O_2$, THF, rt, 20 min; (vi) $NH_4OH$, 37 °C, 16 h; (vii) 2-methylimidazole, $PPh_3$, 2,2′-dipydiyl disulfide, triethylamine, DMF-DMSO, rt, 6 h; (viii) tributylammonium pyrophosphate, tributylamine, anhydrous DMF, rt, 6 h

7. Cool the reaction mixture to room temperature.
8. Dissolve 1.53 g sugar donor (**1**) (3.03 mmol) in 10 mL anhydrous acetonitrile solution, and add the solution to the flask slowly through syringe.
9. Add 1.65 mL TMSOTf (9.09 mmol) slowly through syringe.
10. Stir the reaction mixture at 65 °C for 5 h.
11. Cool the reaction mixture to room temperature, dilute with 100 mL of $CH_2Cl_2$, and pour into a stirring solution of saturated aqueous $NaHCO_3$ (100 mL). The unreacted nucleobase will precipitate as a yellow solid.

12. Isolate the product by vacuum filtration using a Buchner funnel, and wash the yellow filter cake with 20 mL of $CH_2Cl_2$ (*see* **Note 1**).
13. Transfer the filtrate solution to a 500 mL separatory funnel. Collect the organic layer and wash with 100 mL $H_2O$ and then 100 mL brine.
14. Dry the organic solution with $MgSO_4$ and evaporate to dryness.
15. Dissolve the residue in 50 mL EtOAc, and precipitate the nucleoside **3** by the slow addition of 50 mL hexane under stirring.
16. Filter the mixture and collect nucleoside **3** (1.2 g, yield 60%) as a yellow solid, which could be directly used for the next step.

### *3.2 3-(2′-O-Benzoyl-α-L-threofuranosyl)-1,3-diaza-2-oxophenothiazine (4)*

1. Add 1.54 g yellow solid **3** (2.32 mmol) in a 100 mL round-bottom flask.
2. Place a magnetic stir bar into the reaction flask.
3. Add 30 mL THF and cool the solution to 0 °C.
4. Add 2.32 mL tetrabutylammonium fluoride (TBAF, 1 M solution in THF) slowly through syringe.
5. Stir the reaction mixture at 0 °C for 1 h.
6. Evaporate THF to dryness under reduced pressure (*see* **Note 2**).
7. Purify the residue by silica gel column chromatography, eluting with 20:1 (v/v) $CH_2Cl_2$/MeOH to afford the product **4** in 92% yield (0.91 g) as a yellow foam. TLC (DCM/MeOH, 10:1): $R_f$ = 0.48. 1H NMR (500 MHz, $CDCl_3$): δ 4.23–4.25 (m, 1H), 4.33–4.35 (m, 1H), 4.43 (m, 1H), 5.22 (brs, 1H), 5.90 (s, 1H), 6.80 (s, 2H), 6.86–6.89 (m, 2H), 7.37–7.43 (m, 3H), 7.55–7.58 (m, 2H), 7.95–7.96 (d, 2H, J = 4.0 Hz), 9.04 (brs, 1H). 13C NMR (125.8 MHz, CDC13): δ 73.8, 76.7, 82.1, 92.6, 96.8, 116.6, 117.7, 124.6, 125.9, 127.4, 128.6, 129.1, 130.0, 135.5, 135.6, 155.6, 160.5, 166.0. HRMS (ESI-TOF) calculated for $C_{21}H_{17}O_5N_3SNa$ $[M+Na]^+$ 446.0787; observed 446.0777.

### *3.3 3-(2′-O-Benzoyl-α-L-threofuranosyl)-1,3-diaza-2-oxophenothiazine-3′-(2-cyanoethyl)-N,N′-diisopropyl Phosphoramidite (5)*

1. Add 0.34 g compound **4** (0.80 mmol) and 20 mg DMAP (0.16 mmol) in an oven-dried 50 mL round-bottom flask.
2. Place a magnetic stir bar into the reaction flask.
3. Introduce an argon atmosphere into the flask, close with a rubber stopper, and maintain it throughout the reaction.
4. Add 20 mL anhydrous $CH_2Cl_2$.
5. Add 210 μL DIPEA (1.21 mmol) through syringe.

6. Add 215 μL 2-cyanoethyl *N*,*N*-diisopropylchlorophosphoramidite (0.96 mmol).
7. Stir the reaction mixture at room temperature for 40 min.
8. Dilute the mixture with 40 mL $CH_2Cl_2$.
9. Wash sequentially with 40 mL 5% aq. $NaHCO_3$ and 40 mL brine.
10. Dry the organic solution over $Na_2SO_4$, and evaporate solvent under reduced pressure.
11. Purify the residue by silica gel column chromatography, eluting with 6:1 (v/v) $CH_2Cl_2$/acetone to afford nucleoside phosphoramidite **5** in 93% yield (0.47 g) as a yellow foam. $R_f = 0.33$ ($CH_2Cl_2$/acetone, 4:1). $^{31}P$ NMR (162 MHz, $CDCl_3$) δ 152.87, 150.85. HRMS (ESI-TOF) calculated for $C_{30}H_{34}O_6N_5PSNa$ $[M+Na]^+$ 646.1865; observed 646.1862.

### 3.4 3-(2′-O-Benzoyl-α-L-threofuranosyl)-1,3-diaza-2-oxophenothiazine-3′-bis(2-cyanoethyl)-phosphotriester (6)

1. Add 0.47 g compound **5** (0.75 mmol) in an oven-dried 50 mL round-bottom flask.
2. Place a magnetic stir bar into the reaction flask.
3. Introduce an argon atmosphere into the flask and close with a rubber stopper.
4. Add 15 mL anhydrous acetonitrile to dissolve **5**.
5. Add 82 μL 3-hydroxypropionitrile (1.2 mmol).
6. Add 2.67 mL 0.45 M tetrazole in THF (1.2 mmol), and stir the reaction mixture at room temperature for 3 h.
7. Add 153 μL 30% $H_2O_2$ (1.5 mmol), and stir at room temperature for 20 min.
8. Dilute the reaction mixture with 50 mL $CH_2Cl_2$, and wash with 40 mL brine.
9. Dry the organic solution over $Na_2SO_4$, and evaporate to dryness under reduced pressure.
10. Purify the residue by silica gel column chromatography, eluting with 1:1 (v/v) $CH_2Cl_2$/acetone to afford free nucleoside **6** in 60% yield (275 mg) as a yellow foam. $R_f = 0.28$ ($CH_2Cl_2$/acetone, 1:1). $^1H$ NMR (500 MHz, $CDCl_3$): $\delta$ 2.75–2.82 (m, 4H), 4.33–4.38 (m, 5H), 4.58–4.61 (d, J = 11.5 Hz, 1H), 5.02–5.04 (m, 1H), 5.73 (s, 1H), 6.14 (s, 1H), 6.84–6.85 (m, 2H), 6.99–7.00 (m, 1H), 7.09–7.11 (d, J = 7.5 Hz, 1H), 7.35 (s, 1H), 7.42–7.60 (m, 3H), 8.01 (d, 2H, J = 8.0 Hz), 9.71 (brs, 1H). $^{13}C$ NMR (125.8 MHz, $CDCl_3$) δ 19.6, 19.7, 63.1, 63.2, 74.5, 78.9, 79.9, 90.7, 96.6, 116.6, 118.2, 124.6, 125.8, 127.6, 128.4, 128.5, 130.0, 133.4, 133.9, 135.4, 154.5, 160.5, 164.9. $^{31}P$ NMR (162 MHz, $CDCl_3$): $\delta - 3.80$. HRMS (ESI-TOF) calculated for $C_{27}H_{24}O_8N_5PSNa$ $[M+Na]^+$ 632.0981; observed 632.0955.

### 3.5 3-α-L-Threofuranosyl-1,3-diaza-2-oxophenothiazine 3′-monophosphate (7)

1. Add 185 mg compound **6** (0.30 mmol) in a sealed 50 mL tube.
2. Place a magnetic stir bar into the tube.
3. Add 5 mL ammonium hydroxide and close with a stopper.
4. Stir the reaction mixture at 37 °C for 16 h.
5. Cool the mixture to room temperature, dilute with 15 mL water, and wash with $CH_2Cl_2$ (3 × 40 mL).
6. Freeze-dry the aqueous solution to afford monophosphate 7 (119 mg) in 91% yield as a yellow solid. $^1$H NMR (400 MHz, $D_2O$): δ 4.31–4.35 (dd, J = 10.0, 4.0 Hz, 1H), 4.42–4.46 (m, 2H), 4.55–4.58 (m, 1H), 5.74 (s, 1H), 6.83 (d, J = 6.0 Hz, 1H), 7.00–7.10 (m, 3H), 7.56 (s, 1H). $^{31}$P NMR (162 MHz, $D_2O$): *δ* 2.55. HRMS (ESI-TOF) calculated for $C_{14}H_{13}O_7N_3PS$ $[M-H]^-$ 398.0212; observed 398.0219 (*see* **Note 3**).

### 3.6 3′-phosphor-(2-methyl)imidazolide (8)

1. Add 118 mg monophosphate 7 (0.27 mmol), 200 mg 2-methylimidazole (2.44 mmol), 140 mg triphenylphosphine (0.53 mmol), and 160 mg 2,2′-dipyridyl disulfide (0.73 mmol) in an oven-dried 10 mL round-bottom flask.
2. Place a magnetic stir bar into the reaction flask.
3. Introduce an argon atmosphere into the flask, and close with a rubber stopper.
4. Add 2 mL anhydrous DMSO, 0.1 mL triethylamine and 2 mL anhydrous DMF.
5. Stir the reaction mixture at room temperature for 6 h.
6. Prepare a solution of acetone (60 mL), diethyl ether (80 mL), triethylamine (6 mL), and saturated $NaClO_4$ in acetone (0.4 mL) (*see* **Note 4**).
7. Slowly add the reaction mixture to the stirring $NaClO_4$ solution mentioned above.
8. Centrifuge the mixture at 4400 rpm (3000 × *g*) for 15 min, discard the supernatant, and wash the pellet with acetone/diethyl ether (v/v, 1:1) three times.
9. Dry the pellet over vacuum to afford product **8** in 83% yield as sodium salt. $^{31}$P NMR (162 MHz, DMSO-$d_6$): *δ* – 10.04. HRMS (ESI-TOF) calculated for $C_{18}H_{18}O_6N_5PSNa$ $[M+Na]^+$ 486.0613; observed 486.0602.

### 3.7 3-α-L-Threofuranosyl-1,3-diaza-2-oxophenothiazine 3′-triphosphate (9)

1. Add 49 mg compound **8** (0.10 mmol) and 110 mg tributylammonium pyrophosphate (0.20 mmol) in an oven-dried 25 mL round-bottom flask.
2. Place a magnetic stir bar into the reaction flask.
3. Introduce an argon atmosphere into the flask, and close with a rubber stopper.

4. Add 2 mL anhydrous DMF.
5. Add 47 μL tributylamine (0.20 mmol).
6. Stir the reaction mixture at room temperature for 6 h.
7. Add 2 mL saturated $NaClO_4$ in acetone into 10 mL acetone.
8. Add the reaction mixture dropwise to the stirring $NaClO_4$ solution mentioned above.
9. Centrifuge the mixture at 4400 rpm (3000 × *g*) for 15 min, discard the supernatant, and wash the pellet with 10 mL of acetone three times.
10. Dry the solid under vacuum.
11. Redissolve the solid in 5 mL 0.1 M TEAA buffer.
12. Perform HPLC purification with a C18 reverse phase 250 × 21.2 mm HPLC column using a gradient of 15% acetonitrile in 0.1 M TEAA buffer (pH 7.0) (*see* **Note 5**).
13. Collect the fractions containing triphosphate (peak II), and concentrate to ~10 mL.
14. Adjust the pH to ~8 by triethylamine, and lyophilize to afford the product as a yellow solid (*see* **Note 6**).
15. Suspend the solid in 2 mL methanol, and slowly add to a solution containing 40 mL of acetone and 2 mL of saturated $NaClO_4$ in acetone.
16. Centrifuge the mixture at 4400 rpm (3000 × *g*) for 15 min, discard the supernatant, and wash the pellet with 10 mL of acetone three times.
17. Dry the yellow solid under vacuum at room temperature. $^1H$ NMR (400 MHz, $D_2O$): δ 4.33–4.37 (m, 1H), 4.47–4.51 (m, 2H), 4.81–4.83 (m, 1H), 5.74 (s, 1H), 6.84 (d, J = 8.0 Hz, 1H), 6.98 (t, J = 8.0 Hz, 1H), 7.05–7.11 (m, 2H), 7.55 (s, 1H). $^{31}P$ NMR (162 MHz, $D_2O$): $\delta$ – 8.49 (d, J = 20.0 Hz), −12.50 (d, J = 20.0 Hz), −22.40 (t, J = 20.0 Hz). HRMS (ESI-TOF) calculated for $C_{14}H_{14}O_{13}N_3P_3SK$ $[M-2H+K]^-$ 595.9097; observed 595.9072.
18. Dissolve the solid in RNase-free water containing 10 mM of Tris (pH 8.0), and use it as a stock solution for TNA synthesis (*see* **Note** 7).

## 4 Notes

1. The yellow pellet was washed with water and acetone and can be recovered as clean nucleobase **2**.
2. The residue should be purified by flash column chromatography; otherwise degradation of **4** will occur.

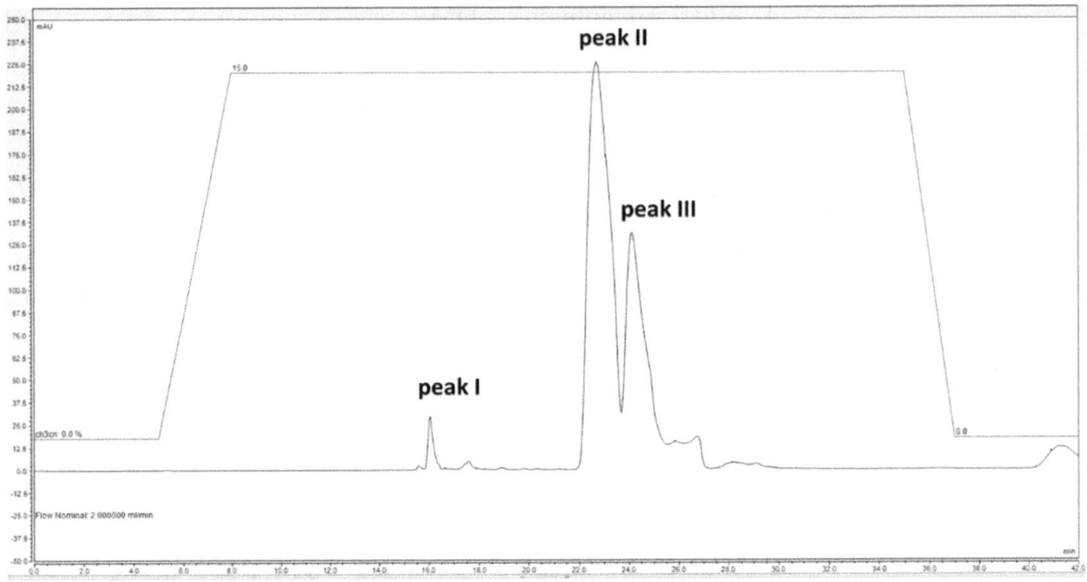


**Fig. 3** 3′-Triphosphate tC$^f$TP **9** was purified on a preparative C18 reverse-phase HPLC column (250 × 21.2 mm) using a gradient of 15% acetonitrile in 0.1 M TEAA buffer (pH 7.0)

3. Monophosphate 7 should be used as ammonium salt directly for next step. If precipitated as sodium salt, imidazolization reaction will not work.
4. Triethylamine is necessary for the precipitation of imidazolide monophosphate **8**; otherwise it cannot react with pyrophosphate to afford 3′-triphosphate (tC$^f$TP).
5. HPLC purification of tC$^f$TP **9** must be careful as 3-′-monophosphate and 3′-triphosphate have similar mobility. Only the main peak (peak II) should be collected as peak III is the 3′-monophosphate (*see* Fig. 3).
6. During the concentrating step, we observed the degradation of tC$^f$TP due to the pH of the solution becoming acidic from a buildup of acetic acid from the ammonium acetate running buffer. This problem is overcomed by monitoring the pH of the concentrated solution and adjusting the pH to ~8 as needed with triethylamine.
7. Triphosphate should be stored in Tris buffer (pH 8.0) for stability reason.

## Acknowledgment

This work was supported by the DARPA Folded Non-Natural Polymers with Biological Function Fold F(x) Program under award number N66001-16-2-4061 and by grants from the National Science Foundation 1615804 and 1607111.

## References


1. Schöning KU, Scholz P, Guntha S, Wu X, Krishnamurthy R, Eschenmoser A (2000) Chemical etiology of nucleic acid structure: the alpha-threofuranosyl-(3′-->2′) oligonucleotide system. Science 290:1347–1351
2. Schöning KU, Scholz P, Guntha S, Delgado G, Krishnamurthy R, Eschenmoser A (2002) The a-L-Threofurnaosyl-(3′-2′)-oligonucleotide system ('TNA'): synthesis and pairing properties. Helv Chim Acta 85:4111–4153
3. Anosova I, Kowal EA, Sisco NJ, Sau SP, Liao JY, Bala S, Rozners E, Egli M, Chaput JC, Van Horn WD (2016) Structural insights into conformational differences between DNA/TNA and RNA/TNA chimeric duplexes. Chembiochem 17:1705–1708
4. Yu H, Zhang S, Chaput JC (2012) Darwinian evolution of an alternative genetic system provides support for TNA as an RNA progenitor. Nat Chem 4:183–187
5. Culbertson MC, Temburnikar KW, Sau SP, Liao JY, Bala S, Chaput JC (2016) Evaluating TNA stability under simulated physiological conditions. Bioorg Med Chem Lett 26:2418–2421
6. Dunn MR, Jimenez RM, Chaput JC (2017) Analysis of aptamer discovery and technology. Nat Rev Chem 1:0076
7. Dunn MR, Otto C, Fenton KE, Chaput JC (2016) Improving polymerase activity with unnatural substrates by sampling mutations in homologous protein architectures. ACS Chem Biol 11:1210–1219
8. Dunn MR, Larsen AC, Zahurancik WJ, Fahmi NE, Meyers M, Suo Z, Chaput JC (2015) DNA polymerase-mediated synthesis of unbiased threose nucleic acid (TNA) polymers requires 7-deazaguanine to suppress G:G mispairing during TNA transcription. J Am Chem Soc 137:4014–4017
9. Lin KY, Jones RJ, Matteucci M (1995) Tricyclic 2-deoxycytidine analogs: syntheses and incorporation into oligodeoxynucleotides which have enhanced binding to complementary RNA. J Am Chem Soc 117:3873–3874
10. Engman KC, Sandin P, Osborne S, Brown T, Billeter M, Lincoln P, Norden B, Albinsson B, Wilhelmsson LM (2004) DNA adopts normal B-form upon incorporation of highly fluorescent DNA base analogue tC: NMR structure and UV-vis spectroscopy characterization. Nucleic Acids Res 32:5087–5095
11. Mei H, Shi C, Jimenez RM, Wang Y, Kardouh M, Chaput JC (2017) Synthesis and polymerase activity of a fluorescent cytidine TNA triphosphate analogue. Nucleic Acids Res 45:5629–5638
12. Sau SP, Fahmi NE, Liao JY, Bala S, Chaput JC (2016) A scalable synthesis of α-L-threose nucleic acid monomers. J Org Chem 81:2302–2307
13. Bala S, Liao J, Mei H, Chaput JC (2017) Synthesis of α-L-threofuranosyl nucleoside 3-′-monophosphates, 3′-phosphoro(2-methyl) imidazolides, and 3′-triphosphates. J Org Chem 82:5910–5916
14. Sandin P, Lincoln P, Brown T, Wilhelmsson LM (2007) Synthesis and oligonucleotide incorporation of fluorescent cytosine analogue tC: a promising nucleic acid probe. Nat Protoc 2:615–623

# Chapter 4

# Synthesis of Base-Modified dNTPs Through Cross-Coupling Reactions and Their Polymerase Incorporation to DNA

**Petra Ménová, Hana Cahová, Milan Vrábel, and Michal Hocek**

## Abstract

Synthesis of base-modified dNTPs through the Suzuki or Sonogashira cross-coupling reactions of halogenated dNTPs with boronic acids or alkynes is reported, as well as the use of these modified dNTPs in polymerase incorporations to oligonucleotides or DNA by primer extension or PCR.

**Key words** Nucleotides, Nucleoside triphosphates, DNA polymerases, Enzymatic synthesis, Modified DNA

## 1 Introduction

Base-modified nucleic acids find a number of promising applications in diagnostics, chemical biology, and even material sciences. Apart from the classical chemical synthesis on solid support by phosphoramidite method, an alternative approach based on polymerase-catalyzed templated synthesis from modified 2′-deoxynucleoside triphosphates (dNTPs) attracts growing attention [1]. 5-Substituted pyrimidine and 7-substituted 7-deazapurine dNTPs are usually good substrates for at least some DNA polymerases [2–4]. Primer extension (PEX) using a template and primer in stoichiometric amounts is a convenient method for the synthesis of shorter DNA duplexes bearing several modifications in one strand. Complementary to PEX, the use of modified dNTPs in polymerase chain reaction (PCR) allows the synthesis of large segments of DNA with high number of modifications. We have successfully prepared diverse redox-labelled DNA probes for electrochemical detection [5–8], some modified DNA for the protection against cleavage by restriction enzymes [9–11], as well as reactive aldehyde-modified DNA for bioconjugations [12].

Nathaniel Shank (ed.), *Non-Natural Nucleic Acids: Methods and Protocols*, Methods in Molecular Biology, vol. 1973, https://doi.org/10.1007/978-1-4939-9216-4_4, © Springer Science+Business Media, LLC, part of Springer Nature 2019

The starting modified dNTPs are usually prepared by triphosphorylation of the corresponding nucleosides [13]. In our laboratory, we use an efficient direct one-step synthesis of base-modified dNTPs through the Suzuki or Sonogashira cross-coupling reactions of halogenated dNTPs with modified arylboronic acids or with terminal alkynes. The reactions are performed in water/acetonitrile mixtures at elevated temperatures (60–120 °C) for limited time (max. 30–60 min) to avoid extensive hydrolysis of the triphosphates. Here follow two representative examples of experimental procedures for cross-coupling reactions of dNTPs and procedures for PEX and PCR synthesis of modified DNA.

## 2 Materials

### 2.1 The Suzuki-Miyaura Cross-Coupling Reaction Components

#### 2.1.1 Chemicals and Reagents

1. 9-(2′-Deoxy-β-D-erythro-pentofuranosyl)-7-iodo-7-deazaadenine 5′-*O*-triphosphate (**1**) (prepared according to Čapek et al. [14], also available from Jena Bioscience).
2. 3-Aminophenylboronic acid hydrochloride (Sigma-Aldrich).
3. Cesium(I) carbonate ($Cs_2CO_3$, Sigma-Aldrich).
4. Palladium(II) acetate ($Pd(OAc)_2$, Fluka).
5. Tris(3-sulfonatophenyl)phosphine hydrate, sodium salt (TPPTS, Strem Chemicals).
6. Argon.
7. Acetonitrile (Sigma-Aldrich).
8. Water (HPLC grade).
9. Methanol (MeOH, HPLC grade).
10. Triethylammonium bicarbonate (TEAB, pH 7.5 at 10–12 °C, 2 M, stored at 4 °C).

#### 2.1.2 Equipments

1. Reaction flasks (2 × 10 mL, 1 × 50 mL).
2. Magnetic stir bars.
3. Rubber septa (PTFE/silicone).
4. 21-G needles.
5. Argon/vacuum line (oil pump with manifold and cold trap).
6. Single-use syringes 1 mL.
7. Ultrasonic bath.
8. Oil bath.
9. Magnetic stirrer with heating MR Hei-Standard (Heidolph).
10. 17 mm nylon syringe filter 0.2 μm (National Scientific).
11. Waters 600 HPLC system.
12. Column packed with 10 μm C18 reversed phase (Phenomenex, Luna 10 μm C18, 100A HPLC Column 250 × 21.2 mm).

13. Nylon-66 membranes 0.2 μm × 47 mm (Supelco).
14. Solvent filtration apparatus (Supelco).
15. Rotary evaporator (Heidolph) equipped with a vacuum system (membrane pump, Vacuubrand).
16. Dowex 50WX8 in $Na^+$ cycle (Fluka).
17. Glass column (125 × 15 mm).
18. Freeze-dryer (Labconco).

#### 2.1.3 Solutions

1. Prepare a water/acetonitrile (2:1) mixture.
2. Prepare mobile phase mixtures:
   Mobile phase solution A: 0.1 M TEAB.

   Mobile phase solution B: 0.1 M TEAB in 50% MeOH.

   Mobile phase solution C: MeOH.

   Filter the mixtures on a solvent filtration apparatus using nylon-66 membranes (*see* **Note 1**).

### 2.2 The Sonogashira Cross-Coupling Reaction Components

#### 2.2.1 Chemicals and Reagents

1. 1-(2′-Deoxy-β-D-erythro-pentofuranosyl)-5-iodothymidine 5′-*O*-triphosphate (**3**) (prepared according to Cahová et al. [4], also available from TriLink BioTechnologies).
2. Ethynylferrocene (Chemos CZ, Czech Republic).
3. Palladium(II) acetate (Fluka).
4. Tris(3-sulfonatophenyl)phosphine hydrate, sodium salt (TPPTS, Strem Chemicals).
5. Copper(I) iodide, 99.999% (CuI, Sigma-Aldrich).
6. Triethylamine (TEA; Fluka).
7. Argon.
8. Acetonitrile (Sigma-Aldrich).
9. Water (HPLC grade).
10. Methanol (MeOH, HPLC grade).
11. Triethylammonium bicarbonate (TEAB, pH 7.5 at 10–12 °C, 2 M, stored at 4 °C).

#### 2.2.2 Equipments

1. Reaction flasks (2 × 10 mL, 1 × 50 mL).
2. Magnetic stir bars.
3. Rubber septa (PTFE/silicone).
4. 21-G needles.
5. Argon/vacuum line (oil pump with manifold and cold trap).
6. Single-use syringes 1 mL.
7. Ultrasonic bath.
8. Oil bath.

9. Magnetic stirrer with heating MR Hei-Standard (Heidolph).
10. 17 mm nylon syringe filter 0.2 μm (National Scientific).
11. Waters 600 HPLC system.
12. Column packed with 10 μm C18 reversed phase (Phenomenex, Luna 10 μm C18, 100A HPLC Column 250 × 21.2 mm).
13. Nylon 66 membranes 0.2 μm × 47 mm (Supelco).
14. Glass filter system (Whatman).
15. Rotary evaporator (Heidolph) equipped with a vacuum system (membrane pump Vacuubrand).
16. Freeze-dryer (Labconco).

#### *2.2.3 Solutions*

1. Prepare a water/acetonitrile (2:1) mixture.
2. Prepare mobile phase mixtures:
   Mobile phase solution A: 0.1 M TEAB.
   Mobile phase solution B: 0.1 M TEAB in 50% MeOH.
   Mobile phase solution C: MeOH.
   Filter the mixtures on a solvent filtration apparatus using nylon-66 membranes (*see* **Note 1**).

### 2.3 Primer Extension Experiment with Modified dNTPs

#### *2.3.1 Chemicals and Reagents*

1. Template: 5′-CTAGCATGAGCTCAGTCCCATGCCGCCC ATG-3′ (Sigma-Aldrich).
2. Primer 5′-CATGGGCGGCATGGG-3′ (Sigma-Aldrich).
3. Adenosine 5′-[$\gamma$-$^{32}$P] triphosphate, tetrakis(triethylammonium) salt (M.G.P., Czech Republic).
4. T4 polynucleotide kinase 10 U/μL (New England Biolabs).
5. T4 polynucleotide kinase reaction buffer (10×) (New England Biolabs).
6. PCR Ultra $H_2O$ (Top-Bio, Czech Republic).
7. Vent(exo-) DNA polymerase 2 U/μL (New England Biolabs).
8. ThermoPol reaction buffer (10×) (New England Biolabs).
9. Natural deoxynucleotide triphosphates: dATP, dCTP, dGTP, dTTP (Fermentas).
10. Bromophenol blue sodium salt, for molecular biology (Sigma-Aldrich).
11. Xylene cyanol FF, for molecular biology (Sigma-Aldrich).
12. Formamide (Sigma-Aldrich).
13. Ethylenediaminetetraacetic acid disodium salt dihydrate (EDTA, Sigma-Aldrich).
14. Tris ultrapure (Tris(hydroxymethyl)aminomethane, Duchefa Biochemie).
15. Boric acid (Duchefa Biochemie).

16. Ammonium persulfate, for electrophoresis (APS, Sigma-Aldrich).
17. *N*,*N*,*N′*,*N′*-Tetramethylethylenediamine, for electrophoresis (TEMED, Sigma-Aldrich).
18. Rotiporese sequencing gel concentrate (Carl Roth).
19. $H_2O$ (HPLC grade and deionized grade).

#### 2.3.2 Equipments

1. Microcentrifuge tubes, 1.5 mL (Axygen).
2. MicroSpin G-25 columns (GE Healthcare).
3. Pipettes of various volumes (Eppendorf).
4. Tips for pipettes (Eppendorf).
5. Erlenmeyer flask 100 mL.
6. Magnetic stir bar.
7. Sekuroka radiation protection screens (Carl Roth).
8. Sekuroka radiation protection boxes (Carl Roth).
9. Sekuroka radiation protection waste boxes (Carl Roth).
10. Sequi-Gen electrophoresis apparatus, 21 × 40 cm (Bio-Rad).
11. PowerPac HV electrophoresis power supply (Bio-Rad).
12. Chromatography paper 3 mm (Whatman).
13. Gel cutter (UVP).
14. Plastic wrap (e.g., Saran Wrap or cling film).
15. Storage phosphor screen and cassette (Amersham Biosciences).
16. Gel Dryer Model 583 (Bio-Rad).
17. Typhoon 9410 Gel Imager (Amersham Biosciences).
18. Thermomixer comfort 1.5 mL (Eppendorf).
19. Mini Spin centrifuge (Eppendorf).
20. Vortex Reax control (Heidolph).

#### 2.3.3 Solutions

1. Prepare PAGE stop solution by dissolving 0.37 g disodium EDTA, 0.10 g xylene cyanol, 0.10 g bromophenol blue, and 80 mL formamide in approximately 10 mL deionized water. Adjust the solution to a final volume of 100 mL. Store up to 1 year at 4 °C.
2. Prepare a stock solution of EDTA (0.5 M, pH 8.0) by dissolving 93.1 g EDTA disodium salt in 400 mL deionized water, and adjust the pH with NaOH to 8.0. Top up the solution to a final volume of 500 mL.
3. Make a concentrated (5×) stock solution of TBE by weighing 54.0 g Tris base and 27.5 g boric acid and dissolving both in approximately 900 mL deionized water. Add 20 mL of 0.5 M EDTA (pH 8.0), and adjust the solution to a final volume of

1 L. Store the buffer in glass bottles at room temperature. If a precipitate forms (usually in older solutions), discard and prepare a fresh stock solution.

4. Prepare 2 × TBE buffer (for the preparation of polyacrylamide gel) by mixing 200 mL 5 × TBE and 300 mL deionized $H_2O$. Store up to 1 month at 4 °C.
5. Prepare 1 × TBE buffer (for running the polyacrylamide gel) by mixing 100 mL 5 × TBE and 400 mL deionized $H_2O$. Store up to 1 month at 4 °C.
6. Prepare 4 mM solutions of dNTPs (use PCR Ultra $H_2O$):
   4 mM mixture of dATP, dCTP, dGTP, and dTTP.
   
   4 mM mixture of dCTP, dGTP, and dTTP.
   
   4 mM mixture of dATP, dCTP, and dGTP.
   
   4 mM solution of **2**.
   
   4 mM solution of **4**.
   
   Store at −20 °C (*see* **Note 2**).
7. Prepare 10% APS by dissolving 1.0 g APS in approximately 8 mL deionized water. Adjust the solution to a final volume of 10 mL. Store up to 2 weeks at 4 °C.

### 2.4 Polymerase Chain Reaction with Modified dNTPs

#### 2.4.1 Chemicals and Reagents

1. Template HIV1-Pr, 63 ng/μL (*see* Table 1 for sequence).
2. Primers: Prim S1-HIV1 and Prim S2-HIV1 (Generi Biotech; *see* Table 1 for sequences).
3. PCR Ultra $H_2O$ (Top-Bio).
4. KOD XL DNA polymerase 2.5 U/μL (Merck).
5. PCR buffer for KOD XL DNA polymerase (10×) (Merck).

**Table 1**
**Sequences of the template and primers employed in PCR**

| | Sequence |
|---|---|
| HIV1-Pr | 5′-CCTCAGATCACTCTTTGGCAGCGACCCCTCGTCACAATAAAGATAGGGG GGCAATTAAAGGAAGCTCTATTAGATACAGGAGCAGATGATACAGTATTAGA AGAAATGAATTTGCCAGGAAGATGGAAACCAAAAATGATAGGGGGAATTGG AGGTTTTATCAAAGTAAGACAGTATGATCAGATACTCATAGAAATCTGCGGA CATAAAGCTATAGGTACAGTATTAGTAGGACCTACACCTGTCAACATAATTG GAAAAATCTGTTGACTCAGATTGGCTGCACTTTAAATTTT-3′ |
| Prim S1-HIV1 | 5′-GATCACTCTTTGGCAGCGACCCCTCGTCAC-3′ |
| Prim S2-HIV1 | 5′-TTAAAGTGCAGCCAATCTGAGTCAACAGAT-3′ |

6. Natural deoxynucleotide triphosphates: dATP, dCTP, dGTP, and dTTP (Fermentas).
7. Ethylenediaminetetraacetic acid disodium salt dihydrate (EDTA, Sigma-Aldrich).
8. Boric acid (Duchefa, Netherland).
9. Tris ultrapure (Tris(hydroxymethyl)aminomethane, Duchefa Biochemie).
10. Agarose for DNA electrophoresis (Serva).
11. GelRed nucleic acid stain (10,000×) (Biotium).
12. Gel loading dye blue (6×) (New England Biolabs).
13. 100 bp DNA ladder (New England Biolabs).

#### 2.4.2 Equipments

1. Microcentrifuge tubes 1.5 mL (Axygen).
2. Microcentrifuge tubes 0.5 mL (Axygen).
3. Erlenmeyer flask 250 mL.
4. Microwave oven ZMC 19 M (Zanussi).
5. $H_2O$ (HPLC grade).
6. Vortex Reax control (Heidolph).
7. Mini Spin centrifuge (Eppendorf).
8. Microcentrifuge (Roth).
9. TGradient 96 thermocycler (Whatman Biometra).
10. PowerPac HV (Bio-Rad).
11. Owl B1 type electrophoresis (Owl Separation System, Inc.).
12. UltraCam 8gD Digital Imaging System (Ultra-Lum).
13. Electronic dual wave transilluminator (Ultra-Lum).

#### 2.4.3 Solutions

1. Prepare 0.5 × TBE buffer by mixing 50 mL 5 × TBE and 450 mL deionized $H_2O$. Store up to 1 month at 4 °C.
2. Prepare 4 mM solutions of dNTPs (use PCR Ultra $H_2O$):
   4 mM mixture of dATP, dCTP, dGTP, and dTTP.

   4 mM mixture of dCTP, dGTP, and dTTP.

   4 mM solution of **2**.

   Store at −20 °C (*see* **Note 2**).

## 3 Methods

### *3.1 The Suzuki-Miyaura Cross-coupling Reaction of 9-(2′-Deoxy-β-D-erythro-pentofuranosyl)-7-iodo-7-deazaadenine 5′-O-triphosphate (1) with 3-Aminophenylboronic Acid*

1. Add 35 mg (0.05 mmol) **1**, 17 mg (0.10 mmol, 2 equiv.) aminophenylboronic acid hydrochloride, and 81 mg (0.25 mmol, 5 equiv.) $Cs_2CO_3$ into one 10 mL flask.
2. Add 1.1 mg (0.005 mmol, 10 mol%) $Pd(OAc)_2$ and 14 mg (0.025 mmol, 5 equiv. to Pd) TPPTS to another 10 mL flask (*see* **Note 3**).
3. Place a magnetic stir bar in each flask and seal with a PTFE/silicone septum. Situate the flasks on a magnetic stirrer, and connect to an argon/vacuum line via a 21-G needle inserted through the septum.
4. Evacuate each flask and then slowly refill with argon. Repeat the evacuation and argon back-filling two more times (*see* **Note 4**).
5. Add, through septum, 0.5 mL of water/acetonitrile (2:1) mixture to the flask prepared in **step 1**. Evacuate the flask and refill with argon three times.
6. Add, through septum, 0.3 mL of water/acetonitrile (2:1) mixture to the flask prepared in **step 2**. Evacuate the flask and refill with argon three times. Place the flask in an ultrasonic bath for 1 min for faster dissolution of the catalytic system.
7. When the catalytic system is dissolved, transfer it by a syringe to the first flask. Evacuate the flask and refill with argon three times.
8. Place the flask in an oil bath on a magnetic stirrer and heat to 120 °C. Stir the reaction mixture for 30 min at 120 °C (*see* **Note 5**).
9. Cool the reaction mixture to ambient temperature.
10. Program the HPLC system to start with 90% mobile phase A and 10% mobile phase B. Use a linear gradient to go to 100% mobile phase B in 60 min and then to 100% mobile phase C in 80 min (*see* **Note 6**).
11. Set the detection wavelength to 254 nm (*see* **Note 7**).
12. Equilibrate the HPLC system with the starting mobile phase composition for at least 20 min or until a flat baseline is achieved at the desired detection wavelength.
13. Filter the reaction mixture through a nylon syringe filter.
14. Inject the filtered sample onto the column.
15. Collect the desired fractions either with an automated fraction collector or by observing the chromatogram in real time and manually collecting the eluate.

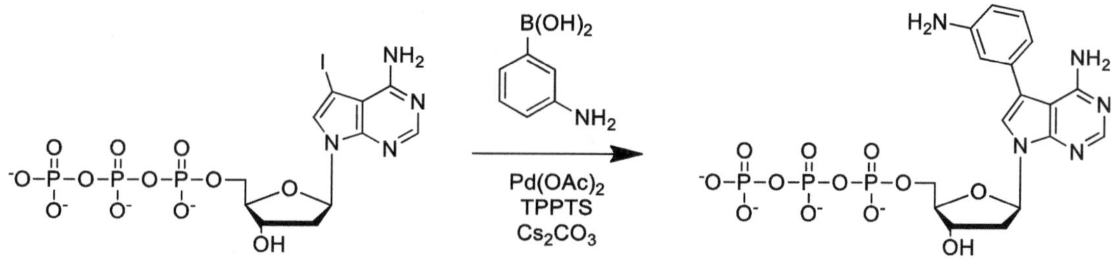


**Fig. 1** Synthesis of 7-(3-aminophenyl)-2′-deoxy-7-deazaadenosine 5′-*O*-triphosphate (**2**)

16. Concentrate the collected fraction on a rotatory evaporator at 40 °C (*see* **Note 8**).
17. Add 3 mL of pure water (HPLC grade) and evaporate to dryness. Repeat this co-evaporation with water until the odor of triethylammonium disappears (at least five times) (*see* **Note 9**).
18. Pour Dowex 50WX8 in $Na^+$ cycle into a glass column, and apply the sample in 5 mL of water. Elute with 25 mL of water.
19. Freeze-dry to obtain a white solid product (*see* **Note 10**).
20. Confirm the structure of the product by NMR, MS (measured in $ESI^-$ mode) and HRMS.
21. *7-(3-Aminophenyl)-2′-deoxy-7-deazaadenosine 5′-O-triphosphate* [4] (**2**) (Fig. 1).
22. Yield 51%. NMR spectra for tetrasodium salt (measured in 50 mM phosphate buffer, pD 7.1): $^1$H NMR (500 MHz, $D_2O$, ref$_{dioxane}$ = 3.75 ppm, pD = 7.1) – 2.45 (ddd, 1H, $J_{gem}$ = 14.2 Hz, $J_{2'b,1'}$ = 6.3 Hz, $J_{2'b,3'}$ = 3.4 Hz, H-2′b); 2.71 (ddd, 1H, $J_{gem}$ = 14.2 Hz, $J_{2'a,1'}$ = 7.8 Hz, $J_{2'a,3'}$ = 6.4 Hz, H-2′a); 4.09–4.20 (m, 2H, H-5′); 4.24 (m, 1H, H-4′); 4.75 (bm, 1H, H-3′); 6.65 (dd, 1H, $J_{1'2'}$ = 7.8, 6.30 Hz H-1′); 6.87 (ddd, 1H, $J_{4,5}$ = 8.0 Hz, $J_{4,2}$ = 2.2 Hz, $J_{4,6}$ = 0.8 Hz, H-4-$C_6H_4NH_2$); 6.94 (bd, 1H, $J_{6,5}$ = 7.5 Hz, H-6-$C_6H_4NH_2$); 6.95 (bs, 1H, H-2-$C_6H_4NH_2$); 7.32 (dd, 1H, $J_{5,4}$ = 8.0 Hz, $J_{5,6}$ = 7.5 Hz, H-5-$C_6H_4NH_2$); 7.50 (s, 1H, H-8); 8.15 (bs, 1H, H-2). $^{13}$C NMR (125.7 MHz, $D_2O$, ref$_{dioxane}$ = 69.3 ppm, pD = 7.1): 46.62 ($CH_2$-2′); 68.19 (d, $J_{C,P}$ = 6.0 Hz, $CH_2$-5′); 73.50 (CH-3′); 86.57 (CH-1′); 88.24 (d, $J_{C,P}$ = 9.0 Hz, CH-4′); 101.87 (C-5); 121.20 (C-7); 124.11 (CH-4-$C_6H_4NH_2$); 124.78 (CH-2-$C_6H_4NH_2$); 126.21 (CH-8); 129.87 (CH-6-$C_6H_4NH_2$); 133.58 (CH-5-$C_6H_4NH_2$); 136.19 (C-1-$C_6H_4NH_2$); 136.39 (C-3-$C_6H_4NH_2$); 145.92 (CH-2); 150.10 (C-4); 153.65 (C-6). $^{31}$P {$^1$H} NMR (202.3 MHz, $D_2O$, ref$_{H3PO4}$ = 0 ppm, pD = 7.1): −23.07 (bdd, $J$ = 19.1, 16.7 Hz, $P_\beta$); −12.04

(d, $J = 19.1$ Hz, $P_\alpha$); −9.15 (d, $J = 16.7$ Hz, $P_\gamma$). MS($ESI^-$): $m/z$ – 580 [M − 1], 500 [M − $PO_3H_2$ − 1]. HRMS: for $C_{17}H_{21}N_5O_{12}P_3$ calculated 580.0400, found 580.0391.

### *3.2 The Sonogashira Cross-Coupling Reaction of 1-(2′-Deoxy-β-D-erythro-pentofuranosyl)-5-iodothymidine 5′-O-triphosphate (3) with Ethynylferrocene*

1. Add 33 mg (0.05 mmol) **3**, 21 mg (0.10 mmol, 2 equiv.) ethynylferrocene, and 2.9 mg (0.015 mmol, 30 mol%) CuI into one 10 mL flask (*see* **Note 11**).
2. Add 1.1 mg (0.005 mmol, 10 mol%) $Pd(OAc)_2$ and 14 mg (0.025 mmol, 5 equiv. to Pd) TPPTS to another 10 mL flask (*see* **Note 3**).
3. Place a magnetic stir bar in each flask and seal with a PTFE/silicone septum. Situate the flasks on a magnetic stirrer and connect to an argon/vacuum line via a 21-G needle inserted through the septum.
4. Evacuate each vial and then slowly refill with argon. Repeat the evacuation and argon back-filling two more times (*see* **Note 4**).
5. Add, through septum, 1.0 mL of water/acetonitrile (2:1) mixture and 56 μL (0.40 mmol, 8 equiv.) TEA to the flask prepared in **step 1**. Evacuate the flask and refill with argon three times.
6. Add, through septum, 0.5 mL of water/acetonitrile (2:1) mixture to the flask prepared in **step 2**. Evacuate the flask and refill with argon three times. Place the flask in an ultrasonic bath for 1 min for faster dissolution of the catalytic system.
7. When the catalytic system is dissolved, transfer it by a syringe to the first flask. Evacuate the flask and refill with argon three times.
8. Place the flask in an oil bath on the magnetic stirrer and heat to 65 °C. Stir the reaction mixture for 60 min at 65 °C (*see* **Note 5**).
9. Cool the reaction mixture to ambient temperature.
10. Program the HPLC system to start with 90% mobile phase A and 10% mobile phase B. Use a linear gradient to go to 100% mobile phase B in 60 min and then to 100% mobile phase C in 80 min (*see* **Note 6**).
11. Set the detection wavelength to 254 nm (*see* **Note** 7).
12. Equilibrate the HPLC system with the starting mobile phase composition for at least 20 min or until a flat baseline is achieved at the desired detection wavelength.
13. Filter the reaction mixture through a nylon syringe filter.
14. Inject the filtered sample onto the column.
15. Collect the desired fractions either with an automated fraction collector or by observing the chromatogram in real time and manually collecting the eluate.

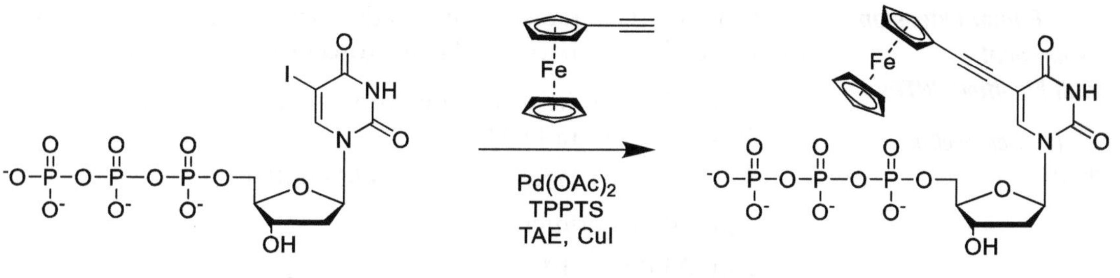


**Fig. 2** Synthesis of 5-(ferrocene-1-yl-ethynyl)-2′-deoxyuridine 5′-*O*-triphosphate (**4**)

16. Concentrate the collected fraction on a rotatory evaporator at 40 °C (*see* **Note 8**).
17. Add 3 mL of pure water (HPLC grade) and evaporate to dryness. Repeat the co-evaporation with water until the odor of triethylammonium disappears (at least five times) (*see* **Note 12**).
18. Freeze-dry to obtain an orange solid product (*see* **Note 10**).
19. Confirm the structure of the product by NMR, MS (measured in ESI$^-$ mode) and UV/vis.
20. *5-(Ferrocene-1-yl-ethynyl)-2′-deoxyuridine 5′-O-triphosphate* [3] (**4**) (Fig. 2).
21. Yield 18%. NMR spectra for tris(triethylammonium) salt (measured in 50 mM phosphate buffer, pD 7.1): $^1$H NMR (500 MHz, $D_2O$, ref$_{dioxane}$ = 3.75 ppm, pD = 7.1) – 2.36 (bdt, 1H, $J_{gem}$ = 13.8 Hz, $J_{2'b,1'}$ = $J_{2'b,3'}$ = 6.9 Hz, H-2′b); 2.65 (bdt, 1H, $J_{gem}$ = 13.8 Hz, $J_{2'a,1'}$ = $J_{2'a,3'}$ = 5.3 Hz, H-2′a); 4.23–4.44 (m, 8H, H-4′,5′ and cp); 4.62 (mbr, 1H, H-3′); 4.67 (mbr, 2H, H-2″,5″); 5.04 (mbr, 2H, H-3″,4″); 6.38 (t br, 1H, $J_{1',2'}$ = 6.9, 5.3 Hz, H-1′); 8.69 ppm (s, 1H, H-6). $^{13}$C NMR (125.7 MHz, $D_2O$, ref$_{dioxane}$ = 69.3 ppm, pD = 7.1): 43.0 ($CH_2$-2′); 67.8 ($CH_2$-5′); 69.4 (CH-3″,4″); 72.7 (CH-3′); 72.9 (CH-cp); 73.2 (CH-2″,5″); 75.1 (C=C-Fc); 88.8 (d, $J_{C,P}$ = 9.0 Hz, CH-4′); 91.0 (CH-1′); 94.3 (C=C-Fc); 113.5 (C-5); 139.0 (CH-6); 158.7 (C-2); 174.1 ppm (C-4); C-1″ not observed. $^{31}$P {$^1$H} NMR (202.3 MHz, $D_2O$, ref$_{H3PO4}$ = 0 ppm, pD = 7.1): −22.61 (t, $J$ = 19.0, 18.0 Hz, $P_\beta$); −10.85 (d, $J$ = 19.0 Hz, $P_\alpha$); −10.32 ppm (d, $J$ = 18.0 Hz, $P_\gamma$). MS (ESI$^-$): *m/z* – 675.0 [M − 1]. UV/vis ($H_2O$): 468 nm ($\varepsilon$ = 7377 $M^{-1}$ $cm^{-1}$), 340 nm ($\varepsilon$ = 58,197 $M^{-1}$ $cm^{-1}$), 282 nm ($\varepsilon$ = 83,607 $M^{-1}$ $cm^{-1}$), 253 nm ($\varepsilon$ = 86,885 $M^{-1}$ $cm^{-1}$).

### 3.3 Primer Extension Experiment with Modified dNTPs

#### 3.3.1 Radiolabelling the Primer

Work on ice. Use radiation protection screens, protection boxes, and protection waste boxes. Wear protection gloves and goggles.

1. To a 1.5 mL microcentrifuge tube on ice, add in this order:
   39 μL PCR ultra $H_2O$.

   5 μL 10 × T4 DNA polynucleotide buffer.

   2 μL 10 μM primer.

   2 μL ATP [$\gamma$-$^{32}$P].

   2 μL (20 U) T4 polynucleotide kinase.
2. Mix using vortex and then briefly spin in a MicroSpin centrifuge to collect the solution at the bottom of the tube.
3. Incubate for 1 h at 37 °C.
4. Incubate for 5 min at 95 °C to inactivate T4 polynucleotide kinase.
5. Resuspend the resin in the MicroSpin column by vortexing, loosen the cap one-quarter turn and snap off the bottom closure, place the column in a supplied collection tube, and pre-spin for 1 min at 735 × *g* at room temperature.
6. Transfer the column to a new microcentrifuge tube. Apply the sample to the center of the resin bed. Spin the column 2 min at 735 × *g*.
7. Withdraw the filtrate from the collection tube, and transfer it to a new 1.5 mL microcentrifuge tube.
8. Add 20 μL 10 μM unlabelled primer to the filtrate to obtain a 3 μM solution of a radiolabelled primer. Vortex and spin (*see* **Note 13**).

#### 3.3.2 Performing PEX with Compounds **2** and **4**

Work on ice. Use radiation protection screens, protection boxes, and protection waste boxes. Wear protection gloves and goggles.

1. Prepare a master mix by adding the following solutions in a 1.5 mL microcentrifuge tube:
   75.0 μL PCR ultra $H_2O$.

   12.0 μL 10 × ThermoPol reaction buffer.

   9.0 μL 3 μM template.

   6.0 μL 3 μM radiolabelled primer.

   6.0 μL (12 U) Vent(exo-) DNA polymerase.

   (*See* **Note 14**).
2. Mix using vortex and then briefly spin in a MicroSpin centrifuge to collect the solution at the bottom of the tube.
3. Prepare five new 1.5 mL microtubes, and label them as positive control (+), negative control for **2** (A−), experiment with **2** ($A^{NH2}$), negative control for **4** (U−), and experiment with **4** ($U^{Fc}$).

4. Transfer 18 μL of the master mix to each microtube.
5. Add 1 μL 4 mM mixture of dATP, dCTP, dGTP, and dTTP and 1 μL PCR Ultra $H_2O$ into the microtube labelled as positive control.
6. Add 1 μL 4 mM mixture of dCTP, dGTP, and dTTP and 1 μL PCR Ultra $H_2O$ into the microtube labelled as negative control for **2**.
7. Add 1 μL 4 mM mixture of dCTP, dGTP, and dTTP and 1 μL 4 mM **2** into the microtube labelled as experiment with **2**.
8. Add 1 μL 4 mM mixture of dATP, dCTP, and dGTP and 1 μL PCR Ultra $H_2O$ into the microtube labelled as negative control for **4**.
9. Add 1 μL 4 mM mixture of dCTP, dGTP, and dTTP and 1 μL 4 mM **4** into the microtube labelled as experiment with **4**.
10. Mix all the samples using vortex and then spin on MicroSpin centrifuge.
11. Incubate for 30 min at 60 °C.
12. Add 40 μL of PAGE stop solution to each microtube.
13. Mix using vortex and then briefly spin in a MicroSpin centrifuge.

#### *3.3.3 Performing Denaturing Polyacrylamide Gel Electrophoresis*

1. Mix the following PAGE gel reagents in a 100 mL Erlenmeyer flask:

   50 mL Rotiphorese sequencing gel concentrate.

   50 mL 2 × TBE buffer.

   40 μL TEMED.

   800 μL 10% APS.
2. Immediately pour the gel mixture in the Sequi-Gen electrophoresis apparatus. Insert a comb (supplied with apparatus) between the glass plates.
3. Let the gel polymerize for about 1 h.
4. Remove the comb and fill the apparatus with 1 × TBE buffer. Remove any air bubbles by gently pipetting buffer into the wells.
5. Preheat the gel to 50 °C using temperature probe (supplied with apparatus) connected to the PowerPac power supply (set up 42 mA).
6. Turn off the power supply and load 2 μL of every sample in neighboring wells. First load primer as a standard (19 μL PCR ultra $H_2O$, 1 μL 3 μM radiolabelled primer, and 40 μL PAGE stop solution). Continue with loading the positive control, the negative control, and the experiment with modified dNTP.

7. Reconnect the electrophoresis apparatus to the power supply, and run the gel at 38 mA for approximately 1 h, until the marker dyes from the PAGE stop solution have migrated the desired distance.
8. Turn off the electric power, discard the electrophoresis buffer, carefully separate the glass plates, cut the desired size of the gel using a gel cutter (DNA is in between the two blue bands used in the stop solution), attach adequate piece of wet chromatography paper, and take down the attached gel from the glass. Then, cover the gel with a plastic wrap.
9. Dry the gel in a gel dryer at 85 °C for 1 h.
10. Place the dry gel in a storage phosphor cassette with screen.
11. After 12–24 h (depending on the specific activity of ATP [$\gamma$-$^{32}$P]), scan the screen on the Typhoon Gel Imager (Fig. 3) (*see* **Note 15**).

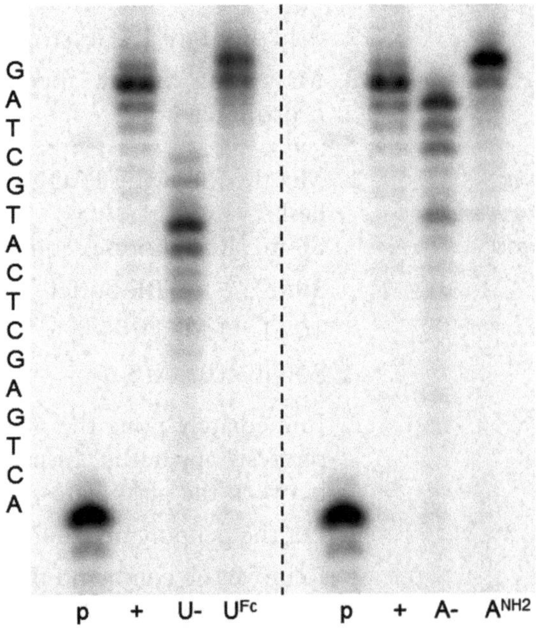


**Fig. 3** 12% Polyacrylamide gel of the PEX products. (+) = positive control, i.e., DNA containing natural nucleotides only; (U−) = negative control for U, i.e., PEX with dATP, dCTP, and dGTP only; ($U^{Fc}$) = DNA containing 5-(ferrocene-1-yl-ethynyl)-2′-deoxyuridine instead of natural deoxythymidine; (A−) = negative control for A, i.e., PEX with dCTP, dGTP, and dTTP only; ($A^{NH2}$) = DNA containing 7-(3-aminophenyl)-2′-deoxy-7-deazaadenosine instead of natural deoxyadenosine

### 3.4 Polymerase Chain Reaction (PCR): Incorporation of 2 into DNA

#### 3.4.1 Performing PCR

1. Prepare a master mix by adding the following solutions in a 1.5 mL microcentrifuge tube (work on ice):
   8 μL 10 × PCR buffer for KOD XL DNA polymerase.
   1.4 μL 63 ng/μL template HIV1-Pr.
   8 μL 20 μM Prim S1-HIV1.
   8 μL 20 μM Prim S2-HIV1.
   6 μL (15 U) KOD XL DNA polymerase.
2. Mix using vortex, and then spin on MicroSpin centrifuge to collect the solution at the bottom of the tube.
3. Prepare three new 0.5 mL microcentrifuge tubes, and label one as positive control (+), one as negative control (−), and one as experiment with **2** ($A^{NH2}$).
4. Transfer 7.85 μL of master mix to each microcentrifuge tube.
5. Add 0.50 μL 4 mM mixture of dATP, dCTP, dGTP, and dTTP and 11.65 μL PCR Ultra $H_2O$ into the microtube labelled as positive control.
6. Add 0.50 μL 4 mM mixture of dCTP, dGTP, and dTTP and 11.65 μL PCR Ultra $H_2O$ into the microtube labelled as negative control.
7. Add 0.05 μL 4 mM mixture of dCTP, dGTP, and dTTP, 2 μL 4 mM **2**, and 9.65 μL PCR Ultra $H_2O$ into the microtube labelled as experiment with **2**.
8. Mix all three samples using vortex and then briefly spin on microcentrifuge.
9. Transfer the tubes to the TGradient thermocycler preheated to 80 °C, and set the following program:

| | | | |
|---|---|---|---|
| 1 cycle | 3 min | 94 °C | (Initial denaturation) |
| 40 cycles | 1 min | 94 °C | (Denaturation) |
| | 1 min | 60 °C | (Annealing) |
| | 2 min | 75 °C | (Extension) |
| 1 cycle | 2 min | 75 °C | (Final extension) |

   (*See* **Note 16**).
10. After the PCR is complete, prepare the samples for agarose gel electrophoresis by adding 1 μL gel loading dye blue (6 ×) to 5 μL of each sample.
11. Mix by vortexing, and then briefly spin on microcentrifuge to collect the solution at the bottom of the tube.

*3.4.2 Performing Agarose Gel Electrophoresis*

1. Add 1.3 g agarose to 100 mL of 0.5 × TBE in a 250 mL Erlenmeyer flask.
2. Put the flask in a microwave oven, and heat at 700 W until the mixture starts boiling. Swirl the flask occasionally (wear protective gloves), put it back in the oven, and let it boil. Repeat this step until the solution is clear and without bubbles.
3. Cool the agarose solution a little under flowing tap water (*see* **Note 17**).
4. Add 10 μL GelRed to the cooled agarose (*see* **Note 18**).
5. Pour the gel in the electrophoresis system, and insert a comb (supplied with the electrophoresis apparatus). If there are any air bubbles in the gel, remove them with a pipette tip.
6. Wait for at least 30 min until the gel solidifies.
7. Remove the comb and then pour 0.5 × TBE buffer on top of the gel.
8. Load 5 μL of every sample in neighboring wells. First, load the solution of 100 bp DNA ladder (1 μL purchased 100 bp DNA ladder, 4 μL 0.5 × TBE, and 1 μL Gel loading dye blue 6×). Next, load the positive control, negative control, and experimental sample, respectively.
9. Connect the electrophoresis apparatus to the PowerPac machine, and run the gel at 145 V for 45 min.
10. Transfer the gel to the electronic dual wave transilluminator, and use UltraCam 8gD Digital Imaging System to image the gel (Fig. 4) (*see* **Note 15**).

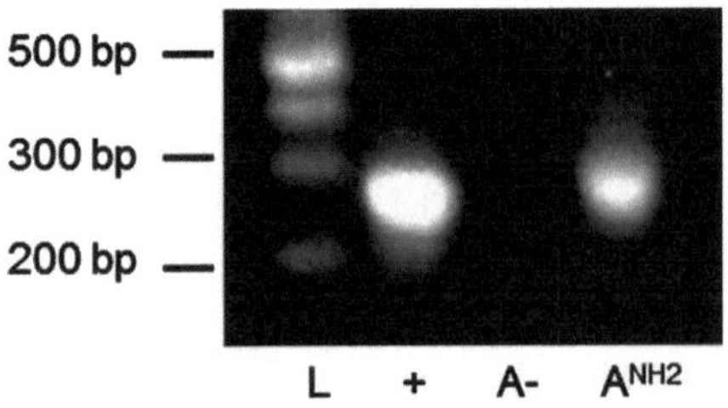


**Fig. 4** 1.3% agarose gel of the PCR products stained with GelRed. (L) = DNA ladder; (+) = positive control, i.e., DNA containing natural nucleotides only; (A−) = negative control for A, i.e., PCR with dCTP, dGTP, and dTTP only; ($A^{NH2}$) = DNA containing 7-(3-aminophenyl)-2′-deoxy-7-deazaadenosine instead of natural deoxyadenosine

## 4 Notes

1. Only HPLC grade solvents should be used. Particles from the mobile phases can accumulate in the column, increasing back pressure and damaging pump seals. Moreover, gas bubbles can form as the components mix. Both air and particles can be removed from the mobile phases by passing them through a nylon filter into a vacuum flask.
2. The solutions of the prepared triphosphates should be portioned into several aliquots, which are thawed and refrozen a maximum of five times. After that, it is advisable to use a new aliquot as frequent thawing and freezing of the solutions cause partial hydrolysis of triphosphates to diphosphates and monophosphates.
3. This flask contains the catalytic system. We find it best to prepare the catalytic complex separately and to transfer the solution to the mixture of the reactants after the dissolution of all the solids.
4. This step exchanges atmospheric air for argon inert atmosphere.
5. 2′-Deoxyribo 5′-*O*-triphosphates are prone toward hydrolysis to diphosphates, especially at higher temperatures and in acidic conditions. Therefore, the reaction should proceed at the lowest possible temperature, and the reaction time should be as short as possible. Prolongation of the reaction time and/or increasing the temperature of the cross-coupling reaction (30 min at 120 °C or 1 h at 65–70 °C) leads to higher conversions but unfortunately also to higher amounts of the diphosphate.
6. For nucleoside triphosphates bearing a different chemical modification, it is recommended to first perform a pilot HPLC separation with a small amount of the reaction mixture to determine a suitable gradient.
7. Most nucleoside triphosphates absorb at 254 nm. However, some compounds may have absorption minima at 254 nm. In such cases, the appropriate detection wavelength should be determined from the UV spectra.
8. In order to prevent hydrolysis of the product to diphosphate, the evaporation temperature should not exceed 40 °C.
9. Certain amount of diphosphate is always formed during the reaction. Sometimes, the triphosphate and the diphosphate have very similar retention times and are not separable by reverse-phase HPLC. In such cases, the mixture of the two compounds can be separated by ion-exchange chromatography using Poros 50 HQ resin.

10. Store the product at −20 °C.
11. The use of metal-free CuI is vital for obtaining the desired product. Trace metals catalyze further cyclization of the ethynyl-bridged ferrocene-uridine conjugate to furopyrimidine [15].
12. We have found that unlike most functionalized nucleoside triphosphates, the ethynylferrocene derivatives are more stable in the form of triethylammonium salt. Therefore, the ion-exchange chromatography is omitted in the standard protocol. However, when performing the Sonogashira cross-coupling reaction with other ethynyl compounds, it would be advisable to exchange the triethylammonium ions for sodium using either Dowex 50WX8 or any other ion-exchange resin in $Na^+$ cycle.
13. Store the radiolabelled primer at −20 °C in a radiation protection box.
14. We always prepare more master mix (usually as for one extra sample) than necessary because of the potential differences in pipetting.
15. Use Adobe Photoshop or a similar software to process the image.
16. The choice of DNA polymerase and PCR additives (e.g., DMSO, TMAC, formamide, Tween, betaine) depends on the type of modification and needs to be optimized. Temperatures and times of annealing, extension, and denaturation depend on the length and sequence of the template and primers. Before using different template, primers, and/or modified dNTPs, optimization of these parameters should be done.
17. Cooling is done to prevent breaking of the electrophoresis system due to rapid temperature changes.
18. GelRed is a very stable and environmentally safe fluorescent nucleic acid dye used for staining DNA on agarose gels instead of the highly toxic ethidium bromide (EB). However, if preferred, you can use EB staining instead. For detailed protocol on using EB, see [16].

## Acknowledgments

This work was supported by the Academy of Sciences of the Czech Republic (Praemium Academiae), by the Czech Science Foundation (18-03305S), and by the European Regional Development Fund; OP RDE (No. CZ.02.1.01/0.0/0.0/16_019/0000729).

## References


1. Hocek M, Fojta M (2008) Cross-coupling reactions of nucleoside triphosphates followed by polymerase incorporation. Construction and applications of base-functionalized nucleic acids. Org Biomol Chem 6:2233–2241
2. Jager S, Rasched G, Kornreich-Leshem H, Engeser M, Thum O, Famulok M (2005) A versatile toolbox for variable DNA functionalization at high density. J Am Chem Soc 127:15071–15082
3. Kielkowski P, Fanfrlík J, Hocek M (2014) 7-Aryl-7-deazaadenine 2-′-deoxyribonucleoside triphosphates (dNTPs): better substrates for DNA polymerases than dATP in competitive incorporations. Angew Chem Int Ed Engl 53:7552–7555
4. Cahová H, Panattoni A, Kielkowski P, Fanfrlík J, Hocek M (2016) 5-substituted pyrimidine and 7-substituted 7-deazapurine dNTPs as substrates for DNA polymerases in competitive primer extension in the presence of natural dNTPs. ACS Chem Biol 11:3165–3171
5. Brázdilová P, Vrábel M, Pohl R, Pivoňková H, Havran L, Hocek M, Fojta M (2007) Ferrocenylethynyl derivatives of nucleoside triphosphates: synthesis, incorporation, electrochemistry, and bioanalytical applications. Chem Eur J 13:9527–9533
6. Cahová H, Havran L, Brázdilová P, Pivoňková H, Pohl R, Fojta M, Hocek M (2008) Aminophenyl- and nitrophenyl-labeled nucleoside triphosphates: synthesis, enzymatic incorporation, and electrochemical detection. Angew Chem Int Ed 47:2059–2062
7. Vrábel M, Horáková P, Pivoňková H, Kalachová L, Černocká H, Cahová H, Pohl R, Šebest P, Havran L, Hocek M, Fojta M (2009) Base-modified DNA labeled by $[Ru(bpy)_3]^{2+}$ and $[Os(bpy)_3]^{2+}$ complexes: construction by polymerase incorporation of modified nucleoside triphosphates, electrochemical and luminescent properties, and applications. Chem Eur J 15:1144–1154
8. Balintová J, Pohl R, Horáková P, Vidláková P, Havran L, Fojta M, Hocek M (2011) Anthraquinone as a redox label for DNA. Synthesis, enzymatic incorporation and electrochemistry of anthraquinone-modified of nucleosides, nucleotides and DNA. Chem Eur J 17:14063–14073
9. Macíčková-Cahová H, Hocek M (2009) Cleavage of adenine-modified functionalized DNA by type II restriction endonucleases. Nucleic Acids Res 37:7612–7622
10. Macíčková-Cahová H, Pohl R, Hocek M (2011) Cleavage of functionalized DNA containing 5-modified pyrimidines by type II restriction endonucleases. Chembiochem 12:431–438
11. Kielkowski P, Macíčková-Cahová H, Pohl R, Hocek M (2011) Transient and switchable (triethylsilyl)ethynyl protection of DNA against cleavage by restriction endonucleases. Angew Chem Int Ed 50:8727–8730
12. Raindlová V, Pohl R, Šanda M, Hocek M (2010) Direct polymerase synthesis of reactive aldehyde-functionalized DNA and its conjugation and staining with hydrazines. Angew Chem Int Ed 49:1064–1066
13. Kovács T, Ötvös L (1988) Simple synthesis of 5-vinyl- and5-ethynyl-2′-deoxyuridine-5-′-triphosphates. Tetrahedron Lett 29:4525–4528
14. Čapek P, Cahová H, Pohl R, Hocek M, Gloeckner C, Marx A (2007) An efficient method for the construction of functionalized DNA bearing amino acid groups through cross-coupling reactions of nucleoside triphosphates followed by primer extension or PCR. Chem Eur J 13:6196–6203
15. Yu CJ, Yowanto H, Wan YJ, Meade TJ, Chong Y, Strong M, Donilon LH, Kayyem JF, Gozin M, Blackburn GF (2000) Uridine-conjugated ferrocene DNA oligonucleotides: unexpected cyclization reaction of the uridine base. J Am Chem Soc 122:6767–6768
16. Macíčková-Cahová H, Vrábel M, Hocek M (2010) Cross-coupling modification of nucleoside triphosphates, PEX, and PCR construction of base-modified DNA. Curr Protoc Chem Biol 2:1–14

Chapter 5

# 2′-*C*,4′-*C*-Ethyleneoxy-Bridged 2′-Deoxyribonucleic Acids (EoDNAs) with Thymine Nucleobases: Synthesis, Duplex-Forming Ability, and Enzymatic Stability

**Takashi Osawa, Satoshi Obika, and Yoshiyuki Hari**

## Abstract

This chapter describes procedures for (1) the synthesis of six 2′-*C*,4′-*C*-ethyleneoxy-bridged thymidine phosphoramidites, i.e., methylene-EoDNA-T, (*R*)-Me-methylene-EoDNA-T, (*S*)-Me-methylene-EoDNA-T, EoDNA-T, (*R*)-Me-EoDNA-T, and (*S*)-Me-EoDNA-T phosphoramidites, (2) the introduction of the phosphoramidites into oligonucleotides, (3) UV-melting experiments of the duplexes of the modified oligonucleotides and complementary RNA, and (4) nuclease degradation experiments of the modified oligonucleotides.

**Key words** EoDNAs, Methylene-EoDNAs, 2′,4′-Bridged nucleic acids, Ethyleneoxy bridge, Oligonucleotides, Duplex-forming ability, Nuclease resistance

## 1 Introduction

Chemically modified oligonucleotides are used in various technologies including oligonucleotide-based therapeutics and diagnostics [1–7]. Among them, oligonucleotides modified by 2′-*O*,4′-*C*-methylene-bridged/locked nucleic acid (2′,4′-BNA or LNA), adopting fixed N-type sugar conformation, have attracted much attention because these oligonucleotides show a dramatic increase in the binding affinity toward complementary RNA and a significant improvement in nuclease resistance [8–10]. Therefore, various 2′,4′-bridged nucleic acid analogs have been developed so far by many groups including us [4, 11–13]. Based on this background, we have succeeded in the development of 2′-*O*,4′-*C*-ethyleneoxy-bridged 5-methyluridine (EoNA-T) possessing a seven-membered bridge and a 6′-oxygen atom as a new class of 2′,4′- bridged nucleic acids [14]. We found that the 6′-oxygen atom in EoNA-T can contribute to increasing both the binding affinity with target nucleic acids and the nuclease resistance. Recently, we designed

Nathaniel Shank (ed.), *Non-Natural Nucleic Acids: Methods and Protocols*, Methods in Molecular Biology, vol. 1973,
https://doi.org/10.1007/978-1-4939-9216-4_5, © Springer Science+Business Media, LLC, part of Springer Nature 2019

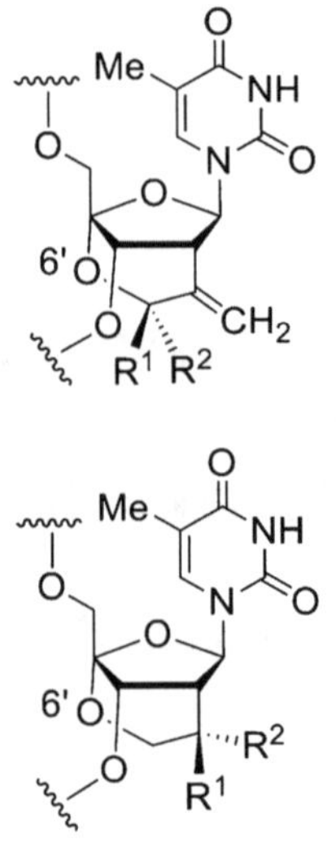

methylene-EoDNA-T ($R^1$ = H, $R^2$ = H)
(*R*)-Me-methylene-EoDNA-T ($R^1$ = Me, $R^2$ = H)
(*S*)-Me-methylene-EoDNA-T ($R^1$ = H, $R^2$ = Me)

EoDNA-T ($R^1$ = H, $R^2$ = H)
(*R*)-Me-EoDNA-T ($R^1$ = Me, $R^2$ = H)
(*S*)-Me-EoDNA-T ($R^1$ = H, $R^2$ = Me)

**Fig. 1** Structures of 2′-*C*,4′-*C*-ethyleneoxy-bridged thymidines

2′-*C*,4′-*C*-ethyleneoxy-bridged 2′-deoxyribonucleic acid (EoDNA) possessing a six-membered bridge and a 6′-oxygen atom, and six EoDNAs with thymine nucleobases were developed, as shown in Fig. 1 [15, 16]. The properties of the oligonucleotides modified by EoDNAs strongly suggest that all EoDNAs are promising candidates for antisense technology. Thus, in this chapter, the procedures for (1) the synthesis of six thymidine phosphoramidites modified by EoDNAs, (2) their introduction into oligonucleotides using an automated DNA synthesizer, (3) UV-melting experiments of duplexes between the modified oligonucleotides and complementary RNA, and (4) enzymatic degradation experiments of the modified oligonucleotides are described.

As shown in Fig. 2, methylene-, (*R*)-Me-methylene-, and (*S*)-Me-methylene-EoDNA-T phosphoramidites **12a–c** can be synthesized from 5-methyluridine (**1**). In the synthesis of **3**, 1,8-diazabicyclo[5.4.0]undec-7-ene (DBU), instead of NaOMe, can also be used for the *E*2 elimination [14]. However, DBU is unsuitable for the large-scale synthesis of olefin **3** because it is difficult to remove the large amount of DBU by silica gel column chromatography. Although the 3′-OH-selective monosilylation of nucleosides using TBSCl, $AgNO_3$, and 1,4-diazabicyclo[2.2.2]octane (DABCO) in THF is reported [17], monosilylation of compound **3** in THF, instead of DMF, does not proceed significantly, possibly because of the lower solubility. Compound **7a** can be directly synthesized in 40% yield by treatment of **4** with mCPBA in 2-propyn-1-ol (*see* **Note 1**). In the synthesis of **7** from **4**, 2-propyn-1-ol, (*R*)-3-butyn-2-ol, and (*S*)-3-butyn-2-ol are used for **12a**, **12b**, and **12c**, respectively.

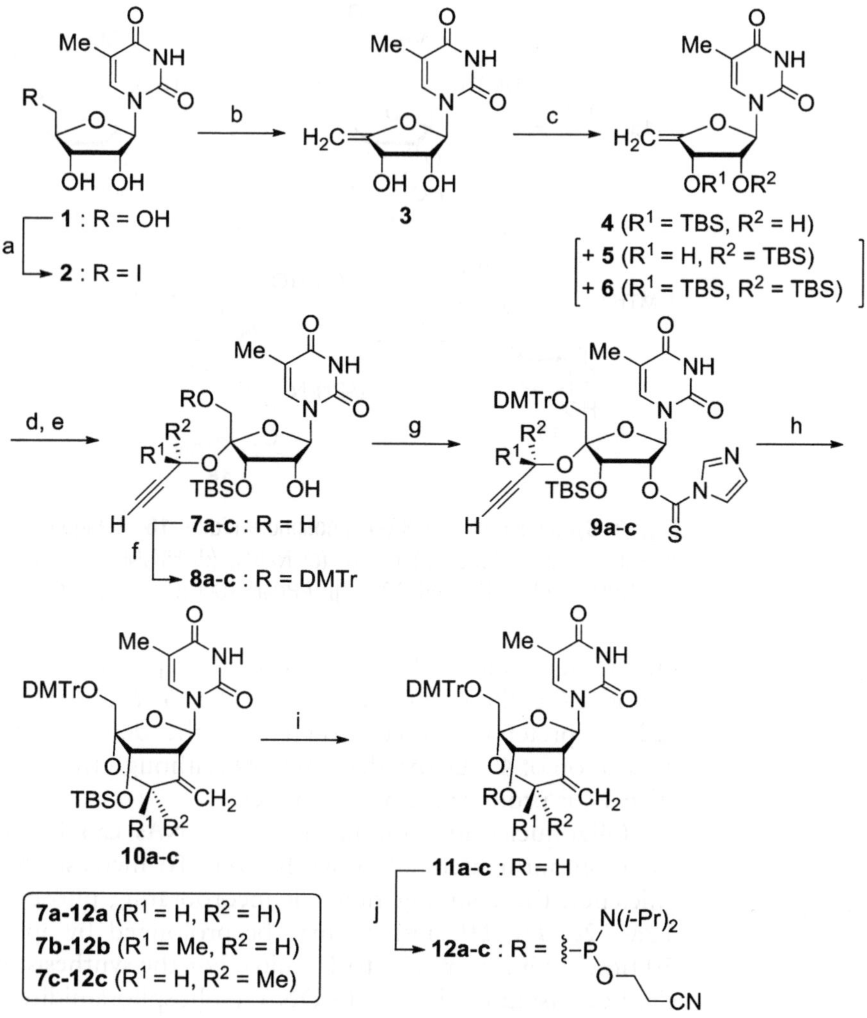


**Fig. 2** Synthesis of methylene-, (*R*)-Me-methylene-, and (*S*)-Me-methylene-EoDNA-T phosphoramidites **12a–c**. Reagents: (a) $Ph_3P$, $I_2$, and imidazole; (b) NaOMe; (c) TBSCl, $AgNO_3$, and DABCO; (d) Oxone®, $NaHCO_3$, and acetone; (e) 2-propyn-1-ol (for **7a**), (*R*)-3-butyn-2-ol (for **7b**), (*S*)-3-butyn-2-ol (for **7c**), and $ZnCl_2$; (f) DMTrOTf and 2,6-lutidine; (g) TCDI; (h) AIBN and $(Me_3Si)_3SiH$; (i) TBAF; (j) $(i\text{-}Pr)_2NP(Cl)O(CH_2)_2CN$ and DIPEA

Figure 3 shows the synthetic route of EoDNA-T phosphoramidite **16**. Three steps are required for the synthesis of **13** from **11a**, because the $OsO_4$-promoted oxidative cleavage of the exocyclic olefin in **10a** (shown in Fig. 2) or **11a** does not proceed at all. The hydrazonation of **13** without $Et_3N$ becomes slow. Furthermore, because the hydrogenation of **14** using $PtO_2$ in the absence of DBU gives an inseparable compound along with compound **15**, **15** cannot be isolated in pure form.

The synthesis of (*R*)- and (*S*)-Me-EoDNA-T phosphoramidites **19** and **24** is shown in Fig. 4. Hydrogenation of the exocyclic

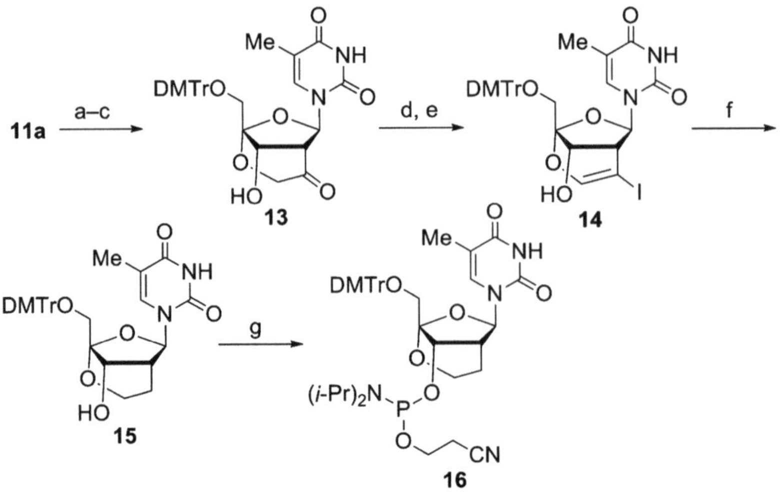

**Fig. 3** Synthesis of EoDNA-T phosphoramidite **16**. Reagents: (a) Oxone®, $NaHCO_3$, and acetone; (b) NaOH; (c) $NaIO_4$; (d) $NH_2NH_2{\cdot}H_2O$ and $Et_3N$; (e) $I_2$ and DBU; (f) $PtO_2$, $H_2$, and DBU; (g) $(i\text{-}Pr)_2NP(Cl)O(CH_2)_2CN$ and DIPEA

methylene group of **10a** stereoselectively produces desired compound **17** as the sole diastereomer. The (*S*)-Me-EoDNA skeleton (**22**) is preferentially constructed by the deoxygenative radical cyclization of 4′-allyloxy derivative **21**, although the (*R*)-Me derivative **17** is produced as a by-product.

Oligonucleotides modified by EoDNAs can be synthesized using an automated DNA synthesizer. To increase the coupling efficiency, the coupling time for incorporating phosphoramidites **12a–12c**, **16**, **19**, and **24** may be prolonged by approximately 20 times (e.g., from 32 s to 10 min). For the synthesis of oligonucleotides using methylene-EoDNA-T phosphoramidites **12a–12c**, 1 M *tert*-BuOOH in *n*-decane/toluene [18] must be used as an oxidizing agent instead of the common 0.02 M iodine solution to avoid decomposing the exocyclic methylene groups.

The UV-melting experiments show that oligonucleotides modified by all EoDNAs can bind to complementary RNA with high affinity. The change in $T_m$ per modification ($\Delta T_m$/mod) compared to natural DNA ranges from 1.7 to 5.0 °C [15, 16]. In addition, the stability of duplexes formed by apoB mRNA antisense oligonucleotides with EoDNAs ($\Delta T_m$/mod = +2.2 to +2.8 °C) except for (*R*)-Me-EoDNA ($\Delta T_m$/mod = +0.6 °C) is comparable to that of LNA-modified oligonucleotide ($\Delta T_m$/mod = +3.0 °C) [16]. Enzymatic degradation experiments using *Crotalus adamanteus* venom phosphodiesterase (CAVP), a 3′-exonuclease, demonstrate that all EoDNAs have excellent resistance to nucleolytic degradation and the nuclease resistance of the oligonucleotides modified by EoDNAs is superior to that of oligonucleotides containing EoNA with a larger seven-membered bridge

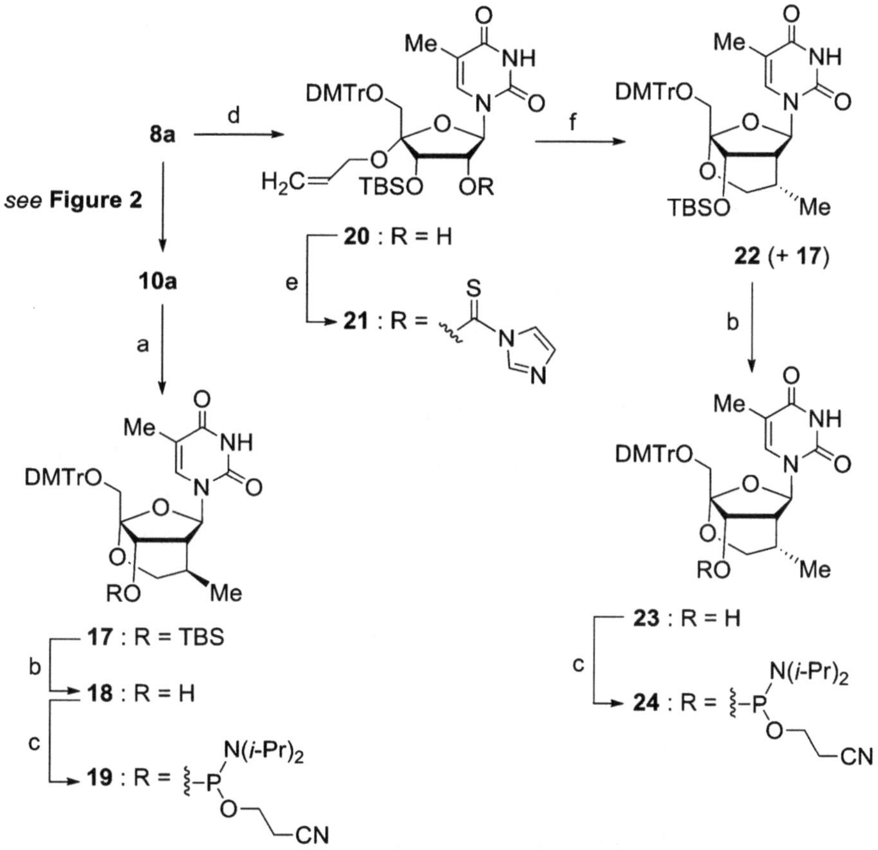


**Fig. 4** Synthesis of (*R*)- and (*S*)-Me-EoDNA-T phosphoramidites **19** and **24**. Reagents: (a) $PtO_2$ and $H_2$; (b) TBAF; (c) $(i\text{-}Pr)_2NP(Cl)O(CH_2)_2CN$ and DIPEA; (d) Lindlar's catalyst and $H_2$; (e) TCDI; (f) AIBN and $(Me_3Si)_3SiH$

[15, 16]. Furthermore, the in vitro gene-silencing potencies of apoB mRNA antisense oligonucleotides modified by all EoDNAs are almost the same as that of the LNA-modified congener [16]. These results strongly suggest that EoDNAs are highly promising building blocks for therapeutic oligonucleotides.

## 2 Materials

Reagents should be purified before use or the proper grade of reagents should be used. Chemical shifts of $^1H$ NMR are reported in parts per million downfield from internal tetramethylsilane (0.00 ppm) and residual DMSO (2.50 ppm) for $^1H$ NMR.

### 2.1 Synthesis of Phosphoramidites

1. Acetone.
2. Acetonitrile (MeCN).
3. Ammonium chloride ($NH_4Cl$).

4. Anisaldehyde solution (*see* **Note 2**): Add 10 mL *p*-anisaldehyde, 5 mL acetic acid, and 15 mL concentrated $H_2SO_4$ in 450 mL ethanol at 0 °C.
5. Argon (Ar) gas.
6. 2,2′-Azobis(isobutyronitrile) (AIBN).
7. Brine.
8. *tert*-Butylchlorodimethylsilane (TBSCl).
9. (*R*)-3-Butyn-2-ol.
10. (*S*)-3-Butyn-2-ol.
11. Celite®.
12. Chloroform ($CHCl_3$).
13. 2-Cyanoethyl-*N,N*-diisopropylchlorophosphoramidite.
14. 1,4-Diazabicyclo[2.2.2]octane (DABCO).
15. 1,8-Diazabicyclo[5.4.0]undec-7-ene (DBU).
16. Dichloromethane ($CH_2Cl_2$).
17. Diethyl ether ($Et_2O$).
18. *N,N*-Diisopropylethylamine (DIPEA).
19. 4,4′-Dimethoxytrityl chloride (DMTrCl).
20. *N,N*-Dimethylformamide (DMF).
21. 1,4-Dioxane.
22. Ethyl acetate (EtOAc).
23. Hexane.
24. Hydrazine monohydrate ($NH_2NH_2 \cdot H_2O$).
25. Hydrogen ($H_2$) gas.
26. Imidazole.
27. Iodine ($I_2$).
28. Lindlar's catalyst.
29. 2,6-Lutidine.
30. Methanol (MeOH).
31. 5-Methyluridine.
32. Nitrogen ($N_2$) gas.
33. Oxone®.
34. Platinum oxide ($PtO_2$).
35. 2-Propyn-1-ol.
36. Saturated aqueous sodium bicarbonate ($NaHCO_3$) solution.
37. Saturated aqueous sodium thiosulfate ($Na_2S_2O_3$) solution.
38. Silica gel.
39. Silver nitrate ($AgNO_3$).

40. Silver trifluoromethanesulfonate (AgOTf).
41. 1 M aqueous sodium hydroxide (NaOH) solution: Weigh 4.0 g (100 mmol) NaOH, and dissolve in 100 mL distilled water.
42. 28% sodium methoxide (NaOMe) in MeOH.
43. Sodium periodate ($NaIO_4$).
44. Sodium sulfate ($Na_2SO_4$).
45. 1 M tetrabutylammonium fluoride (TBAF) in THF.
46. Tetrahydrofuran (THF).
47. Triethylamine ($Et_3N$).
48. 1,1′-Thiocarbonyldiimidazole (TCDI).
49. Toluene.
50. Triphenylphosphine ($Ph_3P$).
51. Tris(trimethylsilyl)silane [$(Me_3Si)_3SiH$].
52. 1 M zinc chloride ($ZnCl_2$) in $Et_2O$.
53. $^1$H NMR spectrometers.
54. MALDI-TOF mass spectrometer.

### 2.2 Synthesis of Oligonucleotides

1. DNA synthesizer.
2. Acetonitrile (MeCN).
3. Thymidine phosphoramidite.
4. *N*$^4$-Acetyl-2′-deoxy-5-methylcytidine phosphoramidite.
5. *N*$^4$-Benzoyl-2′-deoxycytidine phosphoramidite.
6. *N*$^6$-Benzoyl-2′-deoxyadenosine phosphoramidite.
7. *N*$^2$-Isobutyryl-2′-deoxyduanosine phosphoramidite.
8. Controlled pore glass (CPG) column with a 0.2 μmol scale.
9. Deblocking solution: 3% trichloroacetic acid in $CH_2Cl_2$.
10. Activator solution: 0.25 M Activator 42® [5-(3,5-trifluoromethylphenyl)-1*H*-tetrazole] in MeCN.
11. Cap A solution: 10% (v/v) Acetic anhydride in THF.
12. Cap B solution: 8:1:1 (v/v/v) THF/1-methylimidazole/pyridine.
13. Oxidizing solution: 0.02 M $I_2$ in 90.54:9.05:0.41 (v/v/v) THF/$H_2O$/pyridine.
14. 1 M *tert*-Butyl hydroperoxide (*tert*-BuOOH) solution: Dilute 10 mL of 5.5 M *tert*-BuOOH in *n*-decane with 45 mL anhydrous toluene.
15. 28% aqueous ammonia solution.
16. Sep-Pak® C18 plus short cartridges.

17. 0.1 M triethylammonium acetate (TEAA) buffer (pH 7.0): Dilute 100 mL of 1 M TEAA buffer (pH 7.0) with distilled water to a total volume of 1 L.
18. 1% (v/v) trifluoroacetic acid (TFA) solution: Dilute 1.0 mL TFA with distilled water to a total volume of 100 mL.
19. Methanol (MeOH).
20. MALDI-TOF mass spectrometer.
21. 10 mg/mL 3-hydroxypicolinic acid solution: Weigh 10 mg 3-hydroxypicolinic acid, and dissolve in 1 mL distilled water.
22. 1 mg/mL diammonium citrate solution: Weigh 1 mg diammonium citrate, and dissolve in 1 mL distilled water.

### 2.3 UV-Melting Experiments

1. UV spectrophotometers equipped with a $T_m$ analysis accessory.
2. 100 mM sodium cacodylate buffer (pH 7.2): Weigh 2.14 g (10 mmol) sodium cacodylate, and transfer to a glass beaker containing 40–50 mL distilled water. Mix the solution and adjust pH with 0.1 M HCl. Dilute this solution with distilled water to a total volume of 100 mL.
3. 1 M potassium chloride (KCl) solution: Weigh 745 mg (10.0 mmol) KCl, and dissolve in 10 mL distilled water.

### 2.4 Enzymatic Degradation Experiments

1. 1 M Tris–HCl buffer (pH 8.0): Weigh 12.1 g (100 mmol) tris (hydroxymethyl)aminomethane, and transfer to a glass beaker containing 70–80 mL distilled water. Mix and adjust the pH with 6 M HCl. Dilute this solution with distilled water to a total volume of 100 mL.
2. 100 mM magnesium chloride ($MgCl_2$): Weigh 203 mg (1.00 mmol) $MgCl_2 \cdot 6H_2O$, and dissolve in 10 mL distilled water.
3. 10 μg/mL *Crotalus adamanteus* venom phosphodiesterase (CAVP) solution: Dilute 2.0 μL CAVP solution (1 mg/mL) with 198 μL distilled water.

## 3 Methods

### 3.1 Synthesis of Phosphoramidites

**Synthesis of Methylene-, (*R*)-Methylene-, and (*S*)-Methylene-EoDNA-T Phosphoramidites**

#### 3.1.1 Synthesis of Compound 2

1. Add 5-methyluridine (10.0 g, 37.8 mmol) to a round-bottomed flask equipped with a three-way glass stopcock and a balloon filled with Ar. Evacuate the flask using a vacuum line, and flush it with Ar. Repeat this procedure three times.
2. Add anhydrous THF (100 mL), and cool the suspension in an ice bath. Add $Ph_3P$ (11.1 g, 42.6 mmol), imidazole (5.80 g,

85.2 mmol), and $I_2$ (10.8 g, 42.6 mmol) to the mixture, and stir the mixture at room temperature for 2 h.

3. Check the progress of the reaction by TLC on aluminum plates using 7:1 (v/v) $CHCl_3$/MeOH as the eluent. Visualize the spots with a UV lamp and by dipping the TLC plate into the anisaldehyde solution and heating with a heat gun ($R_f = 0.3$).
4. Cool the mixture in an ice bath, and add a saturated aqueous $Na_2S_2O_3$ solution (20 mL) and a saturated aqueous $NaHCO_3$ solution (20 mL). Transfer the resulting solution to a separatory funnel, and extract the product with EtOAc (*see* **Note 3**). Dry over anhydrous $Na_2SO_4$, and filter out the drying agent using a glass funnel plugged with cotton.
5. Concentrate the filtrate under reduced pressure, and purify the residue by silica gel column chromatography (silica gel 200 g) using 30:1 to 10:1 (v/v) $CHCl_3$/MeOH as the eluent. Combine the appropriate fractions, and evaporate the solvent to give compound **2** [19] (12.8 g, 90% yield). White powder. $^1$H NMR (DMSO-$d_6$, 300 MHz): $\delta$ 1.80 (s, 3H), 3.40 (dd, 1H, $J = 6.5, 10.5$ Hz), 3.57 (dd, 1H, $J = 6.0, 10.5$ Hz), 3.80–3.86 (m, 1H), 3.88–3.93 (m, 1H), 4.18 (q, $J = 6.0$ Hz, 1H), 5.36 (d, 1H, $J = 5.0$ Hz), 5.46 (d, 1H, $J = 6.0$ Hz), 5.81 (d, 1H, $J = 6.0$ Hz), 7.52 (s, 1H), 11.37 (s, 1H).

*3.1.2 Synthesis of Compound **3***

1. Add compound **2** (12.8 g, 34.8 mmol) to a round-bottomed flask equipped with a reflux condenser, a three-way glass stopcock, and a balloon filled with Ar. Evacuate the flask using a vacuum line and flush it with Ar. Repeat this procedure three times.
2. Add anhydrous MeOH (70 mL) and 28% (w/v) NaOMe in MeOH (67 mL, 348 mmol) to the flask. Reflux the mixture for 12 h in an oil bath (80 °C).
3. Check the progress of the reaction by TLC on aluminum plates using 7:1 (v/v) $CHCl_3$/MeOH as the eluent. Visualize the spots with a UV lamp and by dipping the TLC plate into the anisaldehyde solution and heating with a heat gun ($R_f = 0.3$).
4. Add a solution of $NH_4Cl$ (18.6 g, 348 mmol) in water (100 mL) to the mixture, and then evaporate the solvent (MeOH). After the addition of water (50 mL) to the residue, collect the precipitate by filtration, and wash it with water (20 mL) and acetone (20 mL). Dry the collected white powder in a vacuum desiccator to give compound **3** (6.85 g, 82% yield). White powder. $^1$H NMR (DMSO-$d_6$, 400 MHz): $\delta$ 1.77 (d, 3H, $J$ = 1.0 Hz), 4.18 (d, 1H, $J$ = 1.0 Hz), 4.23–4.30 (m, 1H), 4.32 (s, 1H), 4.36–4.44 (brs, 1H), 5.40–5.59 (brs, 1H), 5.58–5.65 (brs, 1H), 6.00 (d, 1H, $J$ = 6.5 Hz), 7.49 (d, 1H, $J$ = 1.0 Hz), 11.4 (s, 1H).

High-resolution mass (MALDI-TOF): 263.0638 ($MNa^+$, calcd for $C_{10}H_{12}N_2O_5Na$: 263.0644).

*3.1.3 Synthesis of Compound* ***4***

1. Add compound **3** (9.30 g, 38.7 mmol) to a round-bottomed flask equipped with a three-way glass stopcock and a balloon filled with $N_2$. Evacuate the flask using a vacuum line and flush it with $N_2$. Repeat this procedure three times.
2. Add anhydrous DMF (100 mL), and cool the mixture in an ice bath. Add DABCO (21.7 g, 194 mmol), TBSCl (7.00 g, 46.5 mmol), and $AgNO_3$ (7.90 g, 46.5 mmol), and stir the mixture at room temperature for 12 h.
3. Check the progress of the reaction by TLC on aluminum plates using 1:1 (v/v) *n*-hexane/EtOAc as the eluent. Visualize the spots with a UV lamp and by dipping the TLC plate into the anisaldehyde solution and heating with a heat gun [$R_f = 0.4$ (**4**), 0.5 (**5**), 0.8 (**6**)].
4. Filter the mixture using a glass filter covered with Celite®. Then, dilute the filtrate with $Et_2O$. Transfer the solution to a separatory funnel. Wash the solution with water and brine, dry over $Na_2SO_4$, and filter out the drying agent using a glass funnel plugged with cotton.
5. Concentrate the filtrate under reduced pressure, and purify the residue by silica gel column chromatography (silica gel 500 g) using 3:1 to 3:2 (v/v) hexane/EtOAc as the eluent. Combine the appropriate fractions, and evaporate the solvent to give **4** (5.35 g, 39% yield), **5** (5.90 g, 42% yield), and **6** (1.08 g, 6% yield). Compound **4**. White foam. $^1H$ NMR (500 MHz, $CDCl_3$): $\delta$ 0.20 (s, 6H), 0.95 (s, 9H), 1.95 (s, 3H), 2.97 (d, 1H, $J = 7.5$ Hz), 4.25 (ddd, 1H, $J = 4.5, 5.5, 7.5$ Hz), 4.30 (d, 1H, $J = 2.5$ Hz), 4.61 (d, 1H, $J = 2.5$ Hz), 4.71 (d, 1H, $J = 5.5$ Hz, 1H), 5.93 (d, 1H, $J = 4.5$ Hz), 6.98 (s, 1H), 8.02 (brs, 1H). High-resolution mass (MALDI-TOF): 377.1503 ($MNa^+$, calcd for $C_{16}H_{26}N_2O_5SiNa$: 377.1509). Compound **5**. White foam. $^1H$ NMR (300 MHz, $CDCl_3$): $\delta$ 0.15 (s, 3H), 0.20 (s, 3H), 0.92 (s, 9H), 1.94 (d, 3H, $J = 1.0$ Hz), 2.58 (d, 1H, $J = 7.0$ Hz), 4.30 (dd, 1H, $J = 3.0, 5.5$ Hz), 4.46–4.51 (m, 2H), 4.69 (t, 1H, $J = 1.5$ Hz), 5.92 (d, 1H, $J = 3.0$ Hz), 6.95 (d, 1H, $J = 1.5$ Hz), 9.53 (brs, 1H). High-resolution mass (MALDI-TOF): 355.1684 ($MH^+$, calcd for $C_{16}H_{27}N_2O_5Si$: 355.1689). Compound **6**. White foam. $^1H$ NMR (400 MHz, $CDCl_3$): $\delta$ 0.02 (s, 3H), 0.05 (s, 3H), 0.12 (s, 3H), 0.13 (s, 3H), 0.87 (s, 9H), 0.92 (s, 9H), 1.95 (d, 3H, $J = 1.0$ Hz), 4.23 (dd, 1H, $J = 4.0, 5.5$ Hz), 4.25 (d, 1H, $J = 2.5$ Hz), 4.36 (d, 1H, $J = 4.0$ Hz), 4.53 (d, 1H, $J = 2.5$ Hz), 6.09 (d, 1H, $J = 5.5$ Hz), 6.98 (d, 1H, $J = 1.0$ Hz), 8.88 (brs, 1H). High-resolution mass (MALDI-

TOF): 491.2368 ($MNa^+$, calcd for $C_{22}H_{40}N_2O_5Si_2Na$: 491.2373).

### 3.1.4 Synthesis of Compound 7

1. Add compound **4** (3.00 g, 8.46 mmol), $CH_2Cl_2$ (45 mL), acetone (30 mL), and a saturated aqueous $NaHCO_3$ solution (150 mL) to a round-bottomed flask equipped with a dropping funnel.
2. Cool the mixture in an ice bath, and add dropwise the solution of Oxone® (15.6 g, 25.4 mmol) in water (100 mL) to the mixture over a period of 30 min using the dropping funnel. Stir the mixture at room temperature for 3 h.
3. Check the progress of the reaction by TLC on aluminum plates using 1:1 (v/v) hexane/EtOAc as the eluent. Visualize the spots with a UV lamp and by dipping the TLC plate into the anisaldehyde solution and heating with a heat gun ($R_f = 0.4$).
4. Transfer the mixture to a separatory funnel, and extract the product with $CH_2Cl_2$. Wash the organic extracts with water and brine, dry over $Na_2SO_4$, and filter out the drying agent using a glass funnel plugged with cotton.
5. Evaporate the filtrate, and attach a three-way glass stopcock and a balloon filled with $N_2$ to the flask. Evacuate the flask using a vacuum line and flush it with $N_2$. Repeat this procedure three times.
6. Dissolve the residue in anhydrous $Et_2O$ (40 mL). Cool the mixture in a dry ice/acetonitrile bath (−40 °C), and add 2-propyn-1-ol (5.0 mL, 84.6 mmol) and a solution of 1 M $ZnCl_2$ in $Et_2O$ (8.5 mL, 8.5 mmol). Stir the mixture for 1 h in an ice bath.
7. Check the progress of the reaction by TLC on aluminum plates using 1:1 (v/v) hexane/EtOAc as the eluent. Visualize the spots with a UV lamp and by dipping the TLC plate into the anisaldehyde solution and heating with a heat gun ($R_f = 0.2$).
8. Add a saturated aqueous $NaHCO_3$ solution (100 mL) to the mixture in an ice bath. Filter the mixture using a glass filter covered with Celite®. Transfer the filtrate to a separatory funnel, and extract the product with EtOAc. Wash the organic extracts with brine, dry over $Na_2SO_4$, and filter out the drying agent using a glass funnel plugged with cotton.
9. Concentrate the filtrate under reduced pressure, and purify the residue by silica gel column chromatography (silica gel 50 g) using 2:1 to 2:3 (v/v) hexane/EtOAc as the eluent. Combine the appropriate fractions, and evaporate the solvent to give **7a** (1.89 g, 52% yield over two steps). White foam. $^1H$ NMR (400 MHz, $CDCl_3$): $\delta$ 0.18 (s, 3H), 0.20 (s, 3H), 0.95 (s, 9H), 1.93 (d, 3H, $J = 1.0$ Hz), 2.46 (t, 1H, $J = 2.5$ Hz), 2.63

(dd, 1H, $J$ = 4.5, 8.0 Hz), 3.14 (d, 1H, $J$ = 8.0 Hz), 3.70 (dd, 1H, $J$ = 8.0, 12.0 Hz), 3.86 (dd, 1H, $J$ = 4.5, 12.0 Hz), 4.33 (d, 2H, $J$ = 2.5 Hz), 4.42 (ddd, 1H, $J$ = 4.0, 6.5, 8.0 Hz), 4.69 (d, 1H, $J$ = 6.5 Hz), 5.59 (d, 1H, $J$ = 4.0 Hz), 7.12 (d, 1H, $J$ = 1.0 Hz), 8.84 (brs, 1H). High-resolution mass (MALDI-TOF): 449.1714 ($MNa^+$, calcd for $C_{19}H_{30}N_2O_7$-SiNa: 449.1720).

10. Compound **7b** (1.86 g, 59% yield over two steps) is prepared from **4** (2.53 g, 7.13 mmol) using (*R*)-3-butyn-2-ol instead of 2-propyn-1-ol. The eluent for TLC is 1:1 (v/v) hexane/EtOAc ($R_f$ = 0.2). The eluent for silica gel column chromatography is 2:1 to 2:3 (v/v) hexane/EtOAc. White foam. $^1$H NMR (300 MHz, $CDCl_3$): $\delta$ 0.16 (s, 3H), 0.18 (s, 3H), 0.95 (s, 9H), 1.44 (d, 3H, $J$ = 7.0 Hz), 1.92 (d, 3H, $J$ = 1.0 Hz), 2.34 (dd, 1H, $J$ = 3.0, 9.0 Hz), 2.50 (d, 1H, $J$ = 2.0 Hz), 3.23 (d, 1H, $J$ = 7.5 Hz), 3.57 (dd, 1H, $J$ = 9.0, 11.5 Hz), 3.82 (dd, 1H, $J$ = 3.0, 11.5 Hz), 4.39 (ddd, 1H, $J$ = 3.0, 6.5, 7.5 Hz), 4.59 (dq, 1H, $J$ = 2.0, 7.0 Hz), 4.76 (d, 1H, $J$ = 6.5 Hz), 5.69 (d, 1H, $J$ = 3.0 Hz), 7.06 (d, 1H, $J$ = 1.0 Hz), 8.23 (brs, 1H). High-resolution mass (MALDI-TOF): 463.1871 ($MNa^+$, calcd for $C_{20}H_{32}N_2O_7SiNa$: 463.1876).
11. Compound **7c** (2.42 g, 77% yield over two steps) is prepared from **4** (2.53 g, 7.13 mmol) using (*S*)-3-butyn-2-ol instead of 2-propyn-1-ol. The eluent for TLC is 1:2 (v/v) hexane/EtOAc ($R_f$ = 0.5). The eluent for silica gel column chromatography is 2:1 to 2:3 (v/v) hexane/EtOAc. White foam. $^1$H NMR (300 MHz, $CDCl_3$): $\delta$ 0.18 (s, 3H), 0.20 (s, 3H), 0.95 (s, 9H), 1.46 (d, 3H, $J$ = 6.5 Hz), 1.93 (d, 3H, $J$ = 1.0 Hz), 2.45 (d, 1H, $J$ = 2.0 Hz), 2.59 (dd, 1H, $J$ = 5.5, 7.0 Hz), 3.16 (d, 1H, $J$ = 7.5 Hz), 3.84 (dd, 1H, $J$ = 5.5, 12.0 Hz), 3.89 (dd, 1H, $J$ = 7.0, 12.0 Hz), 4.25–4.35 (m, 1H), 4.62–4.68 (m, 2H), 5.60 (d, 1H, $J$ = 4.5 Hz), 7.12 (d, 1H, $J$ = 1.0 Hz), 8.94 (brs, 1H). High-resolution mass (MALDI-TOF): 463.1873 ($MNa^+$, calcd for $C_{20}H_{32}N_2O_7SiNa$: 463.1876).

#### 3.1.5 Synthesis of Compound 8

1. Add compound **7a** (1.00 g, 2.34 mmol) to a round-bottomed flask equipped with a three-way glass stopcock and a balloon filled with $N_2$. Evacuate the flask using a vacuum line and flush it with $N_2$. Repeat this procedure three times.
2. Add anhydrous $CH_2Cl_2$ (15 mL), and cool the mixture in an ice bath. Add 2,6-lutidine (0.54 mL, 4.68 mmol) and a solution of 1 M 4,4′-dimethoxytrityl trifluoromethanesulfonate (DMTrOTf) in anhydrous $CH_2Cl_2$ (2.6 mL, 2.6 mmol, *see* **Note 4**), and stir the mixture for 1 h at room temperature.

3. Check the progress of the reaction by TLC on aluminum plates using 1:1 (v/v) hexane/EtOAc as the eluent. Visualize the spots with a UV lamp and by dipping the TLC plate into the anisaldehyde solution and heating with a heat gun ($R_f = 0.4$).
4. Cool the mixture in an ice bath and add MeOH (1 mL). Transfer the mixture to a separatory funnel and dilute with $CH_2Cl_2$. Wash the organic layer with water and brine, dry over $Na_2SO_4$, and filter out the drying agent using a glass funnel plugged with cotton.
5. Concentrate the filtrate under reduced pressure, and purify the residue by silica gel column chromatography (silica gel 50 g) using 2:1 to 1:1 (v/v) hexane/EtOAc as the eluent. Combine the appropriate fractions, and evaporate the solvent to give **8a** (1.38 g, 81% yield). White foam. $^1$H NMR (400 MHz, $CDCl_3$): $\delta$ 0.03 (s, 3H), 0.05 (s, 3H), 0.88 (s, 9H), 1.45 (d, 3H, $J$ = 1.0 Hz), 2.34 (t, 1H, $J$ = 2.5 Hz), 3.06 (d, 1H, $J$ = 9.0 Hz), 3.19 (d, 1H, $J$ = 10.0 Hz), 3.58 (d, 1H, $J$ = 10.0 Hz), 3.80 (s, 6H), 4.20 (dd, 1H, $J$ = 2.5, 15.5 Hz), 4.23–4.54 (m, 2H), 4.53 (d, 1H, $J$ = 6.0 Hz), 6.14 (d, 1H, $J$ = 4.5 Hz), 6.83–7.38 (m, 13H), 7.51 (d, 1H, $J$ = 1.0 Hz), 8.43 (brs, 1H). High-resolution mass (MALDI-TOF): 751.3021 ($MNa^+$, calcd for $C_{40}H_{48}N_2O_9SiNa$: 751.3027).
6. Compound **8b** (2.96 g, 96% yield) is prepared from **7b** (1.82 g, 4.13 mmol). The eluent for TLC is 1:1 (v/v) hexane/EtOAc ($R_f = 0.4$). The eluent for silica gel column chromatography is 3:1 to 1:1 (v/v) hexane/EtOAc. White foam. $^1$H NMR (500 MHz, $CDCl_3$): $\delta$ −0.02 (s, 3H), 0.03 (s, 3H), 0.84 (s, 9H), 1.28 (d, 3H, $J$ = 6.5 Hz), 1.47 (s, 3H), 2.50 (d, 1H, $J$ = 2.0 Hz), 3.07 (d, 1H, $J$ = 10.0 Hz), 3.18 (d, 1H, $J$ = 9.5 Hz), 3.49 (d, 1H, $J$ = 10.0 Hz), 3.80 (s, 6H), 4.17–4.20 (m, 1H), 4.47–4.53 (m, 2H), 6.24 (d, 1H, $J$ = 3.5 Hz), 6.83–7.39 (m, 13H), 7.50 (s, 1H), 8.39 (brs, 1H). High-resolution mass (MALDI-TOF): 765.3178 ($MNa^+$, calcd for $C_{41}H_{50}N_2O_9SiNa$: 765.3183).
7. Compound **8c** (3.45 g, 85% yield) is prepared from **7c** (2.40 g, 5.45 mmol). The eluent for TLC is 1:1 (v/v) hexane/EtOAc ($R_f = 0.5$). The eluent for silica gel column chromatography is 3:1 to 3:2 (v/v) hexane/EtOAc. White foam. $^1$H NMR (500 MHz, $CDCl_3$): $\delta$ −0.01 (s, 3H), 0.04 (s, 3H), 0.86 (s, 9H), 1.39 (d, 3H, $J$ = 7.0 Hz), 1.41 (s, 3H), 2.18 (d, 1H, $J$ = 2.0 Hz), 3.11 (d, 1H, $J$ = 8.5 Hz), 3.27 (d, 1H, $J$ = 10.0 Hz), 3.80 (s, 6H), 3.84 (d, 1H, $J$ = 10.0 Hz), 4.18 (ddd, 1H, $J$ = 4.0, 6.5, 8.5 Hz), 4.50 (dq, 1H, $J$ = 2.0, 7.0 Hz), 4.58 (d, 1H, $J$ = 6.5 Hz), 6.07 (d, 1H, $J$ = 4.0 Hz), 6.81–7.38

(m, 13H), 7.55 (s, 1H), 8.13 (brs, 1H). High-resolution mass (MALDI-TOF): 765.3179 ($MNa^+$, calcd for $C_{41}H_{50}N_2O_9$-SiNa: 765.3183).

*3.1.6 Synthesis of Compound **9***

1. Add compound **8a** (2.12 g, 2.91 mmol) to a round-bottomed flask equipped with a reflux condenser, a three-way glass stopcock, and a balloon filled with $N_2$. Evacuate the flask using a vacuum line and flush it with $N_2$. Repeat this procedure three times.
2. Add anhydrous THF (15 mL) and 1,1′-thiocarbonyldiimidazole (TCDI) (1.04 g, 5.82 mmol) at room temperature. Reflux the mixture for 3 h in an oil bath (80 °C).
3. Check the progress of the reaction by TLC on aluminum plates using 1:1 (v/v) hexane/EtOAc as the eluent. Visualize the spots with a UV lamp and by dipping the TLC plate into the anisaldehyde solution and heating with a heat gun ($R_f = 0.3$).
4. Concentrate the mixture under reduced pressure, and purify the residue by silica gel column chromatography (silica gel 100 g) using 1:1 to 1:2 (v/v) hexane/EtOAc as the eluent. Combine the appropriate fractions, and evaporate the solvent to give **9a** (2.20 g, 90% yield). White foam. $^1H$ NMR (400 MHz, $CDCl_3$): $\delta$ −0.09 (s, 3H), −0.08 (s, 3H), 0.72 (s, 9H), 1.51 (s, 3H), 2.36 (t, 1H, $J = 2.5$ Hz), 3.26 (d, 1H, $J = 10.0$ Hz), 3.60 (d, 1H, $J = 10.0$ Hz), 3.80 (s, 6H), 4.26 (dd, 1H, $J = 2.5, 16.0$ Hz), 4.34 (dd, 1H, $J = 2.5, 16.0$ Hz), 4.95 (d, 1H, $J = 6.5$ Hz), 6.06 (dd, 1H, $J = 4.5, 6.5$ Hz), 6.42 (d, 1H, $J = 4.5$ Hz), 6.83–6.86 (m, 4H), 7.04 (t, 1H, $J = 1.0$ Hz), 7.26–7.44 (m, 9H), 7.47 (d, 1H, $J = 1.0$ Hz), 7.69 (t, 1H, $J = 1.0$ Hz), 8.43 (s, 1H), 8.73 (brs, 1H). High-resolution mass (MALDI-TOF): 861.2960 ($MNa^+$, calcd for $C_{44}H_{50}N_2O_9SiSNa$: 861.2965).
5. Compound **9b** (2.68 g, 80% yield) is prepared from **8b** (2.92 g, 3.93 mmol). The eluent for TLC is 1:1 (v/v) hexane/EtOAc ($R_f = 0.3$). The eluent for silica gel column chromatography is 3:2 to 2:3 (v/v) hexane/EtOAc. White foam. $^1H$ NMR (300 MHz, $CDCl_3$): $\delta$ −0.18 (s, 3H), −0.10 (s, 3H), 0.66 (s, 9H), 1.35 (d, 3H, $J = 7.0$ Hz), 1.54 (d, 3H, $J = 1.0$ Hz), 2.41 (d, 1H, $J = 2.0$ Hz), 3.12 (d, 1H, $J = 10.0$ Hz), 3.56 (d, 1H, $J = 10.0$ Hz), 3.80 (s, 6H), 4.56 (dq, 2H, $J = 2.0, 7.0$ Hz), 4.95 (d, 1H, $J = 6.5$ Hz), 6.08 (dd, 1H, $J = 3.0, 6.5$ Hz), 6.38 (d, 1H, $J = 3.0$ Hz), 6.82–6.86 (m, 4H), 7.03 (dd, 1H, $J = 1.0, 1.5$ Hz), 7.26–7.42 (m, 10H), 7.73 (t, 1H, $J = 1.5$ Hz), 8.28 (brs, 1H). 8.47 (s, 1H). High-resolution mass (MALDI-TOF): 875.3116 ($MNa^+$, calcd for $C_{45}H_{52}N_4O_9SiSNa$: 875.3122).

6. Compound **9c** (3.06 g, 78% yield) is prepared from **8c** (3.40 g, 4.58 mmol). The eluent for TLC is 1:1 (v/v) hexane/EtOAc ($R_f = 0.3$). The eluent for silica gel column chromatography is 3:2 to 2:3 (v/v) hexane/EtOAc. White foam. $^1$H NMR (300 MHz, $CDCl_3$): δ −0.10 (s, 3H), −0.09 (s, 3H), 0.71 (s, 9H), 1.41 (d, 3H, *J* = 6.5 Hz), 1.44 (d, 3H, *J* = 1.0 Hz), 2.21 (d, 1H, *J* = 2.0 Hz), 3.36 (d, 1H, *J* = 10.5 Hz), 3.80 (s, 6H), 3.87 (d, 1H, *J* = 10.5 Hz), 4.51 (dq, 1H, *J* = 2.0, 6.5 Hz), 4.97 (d, 1H, *J* = 6.5 Hz), 6.00 (dd, 1H, *J* = 5.5, 6.5 Hz), 6.40 (d, 1H, *J* = 5.5 Hz), 6.83–6.86 (m, 4H), 7.04 (dd, 1H, *J* = 0.5, 1.0 Hz), 7.26–7.41 (m, 9H), 7.43 (d, 1H, *J* = 2.0 Hz), 7.51 (d, 1H, *J* = 1.0 Hz), 7.66 (dd, 1H, *J* = 1.0, 2.0 Hz), 8.39 (brs, 1H). High-resolution mass (MALDI-TOF): 875.3120 ($MNa^+$, calcd for $C_{45}H_{52}N_4O_9SiSNa$: 875.3122).

### 3.1.7 Synthesis of Compound 10

1. Add compound **9a** (2.20 g, 2.62 mmol) to a round-bottomed flask equipped with a three-way glass stopcock and a balloon filled with $N_2$. Evacuate the flask using a vacuum line and flush it with $N_2$. Repeat this procedure three times.
2. Add anhydrous toluene (50 mL) and heat the mixture in an oil bath (90 °C). Add $(Me_3Si)_3SiH$ (2.4 mL, 7.86 mmol) and 2,2′-azobis(isobutyronitrile) (AIBN) (86.1 mg, 0.524 mmol), and stir the mixture at 90 °C for 1 h.
3. Check the progress of the reaction by TLC on aluminum plates using 1:1 (v/v) hexane/EtOAc as the eluent. Visualize the spots with a UV lamp and by dipping the TLC plate into the anisaldehyde solution and heating with a heat gun ($R_f = 0.4$).
4. Concentrate the mixture under reduced pressure, and purify the residue by silica gel column chromatography (silica gel 50 g) using 3:1 to 3:2 (v/v) hexane/EtOAc as the eluent. Combine the appropriate fractions, and evaporate the solvent to give **10a** (1.07 g, 57% yield). White foam. $^1$H NMR (400 MHz, $CDCl_3$): δ −0.04 (s, 3H), 0.06 (s, 3H), 0.76 (s, 9H), 1.19 (s, 3H), 3.22 (d, 1H, *J* = 10.0 Hz), 3.23 (d, 1H, *J* = 4.5 Hz), 3.58 (d, 1H, *J* = 10.0 Hz), 3.79 (s, 3H), 3.79 (s, 3H), 4.28 (d, 1H, *J* = 14.0 Hz), 4.44 (d, 1H, *J* = 14.0 Hz), 4.53 (d, 1H, *J* = 4.5 Hz), 5.02 (s, 2H), 5.87 (s, 1H), 6.81–7.40 (m, 13H), 8.08 (s, 1H), 9.03 (brs, 1H). High-resolution mass (MALDI-TOF): 735.3072 ($MNa^+$, calcd for $C_{40}H_{48}N_2O_8SiNa$: 735.3078).
5. Compound **10b** (1.23 g, 55% yield) is prepared from **9b** (2.64 g, 3.09 mmol) in the same way as compound **10a**. The eluent for TLC is 1:1 (v/v) hexane/EtOAc ($R_f = 0.4$). The eluent for silica gel column chromatography is 5:1 to 2:1 (v/v) hexane/EtOAc. White foam. $^1$H NMR (300 MHz, $CDCl_3$): δ

−0.04 (s, 3H), 0.05 (s, 3H), 0.75 (s, 9H), 1.16 (s, 3H), 1.33 (d, 3H, $J$ = 6.5 Hz), 3.19 (d, 1H, $J$ = 5.0 Hz), 3.24 (d, 1H, $J$ = 10.0 Hz), 3.57 (d, 1H, $J$ = 10.0 Hz), 3.79 (s, 6H), 4.44–4.49 (m, 2H), 4.99 (brs, 2H), 5.87 (s, 1H), 6.80–7.40 (m, 13H), 8.07 (s, 1H), 8.28 (brs, 1H). High-resolution mass (MALDI-TOF): 749.3229 ($MNa^+$, calcd for $C_{41}H_{50}N_2O_8$-SiNa: 749.3234).

6. Compound **10c** (1.05 g, 41% yield) is prepared from **9c** (3.00 g, 3.52 mmol) in the same way as compound **10a**. The eluent for TLC is 1:1 (v/v) hexane/EtOAc ($R_f$ = 0.4). The eluent for silica gel column chromatography is 5:1 to 2:1 (v/v) hexane/EtOAc. White foam. $^1$H NMR (300 MHz, $CDCl_3$): $\delta$ −0.05 (s, 3H), 0.04 (s, 3H), 0.75 (s, 9H), 1.18 (d, 3H, $J$ = 1.0 Hz), 1.35 (d, 3H, $J$ = 6.5 Hz), 3.20 (d, 1H, $J$ = 10.0 Hz), 3.23 (d, 1H, $J$ = 5.0 Hz), 3.63 (d, 1H, $J$ = 10.0 Hz), 3.79 (s, 3H), 3.79 (s, 3H), 4.56 (d, 1H, $J$ = 5.0 Hz), 4.67–4.71 (m, 1H), 5.02 (d, 1H, $J$ = 2.0 Hz), 5.18 (d, 1H, $J$ = 2.0 Hz), 5.63 (s, 1H), 6.80–7.40 (m, 13H), 8.06 (brs, 2H). High-resolution mass (MALDI-TOF): 749.3238 ($MNa^+$, calcd for $C_{41}H_{50}N_2O_8SiNa$: 749.3234).

#### *3.1.8 Synthesis of Compound **11***

1. Add compound **10a** (321 mg, 0.450 mmol) to a round-bottomed flask equipped with a three-way glass stopcock and a balloon filled with $N_2$. Evacuate the flask using a vacuum line and flush it with $N_2$. Repeat this procedure three times.
2. Add THF (5.0 mL) and a solution of 1 M tetrabutylammonium fluoride (TBAF) in THF (0.50 mL, 0.50 mmol) at room temperature, and stir the mixture at room temperature for 15 h.
3. Check the progress of the reaction by TLC on aluminum plates using 1:1 (v/v) hexane/EtOAc as the mobile phase. Visualize the spots with a UV lamp and by dipping the TLC plate into the anisaldehyde solution and heating with a heat gun ($R_f$ = 0.1).
4. Concentrate the mixture under reduced pressure, and purify the residue by silica gel column chromatography (silica gel 10 g) using 1:1 to 1:2 (v/v) hexane/EtOAc as the eluent. Combine the appropriate fractions, and evaporate the solvent to give **11a** (234 mg, 87% yield). White foam. $^1$H NMR (400 MHz, $CDCl_3$): $\delta$ 1.27 (d, 3H, $J$ = 1.0 Hz), 2.06 (d, 1H, $J$ = 10.5 Hz), 3.34 (d, 1H, $J$ = 5.0 Hz), 3.40 (d, 1H, $J$ = 10.0 Hz), 3.42 (d, 1H, $J$ = 10.0 Hz), 3.78 (s, 3H), 3.79 (s, 3H), 4.21 (d, 1H, $J$ = 14.0 Hz), 4.43 (d, 1H, $J$ = 14.0 Hz), 4.58 (dd, 1H, $J$ = 5.0, 10.5 Hz), 5.21 (s, 2H), 5.91 (s, 1H), 6.82–7.44 (m, 13H), 7.87 (d, 1H, $J$ = 1.0 Hz), 9.06 (brs, 1H).

High-resolution mass (MALDI-TOF): 621.2207 ($MNa^+$, calcd for $C_{34}H_{34}N_2O_8Na$: 621.2213).

5. Compound **11b** (889 mg, 89% yield) is prepared from **10b** (1.18 g, 1.62 mmol). The eluent for TLC is 1:1 (v/v) hexane/EtOAc ($R_f$ = 0.1). The eluent for silica gel column chromatography is 1:1 to 1:2 (v/v) hexane/EtOAc. White foam. $^1$H NMR (300 MHz, $CDCl_3$): $\delta$ 1.25 (d, 3H, $J$ = 1.0 Hz), 1.36 (d, 3H, $J$ = 6.0 Hz), 1.75 (d, 1H, $J$ = 11.5 Hz), 3.33 (d, 1H, $J$ = 5.5 Hz), 3.39 (d, 1H, $J$ = 10.0 Hz), 3.42 (d, 1H, $J$ = 10.0 Hz), 3.79 (s, 3H), 3.79 (s, 3H), 4.40–4.50 (m, 1H), 4.55 (dd, 1H, $J$ = 5.5, 11.5 Hz), 5.18 (d, 1H, $J$ = 2.0 Hz), 5.23 (d, 1H, $J$ = 1.5 Hz), 5.88 (s, 1H), 6.82–7.45 (m, 13H), 7.89 (d, 1H, $J$ = 1.0 Hz), 8.28 (brs, 1H). High-resolution mass (MALDI-TOF): 635.2364 ($MNa^+$, calcd for $C_{35}H_{36}N_2O_8Na$: 635.2369).
6. Compound **11c** (485 mg, 58% yield) is prepared from **10c** (1.00 g, 1.38 mmol). The eluent for TLC is 1:2 (v/v) hexane/EtOAc ($R_f$ = 0.4). The eluent for silica gel column chromatography is 1:1 to 1:2 (v/v) hexane/EtOAc. White foam. $^1$H NMR (300 MHz, $CDCl_3$): $\delta$ 1.30 (s, 3H), 1.42 (d, 3H, $J$ = 7.0 Hz), 3.34 (d, 1H, $J$ = 5.0 Hz), 3.43 (d, 1H, $J$ = 10.0 Hz), 3.53 (d, 1H, $J$ = 10.0 Hz), 3.79 (s, 3H), 3.79 (s, 3H), 4.45–4.46 (m, 1H), 4.68–4.75 (m, 1H), 5.12 (s, 1H), 5.27 (s, 1H), 5.76 (s, 1H), 6.83–7.42 (m, 13H), 7.81 (s, 1H), 8.01 (brs, 1H). High-resolution mass (MALDI-TOF): 635.2369 ($MNa^+$, calcd for $C_{35}H_{36}N_2O_8Na$: 635.2369).

### 3.1.9 Synthesis of Compound 12

1. Add compound **11a** (324 mg, 0.541 mmol) to a round-bottomed flask equipped with a three-way glass stopcock and a balloon filled with $N_2$. Evacuate the flask using a vacuum line and flush it with $N_2$. Repeat this procedure three times.
2. Add anhydrous $CH_2Cl_2$ (5.0 mL) and cool the mixture in an ice bath. Add *N,N*-diisopropylethylamine (DIPEA) (0.29 mL, 1.62 mmol) and 2-cyanoethyl-*N,N*-diisopropylchlorophosphoramidite (0.14 mL, 0.649 mmol) to the mixture in an ice bath, and stir the mixture for 2 h at room temperature.
3. Check the progress of the reaction by TLC on aluminum plates using 1:1 (v/v) hexane/EtOAc as the eluent. Visualize the spots with a UV lamp and by dipping the TLC plate into the anisaldehyde solution and heating with a heat gun (two diastereomers: $R_f$ = 0.2 and 0.25).
4. Cool the mixture in an ice bath and add a saturated aqueous $NaHCO_3$ solution. Transfer the mixture to a separatory funnel and extract the product with $CH_2Cl_2$. Wash the organic extracts with water and brine, dry over $Na_2SO_4$, and filter out the drying agent using a glass funnel plugged with cotton.

5. Concentrate the filtrate under reduced pressure, and purify the residue by silica gel column chromatography (silica gel 10 g) using 1:1 to 2:3 (v/v) hexane/EtOAc as the eluent. Combine the appropriate fractions, and evaporate the solvent to give **12a** (333 mg, 77% yield). White foam. $^1$H NMR (400 MHz, $CDCl_3$): $\delta$ 0.99 (d, 3.6H, $J$ = 7.0 Hz), 1.09 (d, 3.6H, $J$ = 7.0 Hz), 1.12–1.13 (m, 5.4H), 1.18 (d, 2.4H, $J$= 7.0 Hz), 2.36 (dt, 0.8H, $J$= 6.0, 9.5 Hz), 2.60 (t, 1.2H, $J$ = 6.0 Hz), 3.32–3.85 (m, 13H), 4.26 (d, 0.4H, $J$= 14.0 Hz), 4.27 (d, 0.6H, $J$= 14.0 Hz), 4.42–4.48 (m, 1H), 4.65 (d, 0.6H, $J$= 5.0, 8.0 Hz), 4.72 (d, 0.4H, $J$= 5.0, 9.5 Hz), 5.04–5.10 (m, 2H), 5.88 (s, 0.4H), 5.91 (s, 0.6H), 6.79–7.49 (m, 13H), 7.96–7.99 (m, 2H). High-resolution mass (MALDI-TOF): 821.3286 ($MNa^+$, calcd for $C_{43}H_{51}N_4O_9PNa$: 821.3291).
6. Compound **12b** (656 mg, 71% yield) is prepared from **11b** (700 mg, 1.14 mmol). The eluent for TLC is 1:2 (v/v) hexane/EtOAc (two diastereomers: $R_f$= 0.5 and 0.6). The eluent for silica gel column chromatography is 2:1 to 2:3 (v/v) hexane/EtOAc. White foam. $^1$H NMR (500 MHz, $CDCl_3$): $\delta$ 1.01 (d, 3.6H, $J$ = 6.5 Hz), 1.11–1.35 (m, 14.4H), 2.39–2.42 (m, 0.8H), 2.61 (t, 1.2H, $J$= 6.0 Hz), 3.35–3.81 (m, 13H), 4.49–4.52 (m, 1H), 4.63 (dd, 0.6H, $J$ = 5.0, 7.5 Hz), 4.70 (dd, 0.4H, $J$ = 5.0, 9.0 Hz), 5.02 (s, 1H), 5.09 (s, 0.4H), 5.11 (s, 0.6H), 5.88 (s, 0.4H), 5.91 (s, 0.6H), 6.84–7.47 (m, 13H), 7.98 (s, 0.4H), 7.99 (s, 0.6H), 8.15 (brs, 0.6H), 8.19 (brs, 0.4H). High-resolution mass (MALDI-TOF): 835.3442 ($MNa^+$, calcd for $C_{44}H_{53}N_4O_9PNa$: 835.3448).
7. Compound **12c** (332 mg, 63% yield) is prepared from **11c** (400 mg, 0.653 mmol). The eluent for TLC is 1:2 (v/v) hexane/EtOAc (two diastereomers: $R_f$ = 0.5 and 0.6). The eluent for silica gel column chromatography is 2:1 to 2:3 (v/v) hexane/EtOAc. White foam. $^1$H NMR (500 MHz, $CDCl_3$): $\delta$ 0.98 (d, 3H, $J$= 7.0 Hz), 1.09–1.31 (m, 15H), 2.32–2.36 (m, 1H), 2.58 (t, 1H, $J$ = 6.0 Hz), 3.20–3.80 (m, 13H), 4.67–4.70 (m, 1.5H), 4.76 (dd, 0.5H, $J$ = 5.0, 9.5 Hz), 5.03 (d, 1H, $J$= 1.5 Hz), 5.23 (d, 0.5H, $J$= 1.0 Hz), 5.24 (d, 0.5H, $J$ = 2.0 Hz), 5.68 (s, 0.5H), 5.70 (s, 0.5H), 6.81–7.44 (m, 13H), 7.98 (s, 0.5H), 8.00 (s, 0.5H), 8.06 (brs, 0.5H), 8.07 (brs, 0.5H). High-resolution mass (MALDI-TOF): 835.3440 ($MNa^+$, calcd for $C_{44}H_{53}N_4O_9PNa$: 835.3448).

### 3.2 Synthesis of EoDNA-T Phosphoramidite

#### 3.2.1 Synthesis of Compound 13

1. Add compound **11a** (1.17 g, 1.95 mmol), $CH_2Cl_2$ (20 mL), acetone (10 mL), and a saturated aqueous $NaHCO_3$ solution (400 mL) to a round-bottomed flask equipped with a dropping funnel.
2. Cool the mixture in an ice bath, and add dropwise the solution of Oxone® (18.0 g, 29.3 mmol) in water (100 mL) to the mixture over a period of 2 h using the dropping funnel. Stir the mixture at room temperature for 24 h.
3. Check the progress of the reaction by TLC on aluminum plates using 7:1 (v/v) $CHCl_3$/MeOH as the eluent. Visualize the spots with a UV lamp and by dipping the TLC plate into the anisaldehyde solution and heating with a heat gun ($R_f = 0.5$).
4. Transfer the mixture to a separatory funnel, and extract the product with $CH_2Cl_2$ (100 mL). Wash the organic extracts with water and brine, dry over $Na_2SO_4$, and filter out the drying agent using a glass funnel plugged with cotton.
5. Evaporate the filtrate, and attach a three-way glass stopcock and balloon to the flask.
6. Dissolve the residue in 1,4-dioxane (20 mL), and add 1 M aqueous NaOH solution (20 mL, 20.0 mmol) to the solution at room temperature. Stir the mixture for 12 h at room temperature.
7. Check the progress of the reaction by TLC on aluminum plates using 7:1 (v/v) $CHCl_3$/MeOH as the eluent. Visualize the spots with a UV lamp and by dipping the TLC plate into the anisaldehyde solution and heating with a heat gun ($R_f = 0.3$).
8. Cool the mixture in an ice bath, and neutralize it with a saturated aqueous $NH_4Cl$ solution. Transfer the mixture to a separatory funnel and extract the product with $CH_2Cl_2$. Wash the organic extracts with brine, dry over $Na_2SO_4$, and filter out the drying agent using a glass funnel plugged with cotton.
9. Concentrate the filtrate under reduced pressure, and purify the residue by silica gel column chromatography (silica gel 10 g) using 25:1 to 7:1 (v/v) $CHCl_3$/MeOH as the eluent. Combine the appropriate fractions, and evaporate the solvent to give a mixture of the triol.
10. Dissolve the obtained triol in THF (7.5 mL) and water (2.5 mL), and cool the mixture in an ice bath. Add $NaIO_4$ (285 mg, 1.33 mmol) and stir the mixture for 2 h at room temperature.
11. Check the progress of the reaction by TLC on aluminum plates using 7:1 (v/v) $CHCl_3$/MeOH as the eluent. Visualize the spots with a UV lamp and by dipping the TLC plate into the anisaldehyde solution and heating with a heat gun ($R_f = 0.5$).

12. Cool the mixture in an ice bath and add a saturated aqueous $Na_2S_2O_3$ solution. Transfer the mixture to a separatory funnel and extract the product with EtOAc. Wash the organic extracts with water and brine, dry over $Na_2SO_4$, and filter out the drying agent using a glass funnel plugged with cotton.
13. Concentrate the filtrate under reduced pressure, and purify the residue by silica gel column chromatography (silica gel 20 g) using 1:2 (v/v) hexane/EtOAc as the eluent. Combine the appropriate fractions, and evaporate the solvent to give **13** (346 mg, 30% yield over three steps). White foam. $^1$H NMR (300 MHz, $CDCl_3$): $\delta$ 1.32 (s, 3H), 3.51 (d, 1H, $J$ = 5.0 Hz), 3.57 (d, 1H, $J$ = 10.5 Hz), 3.60 (d, 1H, $J$ = 10.5 Hz), 3.57 (d, 1H, $J$ = 3.0 Hz), 3.76 (s, 3H), 3.77 (s, 3H), 4.25 (s, 2H), 4.95 (dd, 1H, $J$ = 3.0, 5.0 Hz), 5.99 (s, 1H), 6.82–7.43 (m, 13H), 7.71 (s, 1H), 9.31 (brs, 1H). High-resolution mass (MALDI-TOF): 623.2000 ($MNa^+$, calcd for $C_{33}H_{32}N_2O_9Na$: 623.2006).

### 3.2.2 Synthesis of Compound **14**

1. Add compound **13** (50.0 mg, 0.0832 mmol) and MeOH (2.0 mL) to a round-bottomed flask.
2. Add $Et_3N$ (58 μL, 0.416 mmol) and hydrazine monohydrate (8.0 μL, 0.166 mmol) at room temperature, and stir the mixture at room temperature for 4 h.
3. Check the progress of the reaction by TLC on aluminum plates using 1:2 (v/v) hexane/EtOAc as the eluent. Visualize the spots with a UV lamp and by dipping the TLC plate into the anisaldehyde solution and heating with a heat gun ($R_f$ = 0.05).
4. Evaporate the solvent, and attach a three-way glass stopcock and a balloon filled with Ar to the flask. Evacuate the flask using a vacuum line and flush it with Ar. Repeat this procedure three times.
5. Add anhydrous THF (2.0 mL) and cool the mixture in an ice bath. Add DBU (62 μL, 0.416 mmol) and $I_2$ (43.7 mg, 0.166 mmol), and stir the mixture for 2 h at room temperature.
6. Check the progress of the reaction by TLC on aluminum plates using 1:2 (v/v) hexane/EtOAc as the eluent. Visualize the spots with a UV lamp and by dipping the TLC plate into the anisaldehyde solution and heating with a heat gun ($R_f$ = 0.6).
7. Cool the mixture in an ice bath and add a saturated aqueous $Na_2S_2O_3$ solution. Transfer the mixture to a separatory funnel and extract the product with EtOAc. Wash the organic extracts with water and brine, dry over $Na_2SO_4$, and filter out the drying agent using a glass funnel plugged with cotton.
8. Concentrate the filtrate under reduced pressure, and purify the residue by silica gel column chromatography (silica gel 2.0 g)

using 1:1 to 2:3 (v/v) hexane/EtOAc as the eluent. Combine the appropriate fractions, and evaporate the solvent to give **14** (37.8 mg, 64% yield over two steps). White foam. $^{1}$H NMR (300 MHz, $CDCl_3$): δ 1.27 (d, 3H, *J* = 0.5 Hz), 2.27 (d, 1H, *J* = 9.0 Hz), 3.21 (d, 1H, *J* = 4.5 Hz), 3.47 (d, 1H, *J* = 10.5 Hz), 3.51 (d, 1H, *J* = 10.5 Hz), 3.79 (s, 3H), 3.80 (s, 3H), 4.57 (dd, 1H, *J* = 4.5, 9.0 Hz), 5.98 (s, 1H), 6.63 (s, 1H), 6.81–7.43 (m, 13H), 7.69 (d, 1H, *J* = 0.5 Hz), 8.49 (brs, 1H). High-resolution mass (MALDI-TOF): 733.1017 ($MNa^+$, calcd for $C_{33}H_{31}IN_2O_8Na$: 733.1023).

#### 3.2.3 Synthesis of Compound 15

1. Add compound **14** (323 mg, 0.455 mmol), THF (4.0 mL), DBU (0.20 mL, 1.36 mmol), and $PtO_2$ (51.6 mg, 0.228 mmol) to a round-bottomed flask equipped with a three-way glass stopcock. Attach a balloon filled with $H_2$ to the three-way glass stopcock. Then, evacuate the flask using a vacuum line and flush it with $H_2$. Repeat this procedure three times.
2. Stir the mixture at room temperature for 12 h.
3. Check the progress of the reaction by TLC on aluminum plates using 1:2 (v/v) hexane/EtOAc as the eluent. Visualize the spots with a UV lamp and by dipping the TLC plate into the anisaldehyde solution and heating with a heat gun ($R_f = 0.2$).
4. Filter out the catalyst using a glass filter covered with Celite®. Transfer the filtrate to a separatory funnel and dilute with EtOAc. Wash the organic layer with water and brine, dry over $Na_2SO_4$, and filter out the drying agent using a glass funnel plugged with cotton.
5. Concentrate the filtrate under reduced pressure, and purify the residue by silica gel column chromatography (silica gel 5.0 g) using 1:1 to 1:4 (v/v) hexane/EtOAc as the eluent. Combine the appropriate fractions, and evaporate the solvent to give **15** (251 mg, 94% yield). White foam. $^{1}$H NMR (300 MHz, $CDCl_3$): δ 1.30 (s, 3H), 1.70–1.75 (m, 1H), 2.23–2.39 (m, 2H), 2.76 (t, 1H, *J* = 6.0 Hz), 3.35 (d, 1H, *J* = 11.0 Hz), 3.39 (d, 1H, *J* = 11.0 Hz), 3.78 (s, 3H), 3.79 (s, 3H), 3.87–3.90 (m, 2H), 4.42 (t, 1H, *J* = 6.0 Hz), 5.93 (s, 1H), 6.82–7.43 (m, 13H), 7.84 (s, 1H), 8.84 (brs, 1H). High-resolution mass (MALDI-TOF): 609.2207 ($MNa^+$, calcd for $C_{33}H_{34}N_2O_8Na$: 609.2213).

#### 3.2.4 Synthesis of Compound 16

1. Add compound **15** (251 mg, 0.428 mmol) to a round-bottomed flask equipped with a three-way glass stopcock and a balloon filled with Ar. Evacuate the flask using a vacuum line and flush it with Ar. Repeat this procedure three times.

2. Add anhydrous $CH_2Cl_2$ (5.0 mL) and cool the mixture in an ice bath. Add DIPEA (0.23 mL, 1.28 mmol) and 2-cyanoethyl-*N*,*N*-diisopropylchlorophosphoramidite (0.11 mL, 0.513 mmol) to the mixture in an ice bath, and stir the mixture for 3 h at room temperature.
3. Check the progress of the reaction by TLC on aluminum plates using 1:2 (v/v) hexane/EtOAc as the eluent. Visualize the spots with a UV lamp and by dipping the TLC plate into the anisaldehyde solution and heating with a heat gun (two diastereomers: $R_f = 0.5$ and 0.6).
4. Cool the mixture in an ice bath and add a saturated aqueous $NaHCO_3$ solution. Transfer the mixture to a separatory funnel and extract the product with $CH_2Cl_2$. Wash the organic extracts with water and brine, dry over $Na_2SO_4$, and filter out the drying agent using a glass funnel plugged with cotton.
5. Concentrate the filtrate under reduced pressure, and purify the residue by silica gel column chromatography (silica gel 10 g) using 2:1 to 1:2 (v/v) hexane/EtOAc as the eluent. Combine the appropriate fractions, and evaporate the solvent to give **16** (283 mg, 84% yield). White foam. $^1H$ NMR (500 MHz, $CDCl_3$): $\delta$ 1.02–1.21 (m, 15H), 1.65–1.71 (m, 2H), 2.27–2.63 (m, 2H), 2.76–2.80 (m, 0.5H), 2.83–2.86 (m, 0.5H), 3.13 (d, 0.5H, $J$ = 10.0 Hz), 3.24–3.31 (m, 1.5H), 3.53–3.93 (m, 12H), 4.51–4.58 (m, 1H), 5.99 (s, 0.5H), 6.00 (s, 0.5H), 6.74–7.47 (m, 13H), 7.98 (s, 1H), 8.60 (brs, 0.5H), 8.68 (brs, 0.5H). High-resolution mass (MALDI-TOF): 809.3286 ($MNa^+$, calcd for $C_{42}H_{51}N_4O_9PNa$: 809.3291).

### 3.3 Synthesis of (R)-Me-EoDNA-T Phosphoramidite

#### 3.3.1 Synthesis of Compound 17

1. Add compound **10a** (1.30 g, 1.82 mmol), THF (15 mL), and $PtO_2$ (414 mg, 1.82 mmol) to a round-bottomed flask equipped with a three-way glass stopcock. Attach a balloon filled with $H_2$ to the three-way glass stopcock. Then, evacuate the flask using a vacuum line and flush it with $H_2$. Repeat this procedure three times.
2. Stir the mixture at room temperature for 24 h.
3. Check the progress of the reaction by TLC on aluminum plates using 1:1 (v/v) hexane/EtOAc as the eluent. Visualize the spots with a UV lamp and by dipping the TLC plate into the anisaldehyde solution and heating with a heat gun ($R_f = 0.4$).
4. Filter out the catalyst using a glass filter covered with Celite®. Concentrate the filtrate under reduced pressure, and purify the residue by silica gel column chromatography (silica gel 30 g) using 2:1 to 3:2 (v/v) hexane/EtOAc as the eluent. Combine the appropriate fractions, and evaporate the solvent to give **17** (1.13 g, 87% yield). White foam. $^1H$ NMR (300 MHz,

$CDCl_3$): δ −0.01 (s, 3H), 0.07 (s, 3H), 0.75 (s, 9H), 1.26 (s, 3H), 1.26 (d, 3H, $J = 6.5$ Hz), 2.30–2.40 (m, 1H), 2.50–2.53 (m, 1H), 3.17 (d, 1H, $J = 10.5$ Hz), 3.58 (d, 1H, $J = 10.5$ Hz), 3.68 (dd, 1H, $J = 4.5$, 11.0 Hz), 3.79 (s, 6H), 4.01 (dd, 1H, $J = 7.0$, 11.0 Hz), 4.35 (d, 1H, $J = 5.0$ Hz), 5.70 (s, 1H), 6.79–7.41 (m, 13H), 8.08 (s, 1H), 8.50 (brs, 1H). High-resolution mass (MALDI-TOF): 737.3229 ($MNa^+$, calcd for $C_{40}H_{50}N_2O_8SiNa$: 737.3224).

### 3.3.2 Synthesis of Compound **18**

1. Add compound **17** (1.13 g, 1.59 mmol) to a round-bottomed flask equipped with a three-way glass stopcock and a balloon filled with Ar. Evacuate the flask using a vacuum line and flush it with Ar. Repeat this procedure three times.
2. Add THF (10 mL) and a solution of 1 M TBAF in THF (1.8 mL, 1.8 mmol) at room temperature, and stir the mixture at room temperature for 12 h.
3. Check the progress of the reaction by TLC on aluminum plates using 1:1 (v/v) hexane/EtOAc as the eluent. Visualize the spots with a UV lamp and by dipping the TLC plate into the anisaldehyde solution and heating with a heat gun ($R_f = 0.1$).
4. Concentrate the mixture under reduced pressure, and purify the residue by silica gel column chromatography (silica gel 25 g) using 1:1 to 2:3 (v/v) hexane/EtOAc as the eluent. Combine the appropriate fractions, and evaporate the solvent to give **18** (907 mg, 95% yield). White foam. $^1H$ NMR (300 MHz, $CDCl_3$): δ 1.29 (s, 3H), 1.36 (d, 3H, $J = 6.5$ Hz), 2.10 (d, 1H, $J = 7.5$ Hz), 2.19–2.29 (m, 1H), 2.64–2.67 (m, 1H), 3.38 (d, 1H, $J = 10.5$ Hz), 3.44 (d, 1H, $J = 10.5$ Hz), 3.66 (d, 1H, $J = 12.0$ Hz), 3.79 (s, 6H), 4.12 (dd, 1H, $J = 6.0$, 12.0 Hz), 4.40 (dd, 1H, $J = 6.0$, 7.5 Hz), 5.75 (s, 1H), 6.82–7.46 (m, 13H), 7.83 (s, 1H), 8.85 (brs, 1H). High-resolution mass (MALDI-TOF): 623.2364 ($MNa^+$, calcd for $C_{34}H_{36}N_2O_8Na$: 623.2369).

### 3.3.3 Synthesis of Compound **19**

1. Add compound **18** (230 mg, 0.383 mmol) to a round-bottomed flask equipped with a three-way glass stopcock and a balloon filled with Ar. Evacuate the flask using a vacuum line and flush it with Ar. Repeat this procedure three times.
2. Add anhydrous $CH_2Cl_2$ (3.0 mL) and cool the mixture in an ice bath. Add DIPEA (0.21 mL, 1.15 mmol) and 2-cyanoethyl-*N,N*-diisopropylchlorophosphoramidite (0.10 mL, 0.460 mmol) to the mixture in an ice bath, and stir the mixture for 6 h at room temperature.
3. Check the progress of the reaction by TLC on aluminum plates using 1:2 (v/v) hexane/EtOAc as the eluent. Visualize the spots with a UV lamp and by dipping the TLC plate into the

anisaldehyde solution and heating with a heat gun (two diastereomers: $R_f = 0.5$ and 0.6).

4. Cool the mixture in an ice bath and add a saturated aqueous $NaHCO_3$ solution. Transfer the mixture to a separatory funnel and extract the product with $CH_2Cl_2$. Wash the organic extracts with water and brine, dry over $Na_2SO_4$, and filter out the drying agent using a glass funnel plugged with cotton.
5. Concentrate the filtrate under reduced pressure, and purify the residue by silica gel column chromatography (silica gel 5.0 g) using 1:1 to 2:3 (v/v) hexane/EtOAc as the eluent. Combine the appropriate fractions, and evaporate the solvent to give **19** (166 mg, 54% yield). White foam. $^1$H NMR (500 MHz, $CDCl_3$): $\delta$ 0.92 (d, 4.2H, $J$ = 7.0 Hz), 0.99 (d, 2.1H, $J$ = 1.0 Hz), 1.07–1.21 (m, 9.9H), 1.27 (d, 1.8H, $J$ = 7.0 Hz), 1.92–2.00 (m, 0.3H), 2.25–2.45 (m, 1.7H), 2.30–2.65 (m, 3H), 3.23 (d, 0.3H, $J$ = 10.0 Hz), 3.33 (d, 0.7H, $J$ = 10.0 Hz), 3.40–3.67 (m, 6.4H), 3.79–3.90 (m, 8.6H), 4.02–4.10 (m, 1H), 4.47–4.54 (m, 1H), 5.77 (s, 0.3H), 5.78 (s, 0.7H), 6.79–7.49 (m, 13H), 7.97 (d, 0.7H, $J$ = 1.0 Hz), 8.01 (d, 0.3H, $J$ = 1.0 Hz), 8.10 (brs, 0.7H), 8.15 (brs, 0.3H). High-resolution mass (MALDI-TOF): 823.3442 ($MNa^+$, calcd for $C_{43}H_{53}N_4O_9PNa$: 823.3448).

### 3.4 Synthesis of (S)-Me-EoDNA-T Phosphoramidite

#### 3.4.1 Synthesis of Compound 20

1. Add compound **8a** (9.50 g, 13.0 mmol), EtOAc (150 mL), and 5% (w/w) Lindlar's catalyst (555 mg, 0.261 mmol) to a round-bottomed flask equipped with a three-way glass stopcock. Attach a balloon filled with $H_2$ to the three-way glass stopcock. Then, evacuate the flask using a vacuum line and flush it with $H_2$. Repeat this procedure three times.
2. Stir the mixture at room temperature for 3 h.
3. Check the progress of the reaction by TLC on aluminum plates using 1:1 (v/v) hexane/EtOAc as the eluent. Visualize the spots with a UV lamp and by dipping the TLC plate into the anisaldehyde solution and heating with a heat gun ($R_f = 0.45$).
4. Filter out the catalyst using a glass filter covered with Celite®. Concentrate the filtrate under reduced pressure, and purify the residue by silica gel column chromatography (silica gel 250 g) using 3:2 (v/v) hexane/EtOAc as the eluent. Combine the appropriate fractions, and evaporate the solvent to give **20** (8.57 g, 90% yield). White foam. $^1$H NMR (300 MHz, $CDCl_3$): $\delta$ −0.01 (s, 3H), 0.05 (s, 3H), 0.86 (s, 9H), 1.43 (d, 3H, $J$ = 1.0 Hz), 3.13 (d, 1H, $J$ = 9.0 Hz), 3.14 (d, 1H, $J$ = 10.0 Hz), 3.57 (d, 1H, $J$ = 10.0 Hz), 3.80 (s, 6H), 3.94–3.98 (m, 1H), 4.12–4.23 (m, 2H), 4.55 (d, 1H, $J$ = 6.0 Hz), 5.05–5.18 (m, 2H), 5.73–5.83 (m, 1H), 6.09 (d, 1H, $J$ = 4.5 Hz), 6.83–7.39 (m, 13H), 7.54 (d, 1H,

$J = 1.0$ Hz), 8.33 (brs, 1H). High-resolution mass (MALDI-TOF): 753.3178 ($MNa^+$, calcd for $C_{40}H_{50}N_2O_9SiNa$: 753.3183).

*3.4.2 Synthesis of Compound **21***

1. Add compound **20** (2.00 g, 2.74 mmol) to a round-bottomed flask equipped with a reflux condenser, a three-way glass stopcock, and a balloon filled with Ar. Evacuate the flask using a vacuum line and flush it with Ar. Repeat this procedure three times.
2. Add anhydrous THF (20 mL) and TCDI (975 mg, 5.47 mmol) at room temperature, and reflux the mixture for 2 h in an oil bath (80 °C).
3. Check the progress of the reaction by TLC on aluminum plates using 1:1 (v/v) hexane/EtOAc as the eluent. Visualize the spots with a UV lamp and by dipping the TLC plate into the anisaldehyde solution and heating with a heat gun ($R_f = 0.3$).
4. Concentrate the mixture under reduced pressure, and purify the residue by silica gel column chromatography (silica gel 50 g) using 3:2 (v/v) hexane/EtOAc as the eluent. Combine the appropriate fractions, and evaporate the solvent to give **21** (1.92 g, 83% yield). White foam. $^1H$ NMR (300 MHz, $CDCl_3$): $\delta$ −0.11 (s, 3H), −0.09 (s, 3H), 0.71 (s, 9H), 1.48 (d, 3H, $J = 1.0$ Hz), 3.22 (d, 1H, $J = 10.0$ Hz), 3.60 (d, 1H, $J = 10.0$ Hz), 3.80 (s, 6H), 4.02 (ddt, 1H, $J = 1.0$, 6.0, 12.0 Hz), 4.19 (ddt, 1H, $J = 1.0$, 6.0, 12.0 Hz), 4.93 (d, 1H, $J = 6.0$ Hz), 5.07–5.21 (m, 2H), 5.74–5.87 (m, 1H), 6.02 (dd, 1H, $J = 5.0$, 6.0 Hz), 6.42 (d, 1H, $J = 5.0$ Hz), 6.84–6.87 (m, 4H), 7.03 (t, 1H, $J = 1.0$ Hz), 7.26–7.45 (m, 9H), 7.51 (d, 1H, $J = 1.0$ Hz), 7.65 (t, 1H, $J = 1.0$ Hz), 8.38 (d, 1H, $J = 1.0$ Hz), 8.51 (brs, 1H). High-resolution mass (MALDI-TOF): 863.3116 ($MNa^+$, calcd for $C_{44}H_{52}N_4O_8SiSNa$: 863.3122).

*3.4.3 Synthesis of Compound **22***

1. Add compound **21** (1.92 g, 2.28 mmol) to a round-bottomed flask equipped with a three-way glass stopcock and a balloon filled with Ar. Evacuate the flask using a vacuum line and flush it with Ar. Repeat this procedure three times.
2. Add anhydrous toluene (20 mL) and heat the solution in an oil bath (90 °C). Add $(Me_3Si)_3SiH$ (1.1 mL, 3.42 mmol) and AIBN (75.0 mg, 0.457 mmol), and stir the mixture for 1 h in an oil bath (90 °C).
3. Check the progress of the reaction by TLC on aluminum plates using 1:1 (v/v) hexane/EtOAc as the eluent. Visualize the spots with a UV lamp and by dipping the TLC plate into the anisaldehyde solution and heating with a heat gun [$R_f = 0.4$ (**22**, high polar), 0.4 (**17**, less polar)].

4. Concentrate the mixture under reduced pressure, and purify the residue by silica gel column chromatography (silica gel 60 g) using 3:1 to 3:2 (v/v) hexane/EtOAc as the eluent. Combine the appropriate fractions, and evaporate the solvents to give **22** (1.08 g, 66% yield) and **17** (0.13 g, 8% yield). Compound **22**. White foam. $^1$H NMR (300 MHz, $CDCl_3$): δ −0.03 (s, 3H), 0.07 (s, 3H), 0.80 (s, 9H), 1.04 (d, 3H, $J = 6.5$ Hz), 1.22 (s, 3H), 1.39–1.41 (m, 1H), 2.59–2.67 (m, 1H), 3.17 (d, 1H, $J = 10.0$ Hz), 3.43 (t, 1H, $J = 11.5$ Hz), 3.52 (d, 1H, $J = 10.0$ Hz), 3.72–3.79 (m, 7H), 4.40 (d, 1H, $J = 5.0$ Hz), 5.96 (s, 1H), 6.80–7.40 (m, 13H), 8.07 (s, 1H), 8.82 (brs, 1H). High-resolution mass (MALDI-TOF): 737.3229 ($MNa^+$, calcd for $C_{40}H_{50}N_2O_8SiNa$: 737.3234).

*3.4.4 Synthesis of Compound* ***23***

1. Add compound **23** (900 mg, 1.26 mmol) to a round-bottomed flask equipped with a three-way glass stopcock and a balloon filled with Ar. Evacuate the flask using a vacuum line and flush it with Ar. Repeat this procedure three times.
2. Add THF (10 mL) and a solution of 1 M TBAF in THF (1.4 mL, 1.4 mmol) at room temperature, and stir the mixture at room temperature for 9 h.
3. Check the progress of the reaction by TLC on aluminum plates using 1:1 (v/v) hexane/EtOAc as the eluent. Visualize the spots with a UV lamp and by dipping the TLC plate into the anisaldehyde solution and heating with a heat gun ($R_f = 0.1$).
4. Concentrate the mixture under reduced pressure, and purify the residue by silica gel column chromatography (silica gel 30 g) using 1:1 to 1:2 (v/v) hexane/EtOAc as the eluent. Combine the appropriate fractions, and evaporate the solvent to give **23** (612 mg, 81% yield). White foam. $^1$H NMR (300 MHz, DMSO-$d_6$): δ 0.89 (d, 3H, $J = 6.5$ Hz), 1.15 (s, 3H), 2.29–2.30 (m, 1H), 2.52–2.58 (m, 1H), 3.04 (d, 1H, $J = 10.0$ Hz), 3.18 (d, 1H, $J = 10.0$ Hz), 3.31 (t, 1H, $J = 11.5$ Hz), 3.63 (dd, 1H, $J = 8.5, 11.5$ Hz), 3.73 (s, 6H), 4.34 (t, 1H, $J = 6.0$ Hz), 5.52 (d, 1H, $J = 6.0$ Hz), 5.84 (s, 1H), 6.86–7.42 (m, 13H), 7.70 (s, 1H), 11.33 (brs, 1H). High-resolution mass (MALDI-TOF): 623.2364 ($MNa^+$, calcd for $C_{34}H_{36}N_2O_8Na$: 623.2369).

*3.4.5 Synthesis of Compound* ***24***

1. Add compound **23** (372 mg, 0.619 mmol) to a round-bottomed flask equipped with a three-way glass stopcock and a balloon filled with Ar. Evacuate the flask using a vacuum line and flush it with Ar. Repeat this procedure three times.
2. Add anhydrous $CH_2Cl_2$ (5.0 mL) and cool the mixture in an ice bath. Add DIPEA (0.55 mL, 3.10 mmol) and 2-cyanoethyl-*N*,*N*-diisopropylchlorophosphoramidite (0.21 mL,

0.929 mmol) to the mixture in an ice bath, and stir the mixture for 4 h at room temperature.

3. Check the progress of the reaction by TLC on aluminum plates using 1:2 (v/v) hexane/EtOAc as the eluent. Visualize the spots with a UV lamp and by dipping the TLC plate into the anisaldehyde solution and heating with a heat gun (two diastereomers: $R_f = 0.5$ and 0.6).
4. Cool the mixture in an ice bath and add a saturated aqueous $NaHCO_3$ solution. Transfer the mixture to a separatory funnel and extract the product with $CH_2Cl_2$. Wash the organic extracts with water and brine, dry over $Na_2SO_4$, and filter out the drying agent using a glass funnel plugged with cotton.
5. Concentrate the filtrate under reduced pressure, and purify the residue by silica gel column chromatography (silica gel 15 g) using 1:1 to 2:3 (v/v) hexane/EtOAc as the eluent. Combine the appropriate fractions, and evaporate the solvent to give **24** (287 mg, 58% yield). White foam. $^1$H NMR (500 MHz, $CDCl_3$): $\delta$ 1.00 (d, 3.6H, $J = 7.0$ Hz), 1.03–1.14 (m, 12H), 1.27 (d, 2.4H, $J = 7.0$ Hz), 2.27 (dt, 0.4H, $J = 6.0$, 17.0 Hz), 2.40 (dt, 0.4H, $J = 7.0$, 17.0 Hz), 2.48–2.49 (m, 0.4H), 2.58–2.72 (m, 2.8H), 3.24 (d, 0.4H, $J = 10.0$ Hz), 3.23–3.34 (m, 1.4H), 3.44 (t, 1.2H, $J = 12.0$ Hz), 3.53–3.87 (m, 11H), 4.54–4.62 (m, 1H), 6.00 (s, 0.4H), 6.01 (s, 0.6H), 6.80–7.41 (m, 13H), 7.98 (s, 1H), 8.64 (brs, 1H). High-resolution mass (MALDI-TOF): 823.3445 ($MNa^+$, calcd for $C_{43}H_{53}N_4O_9PNa$: 823.3448).

### 3.5 Synthesis of the Oligonucleotides

1. Weigh the proper amount of thymidine phosphoramidite, $N^4$-acetyl-2′-deoxy-5-methylcytidine phosphoramidite, $N^4$-benzoyl-2′-deoxycytidine phosphoramidite, $N^6$-benzoyl-2′-deoxyadenosine phosphoramidite, $N^2$-isobutyryl-2′-deoxyguanosine phosphoramidite, and compounds **12a–c**, **16**, **19**, and **24**. Dilute the phosphoramidites to 0.1 M with anhydrous MeCN.
2. Install all reagents (i.e., deblocking solution, activator solution, cap A solution, cap B solution, oxidizing solution, and acetonitrile) and phosphoramidite solutions to the DNA synthesizer. To synthesize oligonucleotides including methylene-EoDNAs **12a–12c**, install a 1 M *tert*-BuOOH solution [18] as an oxidizing solution to the DNA synthesizer.
3. Prolong the coupling time for the introduction of **12a–c**, **16**, **19**, and **24** from 32 s (standard coupling time) to 10 min (*see* **Notes 5** and **6**).
4. Prime the lines of all reagents and phosphoramidites, and install the controlled pore glass (CPG) column with a 0.2 μmol scale.

5. Perform the synthesis of oligonucleotide in the trityl-on mode (*see* **Note 7**).
6. Remove the CPG column from the instrument.
7. Treat the CPG column with 28% aqueous ammonia solution (1 mL) for 1.5 h at room temperature to cleave the oligonucleotide from the CPG column. Carry out an additional treatment with 28% aqueous ammonia for 16 h at 55 °C to remove the all protecting groups.
8. Transfer the solution into a microcentrifuge tube, and evaporate the solvent using SpeedVac concentrator at room temperature for 1 h. Dilute the residue with 2–4 mL of 0.1 M TEAA buffer (pH 7.0).
9. Perform "quick and easy purification" of the diluted solution using a Sep-Pak® C18 plus short cartridge according to the protocol of the Sep-Pak® C18 cartridge (*see* **Note 8**). Concentrate the appropriate fractions containing oligonucleotides using the SpeedVac concentrator. Dissolve the residues in 50–100 μL of 0.1 M TEAA buffer (pH 7.0), and combine them.
10. Analyze the solution using reversed-phase high-performance liquid chromatography (HPLC). If necessary, purify the solution using reversed-phase HPLC, and after SpeedVac concentration to remove MeCN, lyophilize the residue to obtain a pure oligonucleotide. Confirm the purity of the oligonucleotide by reversed-phase HPLC and the composition of the oligonucleotide by MALDI-TOF mass analysis using 1:1 (v/v) 3-hydroxypicolinic acid (10 mg/mL)/diammonium citrate (1 mg/mL) as a matrix.

    HPLC conditions:

    Column: Waters XBridge® MS $C_{18}$ 2.5 μm, 4.6 × 50 mm (for analysis) and 10 × 50 mm (for purification).

    Temperature: 50 °C.

    Flow rate: 1 mL/min (for analysis) and 4.5 mL/min (for purification).

    Gradient eluent: 5–20% or 10–25% MeCN in 0.1 M TEAA buffer (pH 7.0) for 30 min (*see* **Note 9**), UV detection at 260 nm.
11. Dissolve the lyophilized oligonucleotide in 200–500 μL Milli-Q water to prepare a stock solution. Calculate the concentration of the stock solution and the total quantity of the oligonucleotide by measuring the absorbance at 260 nm.

### 3.6 UV-Melting Experiments

1. Sample preparation [condition: 10 mM sodium cacodylate (pH 7.2), 140 mM KCl, 4 μmol of each oligonucleotide]: Add 13.0 μL of 100 mM sodium cacodylate buffer (pH 7.2), 18.2 μL of 1 M KCl solution, and 520 pmol of each

oligonucleotide (*see* **Note 10**) in a microcentrifuge tube, and dilute the solution with distilled water to a total volume of 130 μL.

2. Put the prepared sample in boiling water, and allow to cool to room temperature slowly.
3. Load the samples including blank into an 8-series microcell for $T_m$ measurement. Record the absorbance at 260 nm from 20 to 95 °C at a scan rate of 0.5 °C/min.
4. Determine $T_m$ values by the two-point average method and the final values by averaging three independent measurements (the errors are within 1 °C).

### 3.7 Enzymatic Degradation Experiments

1. Sample preparation [condition: 2.5 μg/mL CAVP, 10 mM $MgCl_2$, 50 mM Tris-HCl buffer (pH 8.0), 7.5 μmol of an oligonucleotide]: Add 5.0 μL of 1.0 M Tris-HCl buffer (pH 8.0), 10.0 μL of 100 mM $MgCl_2$, and 750 pmol of each oligonucleotide (*see* **Note 10**) in a microcentrifuge tube, and dilute the solution with distilled water to a total volume of 75 μL. Add 25 μL of CAVP (10 μg/mL) to the solution.
2. Vortex the sample, and centrifuge briefly to collect the solution at the bottom of the tube.
3. Incubate the mixture at 37 °C using a block incubator.
4. Quenching of the enzyme reaction: At each predetermined time after starting the incubation, transfer the mixture (10 μL) into a microcentrifuge tube, and quickly put the sample into the block incubator (90 °C). Incubate the sample at 90 °C for 2 min, and then place the sample on ice.
5. Analyze the amount of the intact oligonucleotide using reversed-phase HPLC under the following conditions:

   Column: Waters XBridge® MS $C_{18}$ 2.5 μm, 3.0 × 50 mm.

   Temperature: 50 °C.

   Flow rate: 0.8 mL/min.

   Gradient eluent: 6–12% or 7–13% MeCN in 0.1 M TEAA buffer (pH 7.0) for 15 min (*see* **Note 9**), UV detection at 260 nm.

## 4 Notes

1. The anhydrous mCPBA must be freshly prepared. The presence of water leads to a drastic reduction in the yield of **7a**.
2. The anisaldehyde solution is useful for detecting nucleosides. A phosphomolybdic acid (PMA) solution can be used. To prepare the PMA solution, weigh 5 g phosphomolybdic acid hydrate, and dissolve it in 100 mL EtOH.

3. Do not wash the organic extracts with water because compound **2** is slightly soluble in water.
4. Preparation of 1 M DMTrOTf in anhydrous $CH_2Cl_2$ [20]: Under anhydrous conditions, add DMTrCl (874 mg, 2.58 mmol) and anhydrous $CH_2Cl_2$ (2.6 mL) to a flask. Cool the mixture in an ice bath, and then add AgOTf (663 mg, 2.58 mmol). Stir the mixture for 0.5 h in an ice bath.
5. The coupling yield at the prolonged coupling time is estimated to be over 95% yield by a trityl monitor.
6. When detritylation immediately after coupling of phosphoramidites **12a–12c**, **16**, **19**, and **24** becomes slow, the waiting time for detritylation should be prolonged from 4 s × 2 (standard condition) to 30 s × 2.
7. Synthesis in the trityl-off mode is also possible. In this case, HPLC purification without the quick and easy purification step using a Sep-Pak® C18 cartridge should be performed.
8. In this step, the 5′-DMTr group of the synthesized oligonucleotide is removed.
9. The proper gradient eluent should be used for HPLC analysis and purification of the synthesized oligonucleotides, because the polarity of oligonucleotides generally depends on their sequence and the type of chemical modifications.
10. It is recommended to prepare a diluted solution of approximately 30 μM of the oligonucleotide, which is useful for sample preparation for UV-melting and enzymatic degradation experiments.

## References


1. Yamamoto T, Nakatani M, Narukawa K, Obika S (2011) Antisense drug discovery and development. Future Med Chem 3:339–365
2. Sharma VK, Rungta P, Prasad AK (2014) Nucleic acid therapeutics: basic concepts and recent developments. RSC Adv 4:16618–16631
3. Lam JKW, Chow MYT, Zhang Y, Leung SWS (2015) siRNA versus miRNA as therapeutics for gene silencing. Mol Ther Nucleic Acids 4:e252
4. Wan WB, Seth PP (2016) The medicinal chemistry of therapeutic oligonucleotides. J Med Chem 59:9645–9667
5. Sridharan K, Gogtay NJ (2016) Therapeutic nucleic acids: current clinical status. Br J Clin Pharmacol 82:659–672
6. Morihiro K, Kasahara Y, Obika S (2017) Biological applications of xeno nucleic acids. Mol BioSyst 13:235–245
7. Khvorova A, Watts JK (2017) The chemical evolution of oligonucleotide therapies of clinical utility. Nat Biotechnol 35:238–248
8. Kaur H, Babu BR, Maiti S (2007) Perspectives on chemistry and therapeutic applications of locked nucleic acid (LNA). Chem Rev 107:4672–4697
9. Veedu RN, Wengel J (2010) Locked nucleic acids: promising nucleic acid analogs for therapeutic applications. Chem Biodivers 7:536–542
10. Hagedorn PH, Persson R, Funder ED, Albæk N, Diemer SL, Hansen DJ, Møller MR, Papargyri N, Christiansen H, Hansen BR, Hansen HF, Jensen MA, Koch T (2017) Locked nucleic acid: modality, diversity, and drug discovery. Drug Discov Today 23:101–114
11. Obika S, Rahman SMA, Fujisaka A, Kawada Y, Baba T, Imanishi T (2010) Bridged nucleic

acids: development, synthesis and properties. Heterocycles 81:1347–1392

12. Zhou C, Chattopadhyaya J (2012) Intramolecular free-radical cyclization reactions on pentose sugars for the synthesis of carba-LNA and carba-ENA and the application of their modified oligonucleotides as potential RNA targeted therapeutics. Chem Rev 112:3808–3832
13. Hari Y, Obika S (2016) Synthesis and properties of 2′,4′-bridged nucleic acids containing multiple heteroatoms in the bridges. J Syn Org Chem Jpn 74:141–153. (in Japanese)
14. Hari Y, Morikawa T, Osawa T, Obika S (2013) Synthesis and properties of 2′-*O*,4′-*C*-ethyleneoxy bridged 5-methyluridine. Org Lett 15:3702–3705
15. Osawa T, Obika S, Hari Y (2016) Synthesis and properties of novel 2′-*C*,4′-*C*-ethyleneoxy-bridged 2′-deoxyribonucleic acids with exocyclic methylene groups. Org Biomol Chem 14:9481–9484
16. Osawa T, Sawamura M, Wada F, Yamamoto T, Obika S, Hari Y (2017) Synthesis, duplex-forming ability, enzymatic stability, and *in vitro* antisense potency of oligonucleotides including 2′-*C*,4′-*C*-ethyleneoxy-bridged thymidine derivatives. Org Biomol Chem 15:3955–3963
17. Hakimelahi GH, Proba ZA, Ogilvie KK (1982) New catalysts and procedures for the dimethoxytritylation and selective silylation of ribonucleosides. Can J Chem 60:1106–1113
18. Hayakawa Y, Uchiyama M, Noyori R (1986) Nonaqueous oxidation of nucleoside phosphites to the phosphates. Tetrahedron Lett 27:4191–4194
19. Smyslova P, Poga I, Lyčka A, Tejral G, Hlavac J (2015) Non-catalyzed click reactions of ADIBO derivatives with 5-methyluridine azides and conformational study of the resulting triazoles. PLoS One 10:e0144613
20. Tarköy M, Bolli M, Leumann C (1994) Synthesis, pairing properties, and calorimetric determination of duplex and triplex stability of decanucleotides from [(3′*S*,5′*R*)-2′-deoxy-3′,5′-ethano-β-D-ribofuranosyl]adenine and-thymine. Helv Chim Acta 77:716–744

# Chapter 6

# Synthesis Protocols for Simple Uncharged Glycol Carbamate Nucleic Acids

**Tanaya Bose and Vaijayanti A. Kumar**

## Abstract

Glycol carbamate nucleic acid (GCNA) oligomers can be produced from activated carbonate monomers. The synthesized monomers can be very conveniently characterized employing analytical tools like NMR and HR-MS. Moreover, the activated carbonate monomers do not require coupling agents, and hence excess monomers can be recovered at the end of each coupling. Here we illustrate the synthesis of activated glycol carbonate monomers and their subsequent application in synthesis of carbamate oligomers.

**Key words** Uncharged nucleic acids, L-Serine, Chiral nucleic acid mimics, Glycol nucleic acids

## 1 Introduction

Non-natural nucleic acids play an important role in antisense therapy. Numerous non-natural nucleic acids have been reported till date, where nucleic acids with uncharged backbone occupy an immense volume in the history of antisense therapy. Among other uncharged backbones reported are peptide nucleic acid (PNA) [1], phosphorodiamidate morpholino oligomer (PMO) [2], and carbamate linkages [3]. In PNA, nucleobases are attached to the polyamide backbone through conformationally rigid tertiary acetamide linker [4]. PNA forms a strong and sequence-specific binding to both complementary DNA and RNA [5]. Carbamate linkages were first reported in 1974 by Michael J. Gait et al., where these linkages were utilized to form DNA mimics in a dinucleotide analogue containing the oxyformamido-linkage, thymidinylformamido-[3′ (O) 5′ (C)]-5′-deoxythymidine. A number of attempts have been made since then to synthesize different carbamate analogues as these linkages are stable under physiological conditions and are resistant to nuclease action.

Revolutionary research by Eschenmoser's group proved that α-threofuranosyl nucleic acid (TNA) in which the pentose sugar in

Nathaniel Shank (ed.), *Non-Natural Nucleic Acids: Methods and Protocols*, Methods in Molecular Biology, vol. 1973, https://doi.org/10.1007/978-1-4939-9216-4_6, © Springer Science+Business Media, LLC, part of Springer Nature 2019

DNA/RNA was replaced by an atom-edited cyclic form of tetrose sugar [6] is capable of forming antiparallel duplexes by self-pairing and was also able to cross-pair with cDNA and RNA [7]. Inspired by Eschenmoser's TNA structure, Eric Meggers further structurally simplified TNA to an atom economic novel nucleic acid analogue (an acyclic version of TNA), known as glycol nucleic acid (GNA) [8] where *R* and *S* monomers were synthesized in just three steps.

Earlier, pyrrolidinyl carbamate oligonucleotides [9] were reported by our group. The flexibility in the linker group led to destabilization of the complexes. In another report by our group, the internucleoside polyamide linkages in PNA backbone were replaced by carbamate linkages, and the nucleobase linker was retained as in PNA to get polycarbamate nucleic acids (PCNA). Both *R*-PCNA [10] and *S*-PCNA [11] formed more stable duplex with DNA than RNA. Here, we describe the synthesis of carbamate analogue of GNA named as glycol carbamate nucleic acid (GCNA). Below, we also elaborately describe the synthesis of the monomers required for the synthesis of GCNA oligomers.

## 2 Materials

All the non-aqueous reactions were carried out under the inert atmosphere of nitrogen/argon, and the chemicals used were of laboratory or analytical grade. All solvents used were dried and distilled according to standard protocols. TLCs were carried out on pre-coated silica gel 60 $F_{254}$. Column chromatographic separations were performed using silica gel 60–120 mesh or 200–400 mesh and using the solvent systems ethyl acetate (EtOAc)/petroleum ether and methanol/dichloromethane (DCM). $^{1}$H and $^{13}$C NMR spectra were obtained using Bruker AC-200, AC-400, and AC-500 NMR spectrometers. The chemical shifts are reported in delta ($\delta$) values and referred to internal standard trimethylsilane for $^{1}$H. High-resolution mass spectra were recorded on a Thermo Fisher Scientific Q Exactive mass spectrometer.

### *2.1 Synthesis of Monomer Units*

1. Methanol.
2. Sodium borohydride ($NaBH_4$).
3. Ammonium chloride ($NH_4Cl$).
4. *N*3-Benzoylthymine.
5. Triphenylphosphine.
6. Tetrahydrofuran (THF).
7. $N^6$-Benzyloxycarbonyladenine.
8. Tetra-*n*-butylammonium fluoride (TBAF).
9. Pyridine.

10. Triethylamine (TEA).
11. *p*-Nitrophenylchloroformate.
12. Dioxane.
13. Ammonia.
14. Kromasil 5-Amycoat (4.6 × 250 mm) column for chiral HPLC.

**2.2 Solid-Phase Carbamate**

1. Boc-Lys(Z)-OH.
2. Picric acid.
3. 1,2-ethanedithiol.
4. Trifluoroacetic acid (TFA).
5. Trifluoromethane sulfonic acid (TFMSA).
6. Diethyl ether.
7. Acetonitrile (ACN).
8. Thioanisole.
9. Diisopropylethylamine (DIPEA).
10. Ninhydrin.
11. Phenol.
12. *tert*-butanol.
13. Ethanol.
14. Solid-phase flask.
15. Sintered funnel for resin filtration.
16. Cary Varian spectrophotometer.
17. Waters HPLC system for purifying the synthesized oligomers.

## 3 Methods

### 3.1 Synthesis of Monomer Units

The naturally occurring L-serine was versatile starting material for the synthesis of *R*- monomer. However, *S*-monomer could also be easily obtained by simple maneuver of the protecting groups. The hydroxyl group in L-serine was implemented for activation with *p*-nitrophenylchloroformate to obtain the activated carbonate monomers in case of *R*-, while for *S*-, the acid could be reduced to alcohol to achieve hydroxyl functionality (*see* **Note 1**).

#### 3.1.1 Synthesis of R-Monomer Units

The *R*-monomer units were prepared starting from naturally occurring L-serine as depicted in Fig. 1. Reduction of the ester group in L-serine derivative **1** [10] produced alcohol **2** using $NaBH_4$ in MeOH. Mitsunobu reaction was involved for the attachment of protected nucleobases (*N*3-benzoylthymine and $N^{6-}$Cbz adenine) to yield the monomer precursors **3** and **4**. The TBS group was

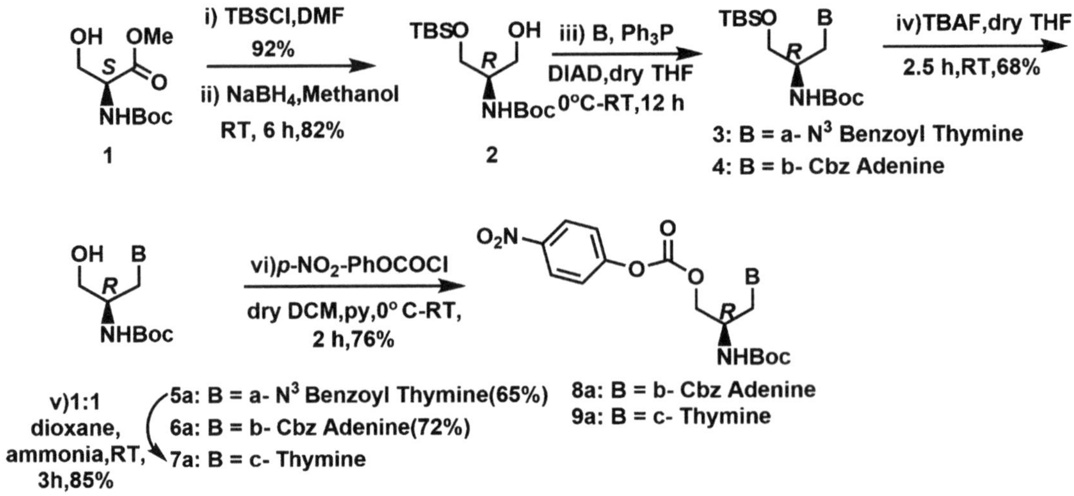


**Fig. 1** Synthesis of *R*-GCNA monomer units

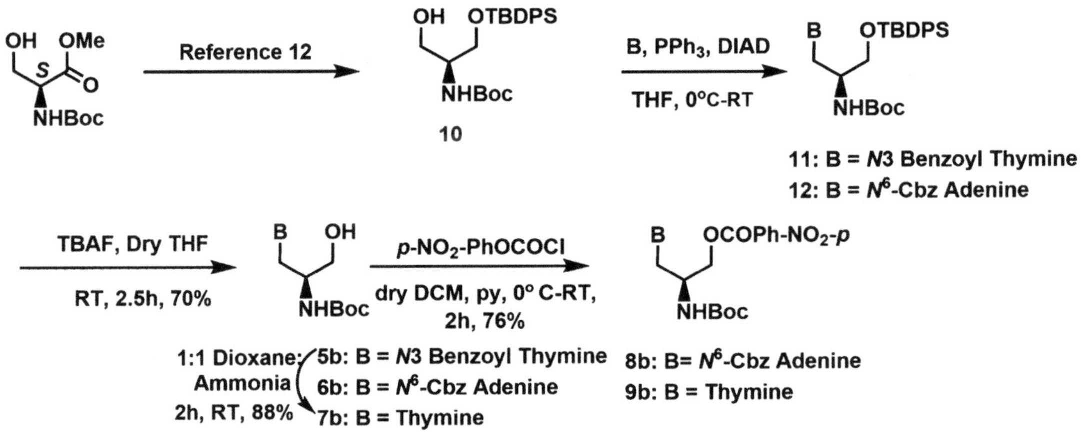


**Fig. 2** Synthesis of *S*-GCNA monomer units

deprotected using TBAF in THF to get the *R*-monomeric units **5a, 6a**. Compound **5a** was debenzoylated using 1:1 dioxane/aq. ammonia to yield **7a**. Monomers **6a** and **7a** were then activated using *p*-nitro-phenyloxycarbonyl chloride yielding **8a** and **9a**, respectively, in good overall yield.

#### *3.1.2 Synthesis of S- monomer units*

Alternatively, the *S*-enantiomers were synthesized from L-serine according to literature reports [12] via Garner's alcohol as in Fig. 2. Mitsunobu reaction with nucleobases (*N*3-benzoylthymine and *N*$^{6-}$Cbz adenine) yielded compounds **11** and **12**, respectively. Removal of TBDPS group was accomplished with TBAF in THF to give *S*-monomer units **5b, 6b**. Debenzylation of **5b** gave **7b** in quantitative yield. Monomers **6b, 7b** were activated using *p*-nitrophenyloxycarbonyl chloride to yield **8b** and **9b**, respectively.

The activated monomers **8a, 9a** and **8b, 9b** could be directly used for solid phase oligocarbamate synthesis. All the new compounds were adequately characterized by $^{1}H$, $^{13}C$ NMR, and HR-MS analysis.

### 3.2 Procedure

#### 3.2.1 (R)-Tert-butyl (1-((tert-butyldimethylsilyl) oxy)-3-hydroxypropan-2-yl) Carbamate (2)

1. A solution of *S*-methyl 2-((*tert*-butoxycarbonyl) amino)-3-((*tert*-butyldimethylsilyl) oxy) propanoate (29 g, 87.1 mmol) in methanol (250 mL) was cooled to 0 °C in an ice bath.
2. Solid $NaBH_4$ (4.9 g, 130.63 mmol) was added in portions for a period of 30 min.
3. The reaction mixture was stirred for a period of 3 h and finally quenched using $NH_4Cl$ solution till pH was neutral.
4. Methanol was removed under reduced pressure, and the residue was extracted with EtOAc (4 × 100 mL), washed with water and brine. The organic layer was dried over anhydrous $Na_2SO_4$, and solvent was removed under reduced pressure.
5. The residue was purified by column chromatography (20% EtOAc in petroleum ether) affording compound **2** (24.1 g, 83%) as colorless oil.
6. $[\alpha]_D^{20}$ + 16.2° (c 0.021, $CHCl_3$); $^{1}H$ NMR (200 MHz, $CDCl_3$), $\delta$ (ppm) 0.08 (s, 6H, Si–($\underline{CH_3}$)$_2$), 0.9 (s, 9H, Si–C ($\underline{CH_3}$)$_3$), 1.45 (s, 9H, *t*Boc), 3.61–3.87 (m, 5H 2–$CH_2$, 1–CH, 2–$CH_2$), 5.15 (br s, OH); $^{13}C$ NMR (200 MHz, $CDCl_3$) $\delta$ (ppm), −5.56 (Si–($\underline{C}H_3$)$_2$), 18.21 (Si–$\underline{C}$($CH_3$)$_3$), 25.84 (Si–C($\underline{C}H_3$)$_3$), 28.38 (O–C($\underline{C}H_3$)$_3$), 52.54($\underline{C}$H), 64.06($CH_2$), 79.64(O–$\underline{C}$($CH_3$)$_3$), 156.06 (NH$\underline{C}$ OO–); HRMS calcd for $C_{14}H_{31}NO_4Si$ Na, 328.1911. Observed mass: 328.1915.

#### 3.2.2 (R)-Tert-butyl (1-(3-benzoyl–thyminyl)-3-((tert-butyldimethylsilyl) oxy) propan-2-yl) Carbamate (3)

(*S*)-*Tert*-butyl (1-(3-benzoyl–thyminyl-)-3-((*tert*-butyldiphenylsilyl) oxy) propan -2-l) Carbamate (**11**)

1. *N*3-Benzoylthymine (0.9 g, 3.92 mmol) and triphenylphosphine (1 g, 3.92 mmol) were dissolved in 40 mL dry THF, and the solution was cooled to 0 °C.
2. At this temperature, compounds **2, 10** (1 g, 3.27 mmol) dissolved in 10 mL dry THF was added to the stirred solution followed by dropwise addition of DIAD (1 mL, 4.91 mmol).
3. The solution was gradually allowed to reach the room temperature, and stirring was continued overnight at room temperature.
4. The solvent was removed under pressure, and the residue was extracted with EtOAc (4 × 100 mL), washed with water and brine.

5. The organic layer was dried over anhydrous $Na_2SO_4$, and solvent was removed under reduced pressure. Compound **3, 11** was used as such for the next reaction without further purification.

*3.2.3 (R)-Tert-butyl (1-(6-benzyloxy-carbonyladeninyl-)-3-((tert-butyl dimethyl silyl) oxy) propan-2-yl)Carbamate* **(4)**

*(S)-Tert-butyl (1-(6-benzyloxy-carbonyladeninyl-)-3-((tert-butyldiphenyl silyl) oxy) propan-2-yl) Carbamate* **(12)**

1. $N^6$-Benzyloxycarbonyladenine (1.1 g, 3.93 mmol) and triphenylphosphine (1 g, 3.92 mmol) were dissolved in 40 mL dry THF, and the solution was cooled to 0 °C.
2. At this temperature compound **2, 10** (1 g, 3.27 mmol) dissolved in 10 mL dry THF was added to the stirred solution followed by dropwise addition of DIAD (1 mL, 4.91 mmol).
3. The solution was gradually allowed to reach the room temperature, and stirring was continued overnight at room temperature.
4. The solvent was removed under pressure, and the residue was extracted with EtOAc (4 × 100 mL), washed with water and brine. The organic layer was dried over anhydrous $Na_2SO_4$, and solvent was removed under reduced pressure.
5. Compound **4, 12** was used as such for the next reaction without further purification.

*3.2.4 (R)-Tert-butyl (1-(3-benzoyl-thyminyl)-3-hydroxypropan-2-yl) Carbamate* **(5a)**

*(S)-Tert-butyl (1-(3-benzoyl-thyminyl)-3-hydroxypropan-2-yl) Carbamate* **(5b)**

1. 0.5 g (0.97 mmol) of the compound **3** and **11** was dissolved in 3 mL of dry THF.
2. To this TBAF (1.5 mL of 1 M TBAF, 1.45 mmol) was added slowly and with constant stirring. The reaction was allowed to stir for 2.5 h. THF was removed under vacuum, and the residue was extracted with DCM (2 × 100 mL), dried over $Na_2SO_4$.
3. The solvent was evaporated under vacuum, and the crude compound was purified by column chromatography (2% MeOH in DCM) furnishing compound **5a, 5b** (0.45 g, 68% from **3** to **5a, 11** to **5b**) as colorless solid.
4. $[\alpha]_D^{20}$ +95.2(c 0.015, MeOH) for **5a**; $[\alpha]_D^{20}$ −95.3(c 0.015, MeOH) for **5b**; $^1$H NMR (200 MHz, $CD_3OD$) $\delta$ (ppm), 1.33 (s, 9H, *t*Boc), 1.80 (s, 3H, =C($\underline{CH_3}$), 3.23–3.50 (m, 3H, $CH_2$, CH), 3.92–4.17 (m, 2H, $CH_2$), 7.41–7.48 (m, =C$\underline{H}$N, Ph-para to CO), 7.59–7.66 (t, Ph-meta to CO, $J$ = 7.2 Hz, $J$ = 7.58 Hz), 8.01–8.05 (d, Ph-ortho to CO, $J$ = 7.57 Hz); $^{13}$C NMR (50 MHz, $CD_3OD$) $\delta$ (ppm), 12.42 (=C(CH3)), 28.83 (O–C($\underline{C}H_3)_3$), 51.76 (N–$\underline{C}H_2$–CHN), 52.55 (N–$CH_2$–$\underline{C}$HN), 62.94 (O–$\underline{C}$H–CHN), 80.42 (O–$\underline{C}$ $(CH_3)_3$), 110.17 (=$\underline{C}$(CH3), 130.28 (Ph-ortho to CO), 131.93 (Ph-meta to CO), 133.11 (Ph to CO), 136.27 (Ph-para to CO), 144.34 ($\underline{C}$HN), 151.60 (N–$\underline{C}$O–N), 158.09 (NH$\underline{CO}$O–), 165.43 (C–$\underline{C}$O–N), 170.56

(N–C̲O–Ph); HRMS calculated for $C_{20}H_{25}O_6N_3Na$, 426.1636. Observed mass: 426.1634.

### 3.2.5 (R)-Tert-butyl (1-(6-benzyloxycarbonyl adeninyl-)-3-hydroxypropan-2-yl) Carbamate (6a)

### (S)-Tert-butyl (1-(6-benzyloxycarbonyl adeninyl-)-3-hydroxypropan-2-yl) Carbamate (6b)

1. 0.6 g (1.08 mmol) of the compound **4** and **12** was dissolved in 3 mL of dry THF. To this TBAF (1.6 mL of 1 M TBAF, 1.62 mmol) was added slowly and with constant stirring.
2. The reaction was allowed to stir for 2.5 h. THF was removed under vacuum, and the residue was extracted with DCM (2 × 100 mL), dried over $Na_2SO_4$.
3. The solvent was evaporated under vacuum, and the crude compound was purified by column chromatography (2% MeOH in DCM) furnishing compound **6a** and **6b** (0.59 g, 73% from **4** to **6a** and **12** to **6b**) as colorless solid.
4. $[\alpha]_D^{20}$ +35.3 (c 0.0102,MeOH) for **6a**; $[\alpha]_D^{20}$ −35.3 (c 0.0103, MeOH) for **6b**;$^1$H NMR (200 MHz, $CD_3OD$) $\delta$ (ppm), 1.21 (s, 9H, *t*Boc), 3.57–3.62 (m, 2H, C$\underline{H_2}$), 4.04–4.08 (m, 1H, C$\underline{H}$), 4.27–4.29 (dd, 1H, $CH_2$), 4.50–4.57 (dd, 1H, $CH_2$), 5.20 (s, 2H, OC$\underline{H_2}$Ph), 7.26–7.38 (m, 5H, Ph), 8.14 (s, 1H, N–C$\underline{H}$=N), 8.50 (s, 1H, N=C$\underline{H}$N); $^{13}$C NMR (125 MHz, DMSO-*d6*) $\delta$ (ppm), 27.86 (O–C($\underline{C}H_3)_3$), 52.02 (N-$CH_2$–$\underline{C}$HN), 62.79 (N–$CH_2$–$\underline{C}$HN), 67.34 (N–$\underline{C}H_2$–CHN), 68.74 (N–$\underline{C}H_2$–CHN), 77.77 (O–$\underline{C}(CH_3)_3$), 123.02 (–N=$\underline{C}$(C)–N),127.47 (–OC$H_2$$\underline{Ph}$-ortho to $CH_2$), 128.00 (–OC$H_2$$\underline{Ph}$-para to $CH_2$), 128.58 (–OC$H_2$$\underline{Ph}$-meta to $CH_2$), 149.22 (–N=C(N)–C), 151.24 (–N$\underline{C}$H=C), 154.27 (–NH$\underline{C}$OOC$H_2$Ph), 154.85 (–NH$\underline{C}$OOC*t*-Bu); HRMS calculated for $C_{21}H_{26}N_6O_5Na$, 465.1857. Observed mass: 465.1857.

### 3.2.6 (R)-Tert-butyl 1-thyminyl-3-hydroxypropan-2-yl Carbamate (7a)

### (S)-Tert-butyl 1-thyminyl-3-hydroxypropan-2-yl Carbamate (7b)

1. 0.89 g (2.21 mmol) of **5a**, **5b** was treated with 20 mL 1:1 ammonia/dioxane and allowed to stir for 3 h.
2. The solvent was then removed completely, and the compound was purified by column chromatography (4% MeOH in DCM) to yield **7a, 7b** as pure white solid (0.63 g, 85%).
3. $[\alpha]_D^{20}$ +42.48 (c 0.025, MeOH) for **7a**; $[\alpha]_D^{20}$ −42.6 (c 0.03, MeOH) for **7b**; $^1$H NMR(200 MHz, $CD_3OD$)$\delta$ (ppm), 1.37 (s, 9H, tBoc), 1.86 (s, 3H, =C($CH_3$)), 3.51–3.59 (m, 3H, $CH_2$, CH), 3.96–4.12 (m, 2H, $CH_2$), 7.35 (s, =CHN); 13C NMR (50 MHz, $CD_3OD$) $\delta$ (ppm), 12.41 (=C($CH_3$)), 28.70 (O–C($CH_3$)3), 51.17 (N–$CH_2$–CHN), 52.33 (N–$CH_2$–CHN), 62.95 (O–$CH_2$–CHN), 80.38 (O–C($CH_3$) 3), 110.48 (=C($CH_3$)), 143.92 (CHN), 153.23 (N–CO–N), 157.94 (NHCOO–), 167.07(C–CO–N); HRMS calculated for $C_{13}H_{21}O_5N_3$ Na, 322.1370. Observed mass: 322.1373.

### 3.2.7 (R)-Tert-butyl (1-(6-benzyloxycarbonyl adeninyl-)-3-(((4-nitrophenoxy) carbonyl) oxy) propan-2-yl) Carbamate (8a)

### (S)-Tert-butyl (1-(6-benzyloxycarbonyl adeninyl-)-3-(((4-nitrophenoxy) carbonyl) oxy) propan-2-l) Carbamate (8b)

1. Compound **6a, 6b** (1 g, 2.26 mmol) was taken in dry DCM (10 mL) and cooled to 0 °C in an ice bath.
2. Dry pyridine (0.8 mL, 6.78 mmol) and dry triethylamine (0.2 mL, 1.12 mmol) were added and allowed to stir for 10–15 min.
3. *p*-Nitrophenylchloroformate (1.14 g, 5.64 mmol) dissolved in DCM (3 mL) was added, and the reaction was allowed to stir at room temperature for 2 h.
4. After completion of the reaction, the reaction mixture was concentrated, and the required compound was purified by column chromatography (70% EtOAc in petroleum ether) to get **8a, 8b** (1.04 g, 76%).
5. $[\alpha]_D^{20}$ +48.4 (c 0.006, $CHCl_3$) for **8a**; $[\alpha]_D^{20}$ −48.4 (c 0.006, $CHCl_3$) for **8b**; $^1H$ NMR(400 MHz, $CDCl_3$) $\delta$ (ppm), 1.26 (s, 9H, *t*Boc), 3.56–3.74 (m, 1H,–N–$\underline{CH_2}$–CH), 4.39–4.71 (m, 4H,–$CH_2$–$\underline{CH}$N,–O$\underline{CH_2}$–CHN, –N–$\underline{CH_2}$–CH), 5.32 (s, 2H,–O$CH_2$Ph), 7.30–7.39 (m, 7H, –O$CH_2\underline{Ph}$,–OCOO-Ph-ortho to –COO), 8.20–8.29 (m, 4H, 2 –N=CN, Ph-meta to –OCOO); $^{13}C$ NMR (125 MHz, $CD_3OCD_3$) $\delta$ (ppm), 29.70 (O–C($\underline{CH_3}$)$_3$), 49.75 (N–$CH_2$–$\underline{C}$HN), 65.35 (N–$\underline{C}H_2$–CHN), 68.07 (–O$\underline{C}H_2$Ph), 69.97 (O–$\underline{C}H_2$–CHN), 81.3 (O–$\underline{C}$($CH_3$)$_3$), 122.35 (Ph-ortho to OCO), 125.94 (–N–$\underline{C}$(C)=C), 126.24 (Ph-meta to OCO), 126.24 (–O$CH_2\underline{Ph}$ ortho to $CH_2$), 127.43 (–O$CH_2\underline{Ph}$ para to $CH_2$), 128.58(–O$CH_2\underline{Ph}$- meta to $CH_2$), 134.27 (–O$CH_2\underline{Ph}$-para to OCO), 139.29(–N=$\underline{C}$HN), 141.02 (Ph-para to OCO), 145.74 (–N–$\underline{C}$(N)=C), 146.20 (–N=$\underline{C}$(C)N), 149.03 (–N$\underline{C}$H=C), 152.37 (–O$\underline{C}$OO),154.74 (–NH$\underline{C}$OOC$H_2$Ph), 155.13 (–NH$\underline{C}$OOC*t*-Bu), 162.69 (Ph–O$\underline{C}$O); HRMS calculated for $C_{28}H_{30}O_9N_7$, 608.2100. Observed mass: 608.2088.

### 3.2.8 (R)-Tert-butyl 1-thyminyl-3-(((4-nitrophenoxy) carbonyl) oxy) propan-2-yl) Carbamate (9a)

### (S)-Tert-butyl 1-thyminyl-3-(((4-nitrophenoxy) carbonyl) oxy) propan-2-yl) Carbamate (9b)

1. Compound **7a, 7b** (1 g, 3.34 mmol) was taken in dry DCM (10 mL) and cooled to 0 °C in an ice bath.
2. Dry pyridine (0.8 mL, 10.02 mmol) was added and allowed to stir for 10–15 min.
3. *p*-Nitrophenylchloroformate (1.7 g, 8.35 mmol) dissolved in DCM (3 mL) was added, and the reaction was allowed to stir at room temperature for 2 h.
4. After completion of the reaction, the reaction mixture was concentrated, and the required compound was purified by column chromatography (80% EtOAc in petroleum ether) to get **9a, 9b** (1.2 g, 76%).
5. $[\alpha]_D^{20}$ +54.9 (c 0.005, $CHCl_3$) for **9a**; $[\alpha]_D^{20}$ −54.75 (c 0.0046, $CHCl_3$) for **9b**; $^1H$ NMR (200 MHz, $CDCl_3$) $\delta$

(ppm), 1.43 (s, 9H, *t*Boc), 1.92 (s, 3H, =C(CH$_3$)), 3.86–4.40 (m, 5H, 2CH$_2$, CH), 5.15 (br s, OH), 7.06 (s, 1H, =CHN) 7.39–7.43 (d, 2H, Ph-ortho to OCO, $J$ = 8.71 Hz), 8.28–8.32 (d, 2H, Ph-meta to OCO, $J$ = 8.84 Hz), 8.76 (s, 1H, CO–NH–CO); $^{13}$C NMR (125 MHz, CD$_3$OCD$_3$) $\delta$ (ppm), 12.36 (=C(CH3)), 28.46 (O–C(CH$_3$)$_3$), 49.47 (N–CH$_2$–CHN), 49.69 (N–CH$_2$–CHN), 69.22 (O–CH$_2$–CHN), 79.53 (O–C(CH$_3$)$_3$), 109.66 (=C(CH3), 123.14 (Ph-ortho to OCO), 126.07 (Ph-meta to OCO), 126.07 (CHN), 142.17 (Ph-para to OCO), 146.44 (N–CO–N), 152.12 (O–CO–O), 153.12 (NHCOO–), 156.66 (Ph–OCO), 164.79 (–C–CO–N–); HRMS calculated for $C_{20}H_{24}N_4O_9Na$, 487.1435. Observed mass: 487.1433.

The chiral purity of the monomers **6a/6b** and **7a/7b** was ascertained by analysis with chiral HPLC (Figs. 3, 4, 5, 6, 7 and 8). Chiral HPLC was accomplished on Kromasil 5-Amycoat (4.6 × 250 mm) column in mobile phase isopropyl alcohol/petroleum ether (50:50). 5 μL of 1 mg/mL solution of the sample in isopropyl alcohol was injected and monitored at 254 nm wavelength.

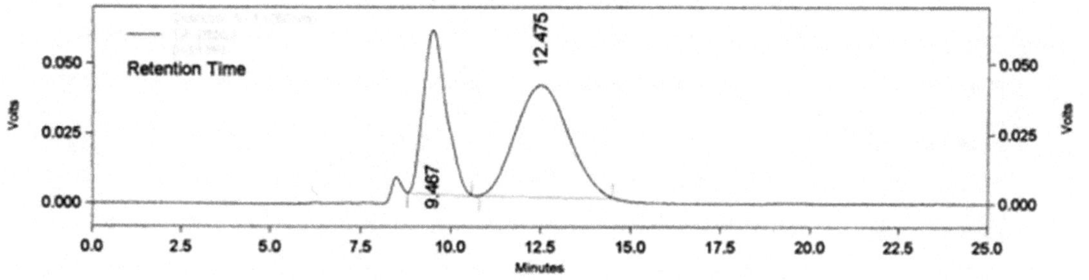


**Fig. 3** Chiral HPLC of racemic compound **6a** + **6b**

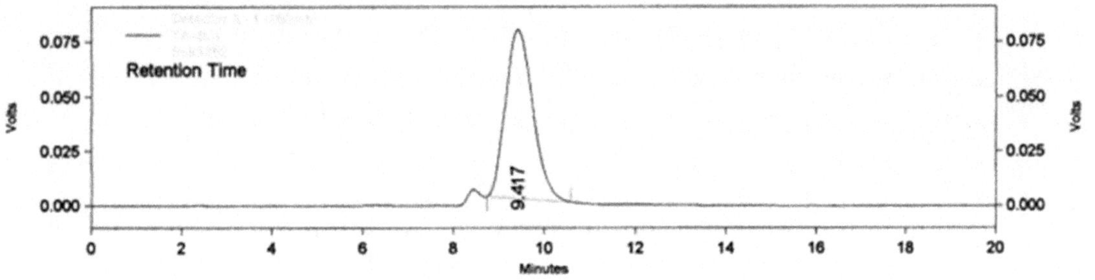


**Fig. 4** Chiral HPLC of compound **6a**

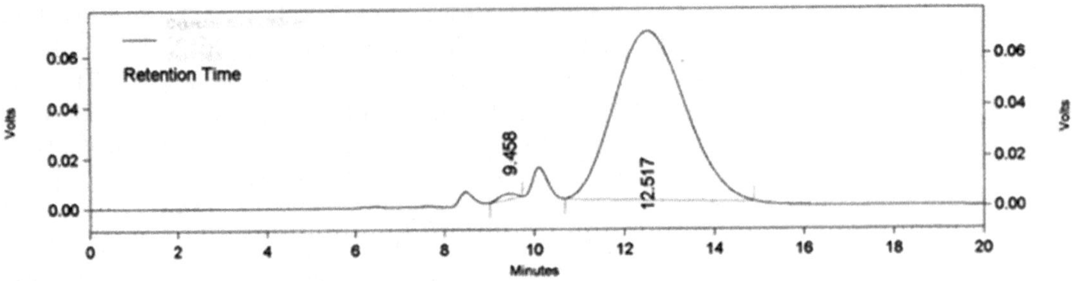


**Fig. 5** Chiral HPLC of compound **6b**

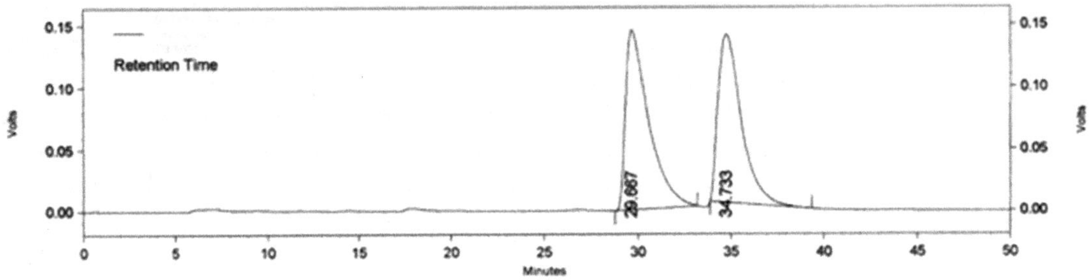


**Fig. 6** Chiral HPLC of racemic compound **7a** + **7b**

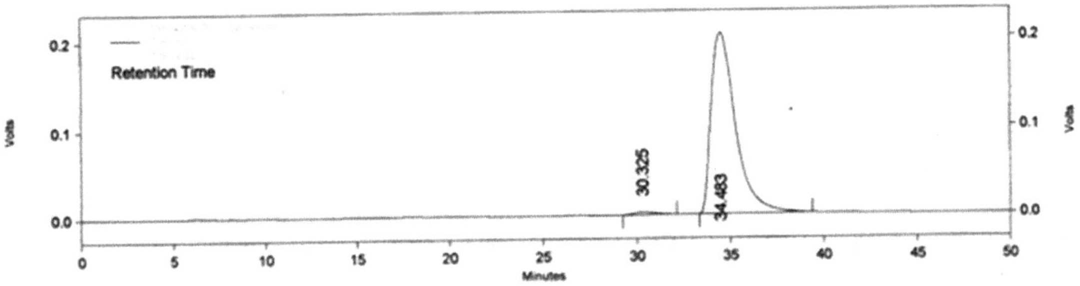


**Fig. 7** Chiral HPLC of compound **7a**

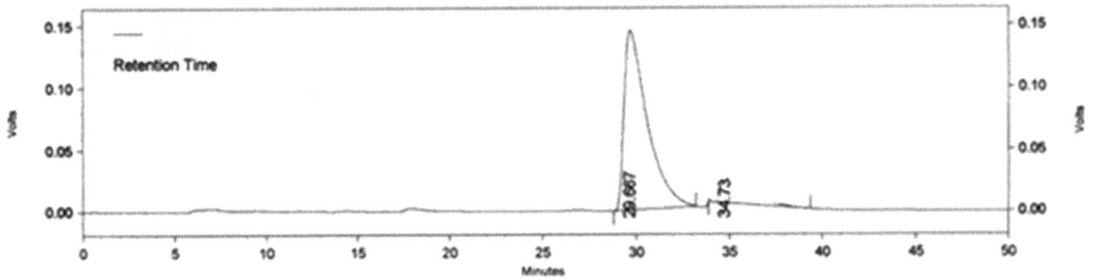


**Fig. 8** Chiral HPLC of compound **7b**

### 3.3 Synthesis, Purification, and Characterization of Oligomers

#### 3.3.1 General Principles of Solid-Phase Carbamate Synthesis (SPCS)

In disparity to the solution phase method, the solid-phase peptide synthesis strategy implemented by Merrifield [13] offers great advantage. In this method, the C-terminal is attached to an insoluble matrix such as polystyrene beads having reactive functional groups, which also act as a permanent protection for the carboxylic acid. The loading value of the resin can be adjusted according as desired. The next carbonate-activated $N^{\alpha}$-protected amino alcohol can be coupled to the resin bound amino acid without the aid of any coupling reagent. The unreacted monomers are then washed out, and the deprotection, coupling reactions, and washing cycles are repeated until the desired peptide is achieved. The need to purify the coupling at every step is obviated. Finally, the resin bound oligomer and the side chain protecting groups are cleaved in one step. Solid-phase carbamate synthesis proposed in the present studies has a major advantage over solid-phase peptide synthesis. As the Boc-amino alcohol is already activated by *p*-nitrophenylchloroformate, expensive coupling reagents are not required for the synthesis of oligomers; therefore, excess monomers can be recovered at the end of each coupling. The resin chosen for the synthesis of carbamate oligomers was MBHA resin (Fig. 9).

The advantages of solid-phase synthesis are (1) all the reactions are performed in a single vessel minimizing the loss due to transfer, (2) large excess of monomer component can be used resulting in high coupling efficiency, and (3) it avoids purification step after each coupling reaction. These properties of solid-phase synthesis make the process of synthesis of oligomers easier and faster in comparison to solution phase synthesis.

#### 3.3.2 Functionalization and Picric Acid Estimation of the MBHA [(4-Methyl Benzhydryl) Amine] Resin

The loading value of MBHA resin was reduced to 0.3 mmol/g, by coupling with calculated amount of di-protected L-lysine carboxylic group ($N^{\alpha}$-amino group protected with Boc- and $N^{\omega}$-amino group by Cl–Cbz) to the free amines on the resin using coupling reagents followed by capping with acetic anhydride [14]. Then the final loading value of the functionalized resin was estimated using picric acid [15].

1. L-lysine-loaded MBHA resin was taken in two different solid-phase vessels (Fig. 8), and their exact weight was recorded.

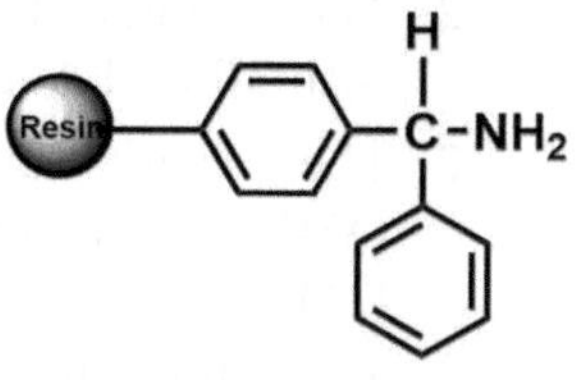


**Fig. 9** Representative structures of MBHA resin in SPCS

2. They were then swelled in DCM for half hour to expose the free amines. The Boc protecting group was deprotected using TFA, and the trifluoroacetate salts of amine were neutralized with 5% DIPEA in DCM.
3. To the resin 0.1 M picric acid solution in DCM was added. After 10 min, the solution was flushed out from solid-phase flask, and the procedure was repeated again. The resin was washed properly with DCM (*see* **Note 2**).
4. The resins were then washed with 5% DIPEA in DCM when the picric acid eluted as amine salt. This process was repeated again, and all the washings were collected in a volumetric flask of 10 mL and the volume adjusted to 10 mL (*see* **Note 3**).
5. This entire process was repeated for another flask containing weighed amount of resin. The absorbances of the picrate solutions were recorded at 358 nm. Loading value of the MBHA resin was calculated from the absorbance applying the formula given below.

$$\text{Loading value} = \frac{\text{Observed absorbance} \times \text{Dilution factor}}{\text{Molar extinction coefficient}}$$

#### *3.3.3 Kaiser Test*

The Kaiser test [16] was used to monitor the *t*-Boc deprotection and coupling steps in the solid-phase synthesis. A few beads of the resin from solid-phase flask were taken in a test tube. To the test tube, 3–4 drops of each of the following solutions (1, 2, and 3) were added.

1. 0.7 mg of KCN in 1 mL of water and 49 mL of pyridine.
2. 1.0 g of ninhydrin in 20 mL of ethanol.
3. 40 g of phenol in 20 mL of n-butanol.

The test tube was heated at 110 °C for 5 min and the color of the beads was noted. A blue color on the beads, which slowly comes into solution, indicated successful deprotection, while colorless beads and the solution confirmed the completion of the coupling reaction.

#### *3.3.4 The Carbamate Synthesis*

1. The synthesis of the oligomers was performed on lysine-loaded MBHA resin in solid-phase flasks.
2. The terminal Boc group was deprotected by treating the resin with 50% TFA in DCM, three times for 10 min each.
3. The solvent was flushed, and the resin was washed with dry DMF (3 times) followed by dry DCM (3 times) and then dry DMF (3 times).
4. The TFA salt of amine was neutralized with 5% DIPEA in DCM (3 times for 5 min each).

5. The resin was then washed with dry DCM (3 times) followed by dry DMF (3 times).
6. The four equivalents of monomer dissolved in dry DMF was added to the resin and allowed to stand in the solid-phase flask for 6–8 h with occasional stirring.
7. After 8 h the coupling cocktail was removed and Kaiser test was performed. The monomer was recovered at the end of each coupling (*see* **Note 4**).
8. If the Kaiser test was negative, we moved forward with deprotection and next coupling, and if negative, we proceed with recoupling.
9. At the end of the synthesis, the resins were cleaved by treating the resin with TFMSA, TFA, ethane 1,2 dithiol, and β-mercaptoethanol cleavage cocktail for 2 h.

#### *3.3.5 Cleavage of the GCNA Oligomers from the Solid Support*

1. The resin-bound GCNA oligomer (5 mg) taken in a glass vial was kept in an ice bath with thioanisole (10 μL) and 1,2-ethanedithiol (4 μL) for 10 min.
2. To it TFA (80 μL) was added and shaken manually and kept for another 10 min (*see* **Note 5**).
3. TFMSA (8 μL) was added and the mixture was allowed to stand for 2 h (*see* **Note 6**)
4. The reaction mixture was filtered through a sintered funnel and collected in 5 mL pear shaped flask (*see* **Note 7**).
5. The residue was washed with TFA (3 × 2 mL), and the combined filtrate and washings were evaporated under vacuum.
6. The residue was precipitated using dry diethyl ether and centrifuged in 1.7 mL centrifuge tubes. The diethyl ether was decanted and the solid residue was redissolved in 1:1 ACN/water.

#### *3.3.6 Reverse-Phase HPLC*

1. The crude GCNAs were purified on a semi-preparative C18 column attached to a waters HPLC system.
2. A gradient elution method contained A = 5% acetonitrile in water +0.1% trifluoroacetic acid and B = 50% acetonitrile in water +0.1% trifluoroacetic acid (A to B = 100% in 20 min with a flow rate of 1.5 mL/min), and the eluent was monitored at 260 nm.
3. The purity of the oligomers was further assessed by an RP-C18 analytical HPLC column. The purity of the purified oligomers was found to be >98%.

**Table 1**
**GCNA sequences synthesized and characterized**

| Code | Sequences | HPLC $t_R$ (min) | MALDI-TOF mass | |
|---|---|---|---|---|
| | | | Calcd. | Obsvd. |
| *R*-GCNA-1 | tttttttt-Lys | 13.1 | 1946.81 | 1951.19 |
| *S*-GCNA-1 | tttttttt-Lys | 13.0 | 1946.81 | 1952.84 |
| *R*-GCNA-2 | atattattaatt-Lys | 11.3 | 2891.07 | 2918.26(M+Na) |
| *S*-GCNA-2 | atattattaatt-Lys | 11.2 | 2891.07 | 2918.14(M+Na) |
| *R*-GCNA-3 | aattaataatat-Lys | 9.4 | 2909.10 | 2929.21(M+Na) |
| *S*-GCNA-3 | aattaataatat-Lys | 9.3 | 2909.10 | 2929.22(M+Na) |

*t* thymine, *a* adenine monomers of R/S-GCNA as indicated in the code

*3.3.7 MALDI-TOF Mass Spectrometry*

Literature reports the analysis of PNA oligomers by MALDI-TOF mass spectrometry in which several matrices have been explored, viz., sinapinic acid (3,5-dimethoxy-4-hydroxycinnamic acid) [17], CHCA (cyano-4-hydroxycinnamic acid), and DHB (2,5-dihydroxybenzoic acid). Out of these, CHCA was found to give the best signal to noise ratio.

For all the MALDI-TOF spectra recorded for the *R/S*-GCNA oligomers reported here, CHCA (α-Cyano-4-hydroxycinnamic acid) was used as the matrix. The MALDI-TOF spectra were recorded on AB SCIEX 5800 MALDI-TOF TOF instrument and are given in Table 1.

## 4 Notes

1. During the synthesis and purification of activated carbonate monomers, the temperature should be maintained below 40 °C as the carbonate monomers are labile due to easy evolution of carbon dioxide.
2. The resins should be washed properly to remove any excess picric acid. The stopcock, cap as well as the vessel should be washed properly to avoid erroneous result.
3. The resins should be washed till the washing is colorless. This should be done in small volumes such that the total volume does not exceed 10 mL. However, the volume can also be made higher but it should be known.
4. The coupling time can also be further increased according to requirement or as the chain elongates.

5. The shaking should be mild otherwise the resins stick to the glass.
6. The mixture should be kept on ice throughout for 2 h.
7. The resin mixture can be poured to the funnel with proper washing to filter the resin. Pear-shaped flask can be used to minimize spreading of the filtrate in the flask as well as facilitate collection of the mixture into a tube for centrifugation in next step.

## Acknowledgments

T.B. acknowledges Senior Research Fellowship from CSIR, New Delhi. V.A.K. acknowledges financial support from CSIR, New Delhi (BSC0123).

### References

1. Nielsen PE, Egholm M, Berg RH, Buchardt O (1991) Sequence-selective recognition of DNA by strand displacement with a thymine-substituted polyamide. Science 254:1497–1500
2. Hendrix C, Rosemeyer H, Verheggen I, Seela F, van Aerschot A, Herdewijn P (1997) 1′,5′-Anhydrohexitol oligonucleotides: synthesis, base pairing and recognition by regular oligodeoxyribonucleotides and oligoribonucleotides. Chem Eur J 3:110–120
3. Gait MJ, Jones AS, Walker RT (1974) Synthetic-analogues of polynucleotides XII. Synthesis of thymidine derivatives containing an oxyacetamido- or an oxyformamido-linkage instead of a phosphodiester group. J Chem Soc Perkin 10:1684–1686
4. Egholm M, Buchardt O, Nielsen PE, Berg RH (1992) Peptide nucleic acids (PNA). Oligonucleotide analogs with an achiral peptide backbone. J Am Chem Soc 114:1895–1897
5. Egholm M, Buchardt O, Christensen L, Behrens C, Freier SM, Driver DA, Berg RH, Kim SK, Norden B, Nielsen PE (1993) PNA hybridizes to complementary oligonucleotides obeying the Watson-crick hydrogen-bonding rules. Nature 365:566–568
6. Schoning KU, Scholz P, Guntha S, Wu X, Krishnamurthy R, Eschenmoser A (2000) Chemical etiology of nucleic acid structure: the α-threofuranosyl-(3′→2′) oligonucleotide system. Science 290:1347–1351
7. Pallan PS, Wilds CJ, Wawrzak Z, Krishnamurthy R, Eschenmoser A, Egli M (2003) Why does TNA cross-pair more strongly with RNA than with DNA? An answer from X-ray analysis. Angew Chem Int Ed Engl 42:5893–5895
8. Zhang L, Peritz A, Meggers E (2005) A Simple Glycol Nucleic Acid. J Am Chem Soc 127:4174–4175
9. Meena KVA (2003) Pyrrolidine carbamate nucleic acids: synthesis and DNA binding studies. Bioorg Med Chem 11:3393–3399
10. Madhuri V, Kumar VA (2010) Design, synthesis and DNA/RNA binding studies of nucleic acids comprising stereoregular and acyclic polycarbamate backbone: polycarbamate nucleic acids (PCNA). Org Biomol Chem 8:3734–3741
11. Kotikam V, Fernandes M, Kumar VA (2012) Comparing the interactions of DNA, polyamide (PNA) and polycarbamate nucleic acid (PCNA) oligomers with graphene oxide (GO). Phys Chem Chem Phys 14:15003–15006
12. Hu F, Zhanga YH, Yaoa ZJ (2007) Parallel synthesis of individual shikimic acid-like molecules using a mixture-operation strategy and ring-closing enyne metathesis. Tetrahedron Lett 48:3511–3515
13. Merrifield RB (1963) Solid phase peptide synthesis. I. The synthesis of a tetrapeptide. J Am Chem Soc 85:2149–2154
14. Dueholm KL, Egholm M, Behrens C, Christensen L, Hansen HF, Vulpius T, Petersen KH, Berg RH, Nielsen PE, Buchardt O (1994) Synthesis of peptide nucleic acid monomers containing the four natural nucleobases: thymine, cytosine, adenine, and guanine and

their Oligomerization. J Org Chem 59:5767–5773

15. Gisin BF (1972) The monitoring of reactions in solid-phase peptide synthesis with picric acid. Anal Chim Acta 58:248–249
16. Kaiser E, Colescott RL, Bossinger CD, Cook PI (1970) Color test for detection of free terminal amino groups in the solid-phase synthesis of peptides. Anal Biochem 34:595–598
17. Beavis RC, Chait BT (1989) Matrix-assisted laser-desorption mass spectrometry using 355 nm radiation. Rapid Commun Mass Spectrom 3(12):436–439

# Chapter 7

# Synthesis of Nucleobase-Functionalized Morpholino Monomers

Bappaditya Nandi, Sankha Pattanayak, Sibasish Paul, Jayanta Kundu, and Surajit Sinha

## Abstract

Morpholino antisense oligonucleotides are used as routine tools in developmental biology to investigate gene function during early embryogenesis. These chemically modified oligos contain morpholine ring connected with phosphorodiamidate linkages as backbone but carry unmodified nucleobases. In this chapter, we describe the methods to further modify the nucleobases using palladium-catalyzed cross-coupling reactions. The key reactions used are halogenations of the nucleobases in suitable position and subsequent Pd-catalyzed Sonogashira and Suzuki reactions. The sequential synthetic steps are described in detail in this chapter, and the examples are shown in tables.

**Key words** Morpholino oligonucleotides, Morpholino monomers, Functionalized nucleobases, Halogenation, Pd-catalyzed cross-coupling

## 1 Introduction

The ability of antisense oligonucleotides to selectively inhibit gene expression has opened a number of applications that include interrogation of gene function by reverse genetic analysis and targeting previously "undruggable" genetic diseases [1]. Phosphorodiamidate morpholino oligonucleotides (PMOs/morpholinos) (*see* Fig. 1) are one such antisense type that has been widely used to block RNA translation or modify pre-mRNA splicing in model organisms such as zebrafish and xenopus [2]. Even in the era of CRISPR/Cas9 technology, morpholinos are valuable research tools in developmental biology because of their high success rate in blocking translation, synthesis, and implementation within days, non-toxicity, and satisfying delivery by microinjection [3]. Morpholinos are commercially available from GeneTools LLC (http://www.gene-tools.com) and were developed by Dr. James Summerton during the late 1990s at Antivirals Inc. (now Sarepta Therapeutics Inc.) [4]. Recently, the first morpholino-based drug eteplirsen

Nathaniel Shank (ed.), *Non-Natural Nucleic Acids: Methods and Protocols*, Methods in Molecular Biology, vol. 1973, https://doi.org/10.1007/978-1-4939-9216-4_7, © Springer Science+Business Media, LLC, part of Springer Nature 2019

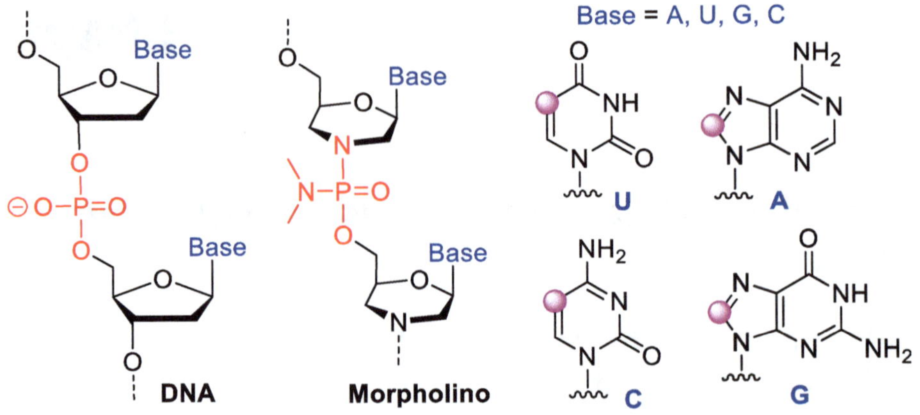


**Fig. 1** Chemical structures of DNA and morpholino oligonucleotide and functionalization sites of nucleobases

has been granted accelerated approval by FDA to treat Duchenne muscular dystrophy (DMD). In comparison to DNA/RNA, morpholinos are synthetically modified 20–25 mer oligonucleotides having morpholine rings in place of the ribose sugars and phosphorodiamidate linkages instead of the phosphate backbone while retaining the same nucleobase structures (adenine, guanine, thymine, cytosine, Fig. 1). However, modifications on the nucleobases in morpholinos are relatively less explored although several types of nucleobase modifications in DNA are known to enhance duplex stability [5, 6] and also employed to attach several labeling (e.g., redox) groups [7, 8]. Our research group has reported the synthesis of nucleobase-functionalized morpholino monomers by using palladium-catalyzed cross-coupling reactions [9, 10] and also used them to prepare cell-penetrating morpholino oligomers [11, 12].

In this chapter, we describe the protocol to synthesize nucleobase-modified morpholino monomers. The key reaction for the modification was halogenation of the suitably protected monomers (position C-5 for pyrimidines and C-8 for purines) and then to use various types of Pd-catalyzed cross-coupling reactions (Sonogashira and Suzuki couplings) to install the desired groups. We will provide representative protocol for the cross-coupling reactions, and examples will be shown in the tables.

## 2 Materials

### *2.1 Reagents for C-5-Substituted Cytidine Morpholino Monomer Synthesis*

1. Trifluoroacetic anhydride [$(CF_3CO)_2O$].
2. Dichloromethane ($CH_2Cl_2$).
3. Chloroform ($CHCl_3$).
4. Iodine ($I_2$).

5. Iodic acid ($HIO_3$).
6. Acetic acid ($CH_3COOH$).
7. Carbon tetrachloride ($CCl_4$).
8. Potassium carbonate ($K_2CO_3$).
9. Methanol (MeOH).
10. Trityl chloride ($Ph_3CCl$).
11. Triethylamine ($Et_3N$).
12. Acetonitrile ($CH_3CN$).
13. Propargylamine.
14. Bis(triphenylphosphine)palladium(II) dichloride [Pd $(PPh_3)_2Cl_2$].
15. Copper(I) iodide (CuI).
16. Anhydrous sodium sulfate ($Na_2SO_4$).
17. [1,1′-Bis(diphenylphosphino)ferrocene]dichloropalladium (II), complex with dichloromethane [Pd(dppf)$Cl_2$. $CH_2Cl_2$, Aldrich].
18. Tripotassium phosphate ($K_3PO_4$).
19. 4-(Trifluoromethoxy)phenylboronic acid.
20. Tetrahydrofuran (THF).
21. Sodium bicarbonate ($NaHCO_3$).
22. Sodium chloride (NaCl).

### 2.2 Reagents (Additional) for C-5-Substituted Uracil Morpholino Monomer Synthesis

1. Iodine monochloride (ICl) solution (1 M in $CH_2Cl_2$, Aldrich).
2. Tetrakis(triphenylphosphine)palladium(0) [$Pd(PPh_3)_4$].
3. Propargyl alcohol.
4. Dimethylformamide (DMF).
5. Ethyl acetate (EtOAc).
6. Sodium thiosulfate ($Na_2S_2O_3$).

### 2.3 Reagents (Additional) for C-8-Substituted Adenosine Morpholino Monomer Synthesis

1. Bromine ($Br_2$).
2. Disodium phosphate ($Na_2HPO_4$).
3. Dioxane.
4. (Triethylsilyl)acetylene.
5. 4-Fluorophenylboronic acid.
6. Sodium sulfite ($Na_2SO_3$).
7. Acetone ($CH_3COCH_3$).

### 2.4 Reagents (Additional) for C-8-Substituted Guanosine Morpholino Monomer Synthesis

1. Tetrabutylammonium fluoride ($n$Bu$_4$NF) solution, 1.0 M in THF.
2. Phenylacetylene.
3. Phenylboronic acid.

## 3 Methods

The base-functionalized morpholino monomers were synthesized from their corresponding halo derivatives using Pd-catalyzed cross-coupling reactions. C-5 position of pyrimidines and C-8 position of purines were chosen for the modification because these are the ideal sites for incorporation of any group in nucleobases without disturbing the Watson-Crick base pair formation [13, 14].

### 3.1 Synthesis of C-5-Substituted Cytidine Morpholino Monomers

To achieve the C-5-substituted cytidine morpholino monomers using Pd-catalyzed cross-coupling reaction, the intermediate 5-iodo cytidine morpholino monomer **5** was synthesized following the reaction sequences presented in Fig. 2. First the morpholino nitrogen in compound **1** [15] was protected with trifluoroacetyl group as it was found to be stable in the iodination conditions. In the next step, iodination of morpholino-modified 7′-*O*-*tert*-butyldiphenylsilyl-*N*-(trifluoroacetyl)cytidine (**2**) was done using

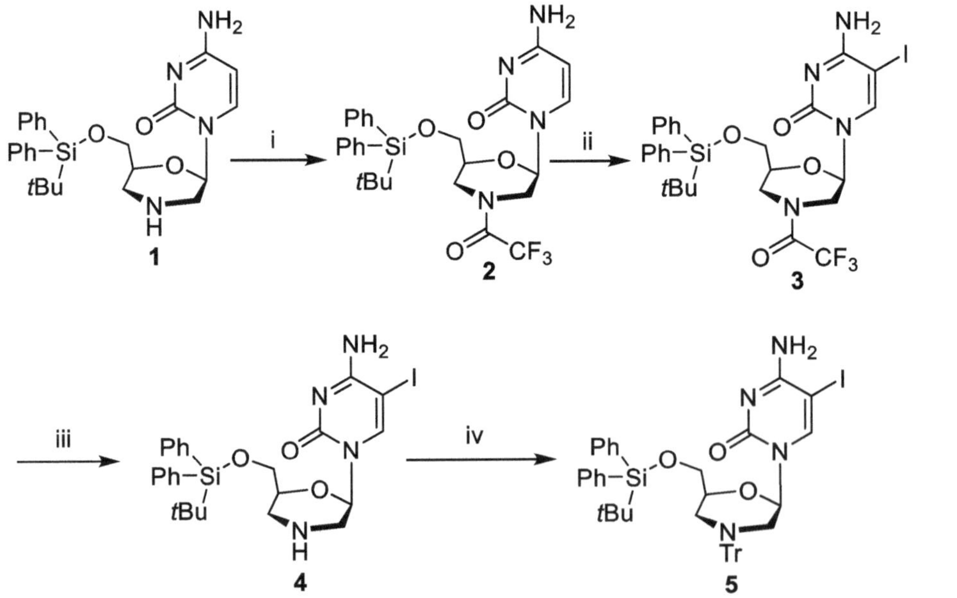


**Fig. 2** Synthesis of *N*-trityl-protected 5-iodocytidine morpholino monomer **5**. Reagents and conditions: (i) $(CF_3CO)_2O$ (1 equiv.), DCM, ice-salt mixture bath temp., 45 min, 95%; (ii) $I_2$ (0.6 equiv.), $HIO_3$ (0.9 equiv.), $CCl_4$–AcOH (1:1), 40 °C, 55 h, 91%; (iii) $K_2CO_3$ (2 equiv.), MeOH, rt, quantitative; (iv) $Ph_3CCl$ (1.1 equiv.), $Et_3N$ (1.5 equiv.), DCM, 5 °C to rt, overnight, 94%. Abbreviation: *Tr* trityl (adapted from Ref. 10)

iodine/iodic acid mixture in $CCl_4$/AcOH (1:1) mixture. The trifluoroacetyl group was subsequently deprotected using potassium carbonate in MeOH to give **4** which was then treated with trityl chloride and triethyl amine in dry $CH_2Cl_2$ to obtain the trityl-protected iodo derivative **5** (*see* Fig. 2). Here we have discussed the synthesis protocol of compounds **2**, **3**, **4**, and **5**.

#### *3.1.1 Morpholino-Modified 7′-O-Tert-butyldiphenylsilyl-N-(trifluoroacetyl)cytidine (**2**) [Ref. 10]*

1. In an oven-dried two-neck round-bottomed flask, compound **1** (8.22 g, 17.7 mmol) was taken in dry $CH_2Cl_2$ (170 mL) under Ar atmosphere. The solution was cooled using ice/salt bath.
2. To this stirred solution of **1**, freshly distilled trifluoroacetic acid anhydride (2.5 mL, 17.7 mmol, diluted in ca. 12 mL $CH_2Cl_2$) was added dropwise at ice/salt bath temperature. The resulting mixture was stirred for 45 min (monitored by TLC) without removal of the ice/salt bath.
3. After completion of the reaction, the reaction mixture was taken up in $CHCl_3$ (200 mL) and transferred to a separating funnel. The organic layer was washed with water (250 mL) and brine (200 mL).
4. Then organic layer was dried with anhydrous $Na_2SO_4$, filtered, and concentrated in vacuo.
5. The solid residue was purified by flash column chromatography on silica gel (230–400 mesh, 5–7% MeOH in $CH_2Cl_2$) to give the product **2** (9.43 g, 95%) as a colorless foamy solid. TLC ($CH_2Cl_2$/MeOH 95:5): $R_f$ 0.34.
6. $^1$H NMR (500 MHz, $CDCl_3$, mixture of two rotamers): $\delta$ 8.84 (br. s, 0.62H), 8.45 (br. s, 0.38H), 7.69–7.64 (comp, 4H), 7.48–7.33 (comp, 7H), 7.15 (br. s, 0.62H), 6.89 (br. s, 0.38H), 6.06 (d, $J$ = 7.5 Hz, 0.62H), 6.00 (d, $J$ = 7.5 Hz, 0.38H), 5.77 (dd, $J$ = 9.5, 2.5 Hz, 1H), 4.67 (d, $J$ = 12.0 Hz, 0.62H), 4.60 (d, $J$ = 12.5 Hz, 0.38H), 4.26 (d, $J$ = 12.5 Hz, 0.38H), 4.18 (d, $J$ = 13.0 Hz, 0.62H), 3.89–3.81 (comp, 2.38H), 3.74–3.70 (m, 0.62H), 3.14 (dd, $J$ = 13.2, 10.8 Hz, 0.62H), 3.02–2.94 (m, 0.76H), 2.65 (app t, $J$ = 11.2 Hz, 0.62H), 1.09 (s, 9H) ppm. $^{13}$C NMR (125 MHz, $CDCl_3$, mixture of two rotamers): $\delta$ 166.4, 166.3, 156.0 (q, $^2J_{C,F}$ = 36.5 Hz), 155.8 (q, $^2J_{C,F}$ = 36.5 Hz), 155.21, 155.18, 139.4, 139.3, 135.54, 135.51, 135.47, 132.9, 132.8, 132.75, 132.6, 130.1, 130.0, 127.93, 127.9. 116.3 (q, $^1J_{C,F}$ = 288.0 Hz), 116.2 (q, $^1J_{C,F}$ = 288.0 Hz), 96.5, 96.1, 80.2, 79.7, 76.4, 76.3, 64.1, 63.6, 48.8, 47.5, 46.3, 44.2, 26.8, 26.7, 19.3, 19.2 ppm. IR (KBr): $\nu$ 3341, 2932, 2859, 1701, 1647, 1487, 1184, 1113 cm$^{-1}$. HRMS (ESI$^+$): *m/z* calcd. for $C_{27}H_{31}N_4O_4F_3Si$ [M+Na] 583.1964; found 583.1960.

#### 3.1.2 Morpholino-Modified 7′-O-Tert-butyldiphenylsilyl-N-trifluoroacetyl-5-iodocytidine (3) [Ref. 10]

1. Iodine (2.52 g, 9.93 mmol) was added to a stirred solution of **2** (9.25 g, 16.5 mmol) in $CCl_4/CH_3COOH$ (1:1, 110 mL) in a round-bottomed flask.
2. Reaction mixture was heated at 40 °C, and then iodic acid (2.62 g, 14.9 mmol) was added to the mixture. The resulting mixture was stirred at 40 °C for 55 h under Ar atmosphere.
3. After completion of the reaction, the mixture was concentrated in vacuo. The residue was dissolved in $CH_2Cl_2/CHCl_3$ (1:1, 300 mL) and transferred to a separating funnel. The organic phase was washed with aqueous $NaHCO_3$ (5% w/v, 100 mL; *see* **Note 1**), water (100 mL), and brine (100 mL), respectively. The organic phase was dried over anhydrous $Na_2SO_4$ and filtered, and the solvents were evaporated.
4. The solid residue was purified by flash column chromatography on silica gel (230–400 mesh, 3–5% MeOH in $CH_2Cl_2$) to give the product **3** (10.31 g, 91%) as a colorless foamy solid. TLC ($CH_2Cl_2$/MeOH 95:5): $R_f$ 0.36.
5. $^1$H NMR (500 MHz, $CDCl_3$, mixture of two rotamers): $\delta$ 9.38 (br. s, 0.53H), 9.26 (br. s, 0.47H), 7.80 (s, 0.53H), 7.72 (s, 0.47H), 7.69–7.63 (comp, 4H), 7.47–7.38 (comp, 6H), 5.86 (br. s, 0.53H), 5.84 (br. s, 0.47H), 5.76 (dd, $J$ = 9.8, 2.8 Hz, 0.47H), 5.70 (dd, $J$ = 10.0, 2.5 Hz, 0.53H), 4.76 (d, $J$ = 13.0 Hz, 0.47H), 4.57 (d, $J$ = 13.5 Hz, 0.53H), 4.34 (d, $J$ = 13.0 Hz, 0.53H), 4.12 (d, $J$ = 13.5 Hz, 0.47H), 3.92–3.83 (comp, 2.53H), 3.76 (dd, $J$ = 10.8, 6.2 Hz, 0.47H), 3.20 (dd, $J$ = 13.5, 11.0 Hz, 0.47H), 3.04–2.94 (m, 1.06H), 2.63 (dd, $J$ = 13.0, 10.5 Hz, 0.47H), 1.10 (s, 4.77H), 1.09 (s, 4.23H) ppm. $^{13}$C NMR (125 MHz, $CDCl_3$, mixture of two rotamers): $\delta$ 164.0, 163.8, 156.1 (q, $^2J_{C,F}$ = 37.7 Hz), 153.6, 153.5, 145.8, 145.6, 135.5, 135.4, 132.8, 132.61, 132.59, 132.5, 130.1, 130.03, 129.99, 127.94, 127.88, 116.2 (q, $^1J_{C,F}$ = 288.0 Hz), 116.1 (q, $^1J_{C,F}$ = 288.0 Hz), 80.3, 79.8, 76.5, 76.4, 63.9, 63.6, 57.5, 48.8, 47.0, 46.2, 43.9, 26.8, 26.7, 19.3, 19.2 ppm. IR (neat): $\nu$ 3314, 2932, 2857, 1703, 1652, 1634, 1472, 1186, 1113 cm$^{-1}$. HRMS (ESI$^+$): *m/z* calcd. for $C_{27}H_{30}N_4O_4F_3SiI$ [M+Na] 709.0931; found 709.0931.

#### 3.1.3 Morpholino-Modified 7′-O-Tert-butyldiphenylsilyl-5-iodocytidine (4)

1. Compound **3** (9.82 g, 14.3 mmol) was taken in MeOH (170 mL) in a round-bottomed flask under Ar atmosphere.
2. $K_2CO_3$ (3.9 g, 28.2 mmol) was added to this solution. The resulting mixture was stirred for 3 h at room temperature.
3. The reaction mixture was concentrated in vacuo. The residue was dissolved in $CH_2Cl_2$ (300 mL) and transferred to a separating funnel. The organic phase was washed carefully with

water (120 mL), in order to avoid the formation of emulsion (*see* **Note 2**).

4. The organic part was dried with anhydrous $Na_2SO_4$ and filtered, and the solvents were evaporated (*see* **Note 3**).
5. The gummy solid residue was co-evaporated with dry $CH_2Cl_2$ two to three to four times in vacuo to remove remaining water content (*see* **Note 4**) and finally dried under high vacuum for a long time to afford compound **4** as white powder. TLC ($CH_2Cl_2$/MeOH 95:5): $R_f$ 0.2.
6. In this step, complete conversion of starting material to product occurred. Compound **4** was used directly for the next tritylation step without further purification.

*3.1.4 Morpholino-Modified 7′-O-Tert-butyldiphenylsilyl-N-trityl-5-iodocytidine (5) [Ref. 10]*

1. Crude compound **4** (14.3 mmol) was taken in dry $CH_2Cl_2$ (220 mL) in a round-bottomed flask under Ar atmosphere (*see* **Note 5**).
2. To this stirred suspension of **4**, triethylamine (3 mL, 21.5 mmol) and trityl chloride (4.38 g, 15.7 mmol) were added at 5 °C. The reaction mixture was allowed to warm slowly to room temperature and stirred overnight at room temperature.
3. The reaction mixture was taken up in $CHCl_3$ (150 mL) and transferred to a separating funnel. The organic layer was washed with water (250 mL) and brine (200 mL). The organic layer was dried with anhydrous $Na_2SO_4$, filtered, and concentrated in vacuo.
4. The solid residue was purified by flash column chromatography on silica gel (230–400 mesh, 2–3% MeOH in $CH_2Cl_2$) to give the product **5** (10.8 g, 90%) together with recovered starting material **4** (338 mg, 4%, recovered in 15–20% MeOH in $CH_2Cl_2$). TLC ($CH_2Cl_2$/MeOH 95:5): $R_f$ 0.4.
5. $^1H$ NMR (500 MHz, $CDCl_3$): $\delta$ 8.71 (br. s, 1H), 7.60–7.28 (comp, 23H), 7.20–7.17 (m, 3H), 6.11 (dd, $J$ = 9.2, 2.2 Hz, 1H), 5.60 (br. s, 1H), 4.27–4.24 (m, 1H), 3.74 (dd, $J$ = 10.5, 4.5 Hz, 1H), 3.60 (dd, $J$ = 10.2, 5.8 Hz, 1H), 3.48 (d, $J$ = 11.5 Hz, 1H), 3.26 (d, $J$ = 12.0 Hz, 1H), 1.50 (app t, $J$ = 11.0 Hz, 1H), 1.19 (dd, $J$ = 11.0, 9.5 Hz, 1H), 0.98 (s, 9H) ppm. $^{13}C$ NMR (125 MHz, $CDCl_3$): $\delta$ 163.7, 154.0, 146.7, 135.6, 133.3, 129.9, 129.8, 129.4, 127.9, 127.84, 127.81, 126.5, 81.9, 77.4, 76.9, 64.7, 56.2, 53.1, 50.0, 26.9, 19.3 ppm. IR (KBr): $\nu$ 3452, 3053, 2857, 1668, 1626, 1489, 1105 $cm^{-1}$. HRMS ($ESI^+$): $m/z$ calcd. for $C_{44}H_{45}N_4O_3SiI$ [M+Na] 855.2203; found 855.2204.

### 3.2 Pd-Catalyzed Cross-Coupling of 5-Iodocytidine Morpholino Monomer 5

The 5-iodocytidine monomer **5** obtained in the previous step was subsequently used for Sonogashira and Suzuki coupling reactions to install a variety of functional groups.

#### 3.2.1 Sonogashira Coupling on 7′-O-silyl-Protected 5-Iodocytidine Morpholino Monomer 5

Sonogashira coupling of compound **5** was achieved using Pd $(PPh_3)_2Cl_2$/CuI catalyst system in dry $CH_2Cl_2$/MeCN (2:1) at room temperature. Using this protocol, several terminal alkyne moieties containing functionalized alkyl, aryl, and heteroaryl groups were incorporated at the C5 position of cytosine base which are summarized in Table 1. Here we have discussed the detailed synthesis protocol of morpholino-modified 7′-*O*-*tert*-butyldiphenylsilyl-*N*-trityl-5-(*N*-trifluoroacetyl-1-aminoprop-2-ynyl)cytidine (**6c**) as, for example. The terminal amine functional group can be further converted to basic guanidine group according to our reported procedure (*see* ref. 12). Other C-5 alkynyl-substituted compounds (**6a–m**) were synthesized adapting similar protocol as compound **6c** (*see* Table 1).

##### Morpholino-Modified 7′-O-Tert-butyldiphenylsilyl-N-trityl-5-(N-trifluoroacetyl-1-aminoprop-2-ynyl) cytidine (6c) [Ref. 10]

1. In an oven-dried round-bottomed flask, compound **5** (3.712 g, 4.46 mmol) was taken in dry $CH_2Cl_2/CH_3CN$ (2:1, 45 mL). The solution was purged with Ar for 5 min.
2. Triethylamine (2.48 mL, 17.8 mmol), *N*-trifluoroacetyl-propargylamine [16] (1.54 g, 10.2 mmol), $Pd(PPh_3)_2Cl_2$ (94 mg, 3 mol%), and CuI (68 mg, 8 mol%) were added, respectively, under Ar atmosphere. The resulting mixture was then stirred for 3.5 h at rt.
3. Reaction mixture was concentrated in vacuo. The gummy solid residue was extracted with $CHCl_3$ (300 mL), washed sequentially with water (100 mL) and brine (100 mL), and dried over anhydrous $Na_2SO_4$.
4. The solvent was removed under reduced pressure, and the solid residue was purified by flash column chromatography on silica gel (230–400 mesh, 2–3% MeOH in $CH_2Cl_2$) to give compound **6c** (3.36 g, 88%) as an off-white foamy solid. TLC ($CH_2Cl_2$/MeOH 95:5): $R_f$ 0.29.
5. $^1$H NMR (500 MHz, $CDCl_3$): $\delta$ 8.04 (br. s, 1H), 7.58–7.23 (comp, 24H), 7.15–7.12 (m, 3H), 6.06 (dd, $J = 9.0, 1.5$ Hz, 1H), 5.97 (br. s, 1H), 4.25–4.20 (m, 2H), 4.05 (dd, $J = 17.8, 4.2$ Hz, 1H), 3.72 (dd, $J = 10.8, 4.8$ Hz, 1H), 3.53 (dd, $J = 10.5, 6.0$ Hz, 1H), 3.43 (d, $J = 11.0$ Hz, 1H), 3.28 (d, $J = 11.5$ Hz, 1 H), 1.41 (app t, $J = 11.2$ Hz, 1H), 1.13 (app t, $J = 10.2$ Hz, 1H), 0.96 (s, 9H) ppm. $^{13}$C NMR (125 MHz, $CDCl_3$): $\delta$ 164.6, 157.3 (q, $^2J_{C,F} = 37.7$ Hz), 153.9, 144.5, 135.6, 135.56, 133.3, 133.2, 129.9, 129.8, 129.3, 127.9,

**Table 1**
**Sonogashira coupling reactions of 5-iodocytidine morpholino monomer 5 with various terminal alkynes**

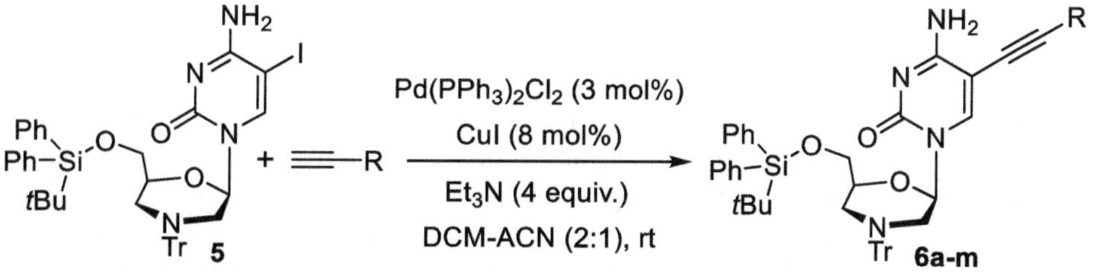


| Entry | -R group (equiv. of alkyne used) | Comp. | Reaction time (h) | Yield (%) | Entry | -R group (equiv. of alkyne used) | Comp. | Reaction time (h) | Yield (%) |
|---|---|---|---|---|---|---|---|---|---|
| 1 |  (2.5) | **6a** | 3 | 70 | 9 | 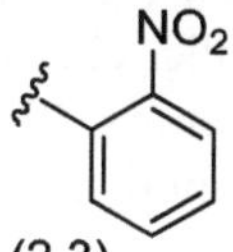 (2.3) | **6i** | 3 | 94 |
| 2 | 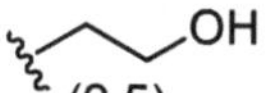 (2.5) | **6b** | 5 | 86 | 10 | 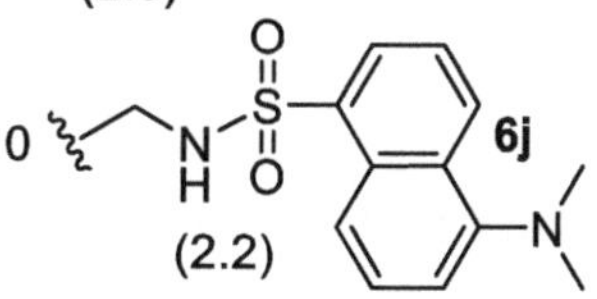 (2.2) | **6j** | 3 | 91 |
| 3 | 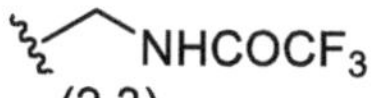 (2.3) | **6c** | 3.5 | 88 | 11 | 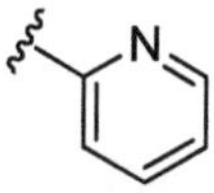 (2.5) | **6k** | 1 | 90 |
| 4 | 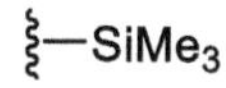 (2.5) | **6d** | 1 | 89 | 12 | 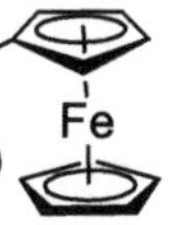 (2.1) | **6l** | 3.5 | 80 |
| 5 | 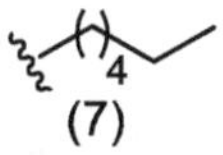 (7) | **6e** | 65 | 67 | 13[(a)] | 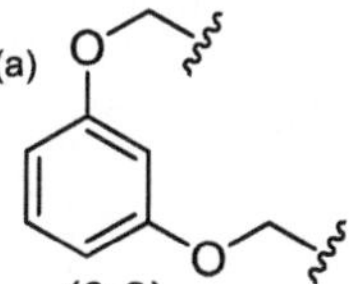 (0.6) | **6m** | 20 | 46[(b)] |
| 6 | 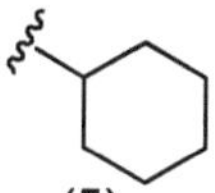 (5) | **6f** | 25 | 68 | | 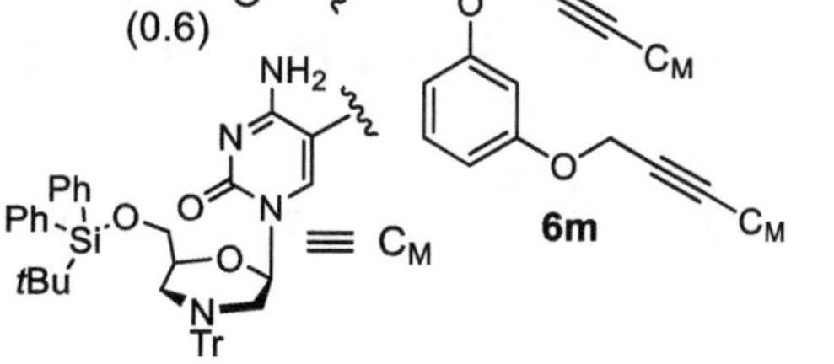 | | | |
| 7 | 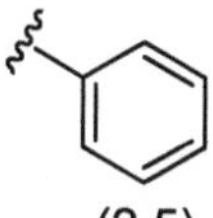 (2.5) | **6g** | 1.5 | 87 | | | | | |
| 8 | 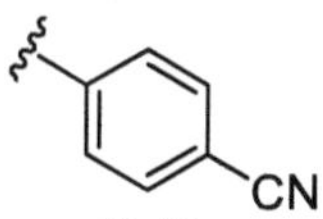 (2.1) | **6h** | 1.5 | 88 | | | | | |

(a) $Et_3N$ (6 equiv.), $Pd(PPh_3)_2Cl_2$ (6 mol%), CuI (16 mol%), and DMF were used as solvent. (b) Yield based on recovered starting material [adapted from ref. 10 with modified representation]

127.86, 126.6, 115.8 (q, ${}^{1}J_{C,F} = 286.8$ Hz), 90.2, 89.9, 82.1, 77.5, 76.9, 75.2, 64.6, 53.2, 50.2, 30.3, 26.9, 19.3 ppm. IR (neat): $\nu$ 3187, 3057, 2859, 2236, 1719, 1651, 1505, 1163 cm$^{-1}$. HRMS (ESI$^{+}$): *m/z* calcd. for $C_{49}H_{48}N_5O_4F_3Si$ [M+Na] 878.3325; found 878.3326.

### *3.2.2 Suzuki Coupling on 5-Iodocytidine Morpholino Monomer* **5**

C-5 aryl-substituted cytosine morpholino monomers (**7a–f**) were synthesized using Suzuki coupling reaction between **5** and boronic acids, employing Pd(dppf)$Cl_2$·$CH_2Cl_2$ catalyst, $K_3PO_4$ base, and THF as solvent in very good yield. The results with various boronic acids are summarized in Table 2. Here, we have discussed synthetic protocol of morpholino-modified 7′-*O*-*tert*-butyldiphenylsilyl-*N*-trityl-5-(4-trifluoromethoxyphenyl)cytidine (**7e**) as, for example. The $-CF_3$ group is biologically important because it increases the lipophilic character in nucleobases. Other C-5 aryl-substituted compounds (**7a–f**) were synthesized following similar protocol mentioned in Table 2 (adapted from ref. 10 with modified representation).

**Table 2**
**Suzuki coupling reactions between 5-iodocytidine morpholino monomer 5 and arylboronic acids**

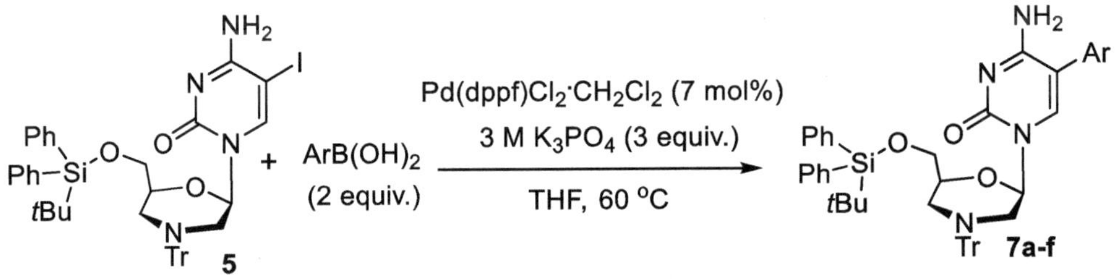


| Entry | -Ar group | Comp. | Reaction time (h) | Yield (%) | Entry | -Ar group | Comp. | Reaction time (h) | Yield (%) |
|---|---|---|---|---|---|---|---|---|---|
| 1 | 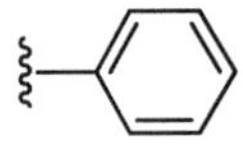 | **7a** | 8 | 81 | 5 | 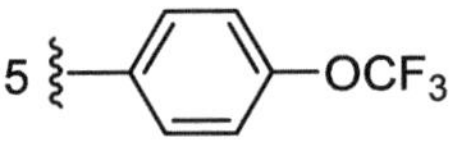  | **7e** | 7 | 81 |
| 2 | 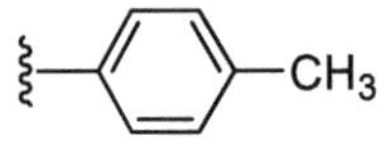  | **7b** | 8 | 75 | 6 | 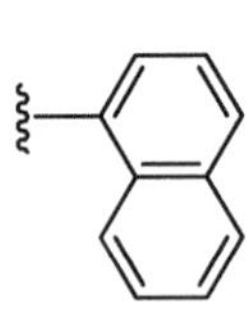 | **7f** | 10 | 87 |
| 3 | 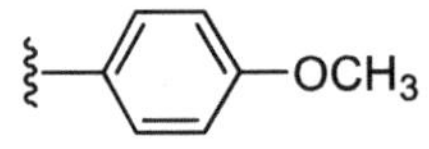  | **7c** | 7 | 78 | | | | | |
| 4 | 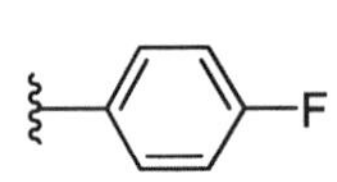  | **7d** | 6 | 80 | 7[(a)] | 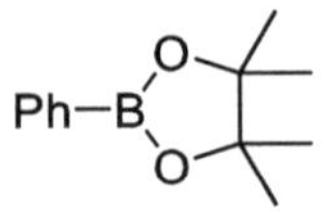  | **7a** | 12 | 79 |

(a) Phenyl boronic ester (2 equiv.) was used as coupling partner [adapted from ref. 10 with modified representation]

#### Morpholino-Modified 7′-*O*-*Tert*-butyldiphenylsilyl-*N*-trityl-5-(4-trifluoromethoxyphenyl) cytidine (7e) [Ref. 10]

1. In an oven-dried round-bottomed flask, compound **5** (300 mg, 0.36 mmol) was taken in dry THF (5 mL) under Ar atmosphere.
2. To this stirred solution of **5**, $K_3PO_4$ (3 M in $H_2O$, 360 μL, 1.08 mmol), 4-(trifluoromethoxy)phenylboronic acid (148 mg, 0.72 mmol), and Pd(dppf)$Cl_2$·$CH_2Cl_2$ (20 mg, 0.025 mmol) were added, respectively. The resulting mixture was heated at 60 °C for 7 h.
3. The reaction mixture was concentrated in vacuo, and the residue was dissolved in $CHCl_3$ (80 mL). The organic phase was washed with water (40 mL) and brine (40 mL) and dried over anhydrous $Na_2SO_4$. The solvent was removed under reduced pressure.
4. The solid residue was purified by flash column chromatography on silica gel (230–400 mesh, 1.5–2.5% MeOH in $CH_2Cl_2$) to give compound (**7e**) (253 mg, 81%) as an off-white foamy solid. TLC ($CH_2Cl_2$/MeOH 95:5): $R_f$ 0.36.
5. $^1$H NMR (500 MHz, $CDCl_3$): δ 8.82 (br. s, 1H), 7.54–7.12 (comp, 30H), 6.17 (d,*J* = 10.0 Hz, 1H), 5.33 (br. s, 1H), 4.26–4.24 (m, 1H), 3.69 (dd, *J* = 10.5, 4.0 Hz, 1H), 3.55 (dd, *J* = 10.4, 5.8 Hz, 1H), 3.47 (d, *J* = 11.0 Hz, 1H), 3.24 (d, *J* = 11.5 Hz, 1H), 1.46 (app t, *J* = 11.0 Hz, 1H), 1.22 (app t, *J* = 10.0 Hz, 1H), 0.90 (s, 9H) ppm. $^{13}$C NMR (125 MHz, $CDCl_3$): δ 164.0, 154.6, 149.2, 139.7, 135.56, 135.5, 133.33, 133.26, 131.9, 130.9, 129.8, 129.7, 129.4, 127.9, 127.74, 127.72, 126.4, 121.9, 120.5 (q, $^1J_{C,F}$ = 286.3 Hz), 107.3, 81.7, 77.2, 77.0, 64.7, 53.0, 50.1, 26.8, 19.3 ppm. IR (neat): ν 3296, 3067, 2930, 1647, 1489, 1256 $cm^{-1}$. HRMS (ESI$^+$): *m/z* calcd. for $C_{51}H_{49}N_4O_4F_3Si$ [M+Na] 889.3373; found 889.3374.

### 3.3 Synthesis of C-5-Substituted Uracil Morpholino Monomers

Uracil morpholino monomer **8** on treatment with ICl/$K_2CO_3$ in methanol gave a mixture of desired iodinated product **9** and addition product **9a** in 5:1 ratio. After workup, this crude reaction mixture on stirring with 10% (v/v) $Et_3N$ in DCM gave **9** in 83% yield [9]. $K_2CO_3$ was added in the reaction mixture to protect acid-sensitive trityl group from HCl which is generated as by-product in the reaction mixture. Sonogashira coupling of **9** was done using Pd $(PPh_3)_4$-CuI catalyst and $Et_3N$ base in DMF at room temperature. Applying this protocol, several alkyne-functionalized uracil morpholino monomers **10** were synthesized (*see* Fig. 3). Here we have discussed the detailed synthesis protocol of morpholino-modified *N*-trityl-5-iodouridine (**9**) and a representative Sonogashira-coupled product **10**.

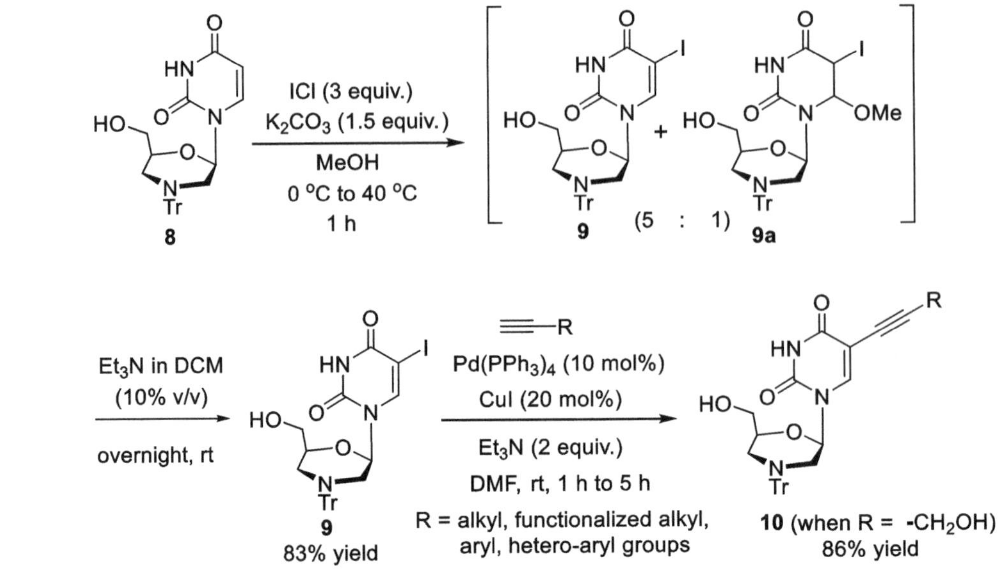

**Fig. 3** Synthesis of C-5 alkyne-functionalized uracil morpholino monomers

*3.3.1 Morpholino-Modified N-Trityl-5-iodouridine (9) [Ref. 9]*

1. To a stirred solution of **8** (1.2 g, 2.55 mmol) in dry MeOH (10.0 mL) in an oven-dried round-bottomed flask, $K_2CO_3$ (530 mg, 3.84 mmol) was added under Ar atmosphere.
2. The reaction mixture was cooled to 0 °C, and ICl solution (7.5 mL, 1.0 M solution in $CH_2Cl_2$) was added in dropwise manner. The reaction mixture was warmed to 40 °C and heated for a period of 1 h in dark. The reaction was monitored by TLC.
3. The reaction mixture was evaporated to dryness and extracted with $CH_2Cl_2$ (2 × 40 mL). The organic layers were washed with $H_2O$ (2 × 20 mL), and the excess iodine was removed by washing with sodium thiosulfate (35% w/v, 20 mL). The organic layer was further washed with water (20 mL) and half-saturated brine (10% w/v, 20 mL; *see* **Note 6**) and dried over $Na_2SO_4$. The organic extract was concentrated in vacuo to obtain a colorless solid.
4. The crude mass obtained was dissolved in $CH_2Cl_2$ (20 mL). To this solution, $Et_3N$ (2.0 mL) was added and stirred for overnight.
5. Reaction mixture was concentrated in vacuo and purified by column chromatography (100–200 mesh silica gel) using 2–3% MeOH in $CH_2Cl_2$ as eluent to get compound **9** (1.26 g, 83%) as colorless solid. TLC ($CH_2Cl_2$/MeOH 95:5): $R_f$ 0.51.
6. $^1H$ NMR (500 MHz, $CDCl_3$): $\delta$ 9.63 (s, 1H), 7.59 (s, 1H), 7.46 (br s, 6H), 7.32–7.29 (m, 6H), 7.21–7.18 (m, 3H), 6.14

(dd, $J = 9.5, 2.0$ Hz, 1H), 4.31–4.27 (m, 1H), 3.64–3.58 (m, 2H), 3.41–3.38 (d, $J = 11.5$ Hz, 1H), 3.11–3.09 (dd, $J = 12.0, 4.0$ Hz, 1H), 2.00 (br s, 1H), 1.49–1.44 (t, $J = 11.0$ Hz, 1H), 1.42–1.38 (t, $J = 9.5$ Hz, 1H) ppm. $^{13}C$ NMR (125 MHz, $CDCl_3$): $\delta$ 159.7, 149.3, 144.4, 129.2, 128.1, 126.8, 92.4, 81.3, 78.2, 68.3, 63.7, 52.5, 48.9 ppm. IR (neat): $\nu$ 3385, 1687, 1681, 1448, 1265, 750 $cm^{-1}$. HRMS ($ESI^+$): *m/z* calcd. for $C_{28}H_{26}N_3O_4I$ [M+Na] 618.0866; found 618.0868.

### *3.3.2 Sonogashira Coupling of Compound 5-Iodouridine Morpholino Monomer* ***9***

Here the synthesis protocol of morpholino-modified *N*-trityl-5-(1-hydroxyprop-2-ynyl)iodouridine (**10**, where R = $-CH_2OH$) has been discussed as, for example, where propargyl alcohol was used as an alkyne counterpart in Sonogashira coupling.

#### Morpholino-Modified *N*-Trityl-5-(1-hydroxyprop-2-ynyl)iodouridine (**10**) [Ref. 9]

1. To a solution of **9** (460 mg, 0.77 mmol) in dry DMF (10 mL) in an oven-dried round-bottomed flask, propargyl alcohol (130 μL, 2.3 mmol) and dry $Et_3N$ (215 μL, 1.55 mmol) were added.
2. The reaction mixture was degassed with argon for 15 min. Next, $Pd(PPh_3)_4$ (88 mg, 0.077 mmol) and CuI (30 mg, 0.157 mmol) were added to the reaction mixture. The resulting solution was stirred under Ar atmosphere for 1.5 h.
3. The reaction mixture was diluted with EtOAc (75 mL). The total organic layer was washed with water (3 × 20 mL) and brine (30 mL), respectively. The organic layer was separated and dried over anhydrous $Na_2SO_4$.
4. The solvent was removed under reduced pressure, and the solid residue was purified by column chromatography (100–200 mesh silica gel) using 2–3% MeOH in $CH_2Cl_2$ as eluent to get compound **10** (347 mg, 86%) as off-white solid. TLC (Pet ether/EtOAc 1:1): $R_f$ 0.38.
5. $^1H$ NMR (500 MHz, $CDCl_3$): $\delta$ 9.53 (s, 1H), 7.52 (s, 1H), 7.45 (m, 6H), 7.29–7.26 (m, 6H), 7.18–7.16 (m, 3H), 6.09 (dd, $J = 8.0, 1.0$ Hz, 1H), 4.28–4.24 (m, 3H), 3.62–3.47 (m, 3H), 3.37–3.35 (d, $J = 11.0$ Hz, 1H), 3.03–3.05 (d, $J = 11.5$ Hz, 1H), 2.11 (br s, 1H), 1.46–1.41 (t, $J = 11.0$ Hz, 1H), 1.38–1.34 (t, $J = 10.5$ Hz, 1H). $^{13}C$ NMR (125 MHz, $CDCl_3$): $\delta$ 162.0, 148.9, 143.1, 129.3, 128.1, 126.7, 99.3, 92.9, 81.2, 77.9, 76.0, 63.6, 52.4, 51.2, 48.7. IR (neat): $\nu$ 3078, 3054, 2217, 1702, 1698, 1510, 1247, 750 $cm^{-1}$. HRMS ($ESI^+$): *m/z* calcd. for $C_{31}H_{29}N_3O_5$ [M+Na] 546.2005; found 546.2003.

### 3.4 Synthesis of C-8-Substituted Adenosine Morpholino Monomers

To achieve 8-substituted morpholino-modified adenosine nucleosides using Pd-catalyzed cross-coupling reactions, morpholino-modified 8-bromo adenosine monomer **13** was used as the starting material.

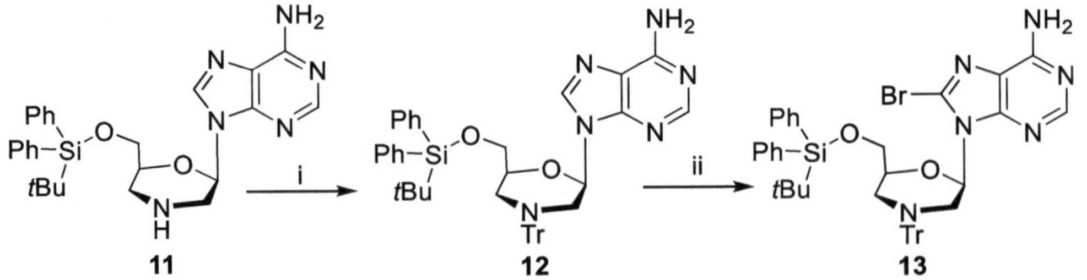

**Fig. 4** Synthesis of 8-bromoadenosine morpholino monomer **13** [Ref. 6]. Reagents and conditions: (i) Trityl chloride (1.1 equiv.), $Et_3N$ (1.5 equiv.), DMF, 0 °C to rt, 1.5 h, 87%; (ii) $Br_2$ in $Na_2HPO_4$ buffer (10% w/v, pH 7), dioxane, 15 °C to rt, overnight, conv. 59%, yield 76% (based on recovered starting material). Abbreviation: *Tr* trityl

Bromination was done instead of iodination in the case of purine nucleobase because direct bromination is possible due to higher reduction potential of bromine than iodine. Bromination was carried out using $Br_2$ in $Na_2HPO_4$ buffer and dioxane biphasic mixture to obtain morpholino-modified 7′-*O*-*tert*-butyldiphenylsilyl-*N*-trityl-8-bromoadenosine **13** (*see* Fig. 4). Bromination was done on *N*-trityl adenosine compound **12**, which was obtained after tritylation of compound **11**. Here we have discussed the synthesis protocol of compounds **12** and **13**.

*3.4.1 Morpholino-Modified 7′-O-Tert-butyldiphenylsilyl-N-trityl adenosine (**12**) [Ref. 15]*

1. In an oven-dried round-bottomed flask, compound **11** (3.0 g, 6.14 mmol) was taken in dry DMF (25 mL) under Ar atmosphere.
2. The solution was cooled to in an ice bath. Dry $Et_3N$ (1.3 mL, 9.33 mmol) was added to it.
3. Then, to this stirred solution, trityl chloride (1.88 g, 6.75 mmol) was added portion-wise (in 4 to 5 times at ~ 5 min gap) at 0 °C. After 30 min, ice bath was removed and stirred for another 1.5 h at room temperature.
4. The reaction mixture was quenched with 1 mL MeOH and concentrated under reduced pressure. This was diluted with EtOAc (250 mL) and transferred to a separating funnel. The organic layers were washed with water (2 × 50 mL) and brine (80 mL). The organic phase was dried over anhydrous $Na_2SO_4$, filtered, and concentrated in vacuo.
5. The solid residue was purified by flash column chromatography on silica gel (230–400 mesh, 2–3% MeOH in $CH_2Cl_2$) to give the product **12** (3.9 g, 87%) as a colorless foamy solid. TLC ($CH_2Cl_2$/MeOH 95:5): $R_f$ 0.44.
6. $^1H$ NMR (500 MHz, $CDCl_3$): $\delta$ 8.35 (s, 1H), 7.62 (s, 1H), 7.26–7.56 (m, 22H), 7.18 (t, $J$ = 6.8 Hz, 3H), 6.30 (d, $J$ = 9.5 Hz, 1H), 5.91–5.96 (m, 2H), 4.32–4.34 (m, 1H), 3.75 (dd, $J$ = 9.8, 4.3 Hz, 1H), 3.58 (dd, $J$ = 11.0, 6.0 Hz,

1H), 3.41 (d, $J$ = 11.5 Hz, 1H), 3.34 (d, $J$ = 12.0 Hz, 1H), 1.75 (t, $J$ = 10.5 Hz, 1H), 1.59 (t, $J$ = 11.0 Hz, 1H), 0.94 (s, 9H) ppm. $^{13}$C NMR (125 MHz, $CDCl_3$): $\delta$ 155.5, 153.3, 149.4, 138.1, 135.6, 133.4, 133.2, 129.8, 129.7, 129.3, 127.9, 127.7, 126.5, 119.2, 79.9, 77.1, 77.0, 64.6, 53.3, 50.0, 26.8, 19.3 ppm. HRMS (ESI$^+$): $m/z$ calcd. for $C_{45}H_{46}N_6O_2Si$ [M+H] 731.3530, found 731.3522.

#### 3.4.2 Morpholino-Modified 7′-O-Tert-butyldiphenylsilyl-N-trityl-8-bromoadenosine (13) [Ref. 10]

1. Bromine (1.05 mL) was added to an aqueous $Na_2HPO_4$ solution (10% w/v, 125 mL), and the mixture was stirred vigorously for 15 min until most of the bromine was dissolved (*see* **Note** 7).
2. The decanted bromine solution (75 mL) was then added dropwise to a stirred solution of **12** (3.5 g, 4.75 mmol) in dioxane (100 mL) at 15 °C. Then, the resulting mixture was stirred for overnight at room temperature.
3. The reaction mixture was cooled again (ice bath), and aqueous $Na_2SO_3$ (60 mL, 1.5 M) was added dropwise. The resulting mixture was extracted with EtOAc (2 × 150 mL). The combined organic layers were washed with brine (100 mL) and dried over anhydrous $Na_2SO_4$. The solvent was removed under reduced pressure.
4. The solid residue was purified by flash column chromatography on silica gel (230–400 mesh, 5–7% acetone in $CH_2Cl_2$) to give bromo compound **13** (2.27 g, conv. 59%; yield 76%, based on recovered starting material) as a colorless foamy solid and unreacted starting material **12**. TLC ($CH_2Cl_2$/MeOH 95:5): $R_f$ 0.46.
5. $^1$H NMR (500 MHz, $CDCl_3$): $\delta$ 8.21 (s, 1H), 7.56–7.25 (comp, 22H), 7.20–7.17 (m,3H), 6.32 (dd, $J$ = 10.2, 2.2 Hz, 1H), 5.62 (br. s, 2H), 4.31–4.28 (m, 1 H), 3.75 (dd, $J$ = 10.8, 4.8 Hz, 1H), 3.58 (dd, $J$ = 10.5, 6.0 Hz, 1H), 3.29 (d, $J$ = 12.0 Hz, 1H), 3.16 (d, $J$ = 11.5 Hz, 1H), 2.81 (app t, $J$ = 10.8 Hz, 1H), 1.62 (app t, $J$ = 11.2 Hz, 1H), 0.95 (s, 9H) ppm.$^{13}$C NMR (125 MHz, $CDCl_3$): $\delta$ 154.4, 153.0,151.3, 135.61, 135.59, 133.4, 133.3, 129.74, 129.68, 129.3, 127.9, 127.7, 127.66, 126.5, 125.6, 119.8, 82.2, 77.6, 77.1, 64.6, 49.9, 26.9, 19.3 ppm. IR (KBr): $\nu$ 3391, 3320, 2930, 2857, 1634, 1597, 1449, 1113, 710 cm$^{-1}$. HRMS (ESI$^+$): $m/z$ calcd. for $C_{45}H_{45}N_6O_2SiBr$ [M+Na] 831.2454; found 831.2426.

### 3.5 Pd-Catalyzed Cross-Coupling of 8-Bromoadenosine Morpholino Monomer 13

The 8-bromoadenosine monomer **13** synthesized in the previous step was subsequently used for Sonogashira and Suzuki coupling reactions to install a variety of functional groups.

**Table 3**
**Sonogashira coupling reactions of 8-bromoadenosine morpholino monomer 13 [adapted from Ref. 10 with modified representation]**

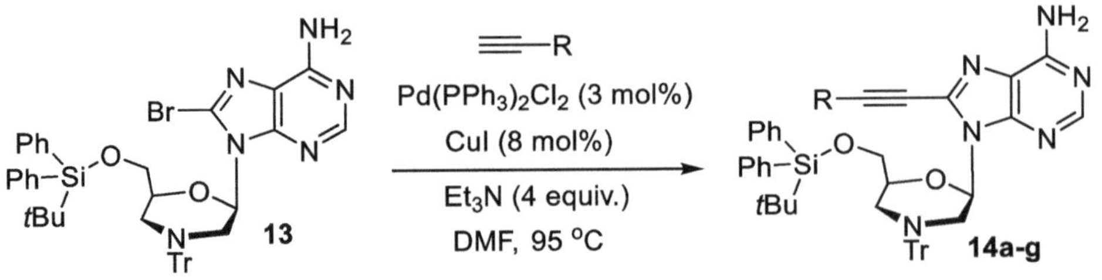


| Entry | -R group (equiv. of alkyne used) | Comp. | Reaction time (h) | Yield (%) | Entry | -R group (equiv. of alkyne used) | Comp. | Reaction time (h) | Yield (%) |
|---|---|---|---|---|---|---|---|---|---|
| 1 | OH (2.5) | **14a** | 4 | 62 | 5 | (2.5) | **14e** | 4.5 | 78 |
| 2 | NHCOCF$_3$ (2.5) | **14b** | 5 | 63 | 6 | CN (2.1) | **14f** | 5 | 74 |
| 3 | –SiEt$_3$ (2.5) | **14c** | 6 | 67 | 7 | N (2.5) | **14g** | 3.5 | 65 |
| 4 | (8) | **14d** | 16 | 68 | | | | | |

*3.5.1 Sonogashira Coupling on 7′-O-Silyl-Protected Adenosine Morpholino Monomer 13*

Several C-8 alkyne-functionalized adenosine morpholino monomers (**14a–g**) were synthesized through Sonogashira coupling reaction in the presence of Pd(PPh$_3$)$_2$Cl$_2$-CuI catalyst and Et$_3$N base in DMF under heating condition. The results with various terminal alkynes having different functional group have been summarized in Table 3. Here we have discussed the synthetic protocol of morpholino-modified 7′-*O*-*tert*-butyldiphenylsilyl-*N*-trityl-8-(triethylsilylethynyl)adenosine (**14c**) as an example. After removal of triethylsilyl group, free 8-ethyne-substituted adenosine morpholino monomer can be synthesized. This terminal acetylene group may be used to label oligonucleotides with fluorescent dyes, sugars, peptides, and other reporter groups using copper-catalyzed alkyne-azide cycloaddition (CuAAC) reaction [17]. Other C-8 alkynyl-substituted compounds (**14a–g**) were synthesized adapting similar protocol as compound **14c** (*see* Table 3).

Morpholino-Modified 7′-*O*-*Tert*-butyldiphenylsilyl-*N*-trityl-8-(triethylsilylethynyl) adenosine(**14c**) [Ref. 10]

1. Compound **13** (500 mg, 0.62 mmol) in dry DMF (5 mL) was taken in an oven-dried round-bottomed flask. The solution was purged with argon gas for 5–10 min.

2. To the stirred solution of **13**, triethylamine (350 μL, 2.5 mmol), (triethylsilyl)acetylene (280 μL, 1.55 mmol), Pd$(PPh_3)_2Cl_2$ (13 mg, 3 mol%), and CuI (10 mg, 8 mol%) were added under Ar atmosphere. The resulting mixture was heated at 95 °C for 6 h.
3. The reaction mixture was concentrated in vacuo, and the residue was dissolved in $CHCl_3$ (150 mL). The organic layer was washed with water (75 mL) and brine (75 mL) and dried over anhydrous $Na_2SO_4$. The solvent was removed under reduced pressure.
4. The crude solid was purified by flash column chromatography on silica gel (230–400 mesh, 4–5% acetone in $CH_2Cl_2$) to obtain compound **14c** (361 mg, 67%) as an off-white foamy solid. TLC ($CH_2Cl_2$/MeOH 95:5): $R_f$ 0.41. (*see* **Note 8**)
5. $^1$H NMR (500 MHz, $CDCl_3$): $\delta$ 8.26 (s, 1H), 7.56–6.16 (comp, 25H), 6.44 (dd, $J$ = 10.0, 2.0 Hz, 1H), 5.92 (br. s, 2H), 4.36–4.34 (m, 1H), 3.80 (dd, $J$ = 10.2, 4.8 Hz, 1H), 3.53 (dd, $J$ = 10.0, 8.0 Hz, 1H), 3.41 (d, $J$ = 11.5 Hz, 1H), 3.18 (d, $J$ = 11.0 Hz, 1H), 2.75 (app t, $J$ = 10.8 Hz, 1H), 1.56 (app t, $J$ = 11.0 Hz, 1H), 1.00 (t, $J$ = 7.8 Hz, 9H), 0.96 (s, 9H), 0.61 (q, $J$ = 8.0 Hz, 6H) ppm. $^{13}$C NMR (125 MHz, $CDCl_3$): $\delta$ 155.3, 154.0, 149.5, 135.59, 135.56, 133.6, 133.5, 133.3, 129.74, 129.7, 129.4, 127.9, 127.73, 127.7, 126.5, 119.2, 100.8, 94.4, 81.9, 77.8, 76.9, 64.6, 50.6, 50.5, 26.9, 19.3, 7.6, 4.1 ppm. IR (neat): $\nu$ 3315, 3183, 2957, 2874, 2166, 1643, 1113 cm$^{-1}$. HRMS (ESI$^+$): $m/z$ calcd. for $C_{53}H_{60}N_6O_2Si_2$ [M+Na] 891.4214; found 891.4214.

#### *3.5.2 Suzuki Coupling on Protected Adenosine Morpholino Monomer **13***

C-8 aryl-substituted adenosine morpholino monomers (**15a–e**) were synthesized using Suzuki coupling reaction between **13** and boronic acids. The results with various boronic acids are summarized in Table 4.

##### Morpholino-Modified 7′-*O*-*Tert*-butyldiphenylsilyl-*N*-trityl-8-(4-fluorophenyl) adenosine (**15b**) [Ref. 10]

1. In an oven-dried round-bottomed flask, compound **13** (500 mg, 0.62 mmol) was taken in dry dioxane (8 mL) under Ar atmosphere.
2. To the stirred solution of **13**, $K_3PO_4$ (3 M in $H_2O$, 620 μL, 1.86 mmol), 4-fluorophenylboronic acid (191 mg, 1.36 mmol), and Pd(dppf)$Cl_2$·$CH_2Cl_2$ (35 mg, 0.043 mmol) were added, respectively, under Ar atmosphere. The resulting mixture was heated at 90 °C for 12 h.
3. The reaction mixture was concentrated in vacuo, and the residue was dissolved in EtOAc (150 mL). The organic phase was washed with water (60 mL) and brine (60 mL), respectively, and dried with anhydrous $Na_2SO_4$. The solvent was removed under reduced pressure.

**Table 4**
**Suzuki coupling reactions of 13 with arylboronic acids [adapted from Ref. 10]**

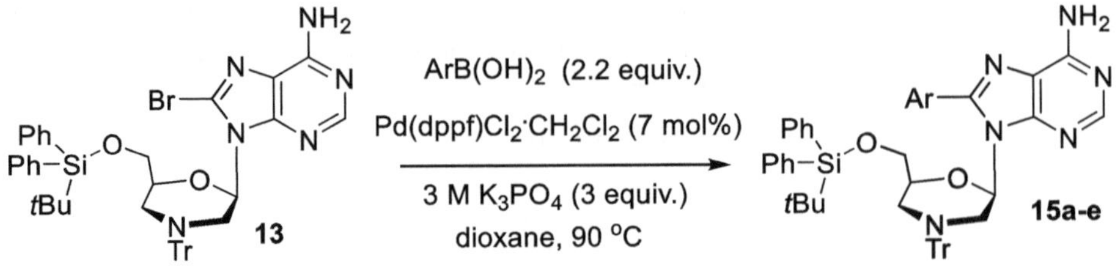


| Entry | -Ar group | Comp. | Reaction time (h) | Yield (%) | Entry | -Ar group | Comp. | Reaction time (h) | Yield (%) |
|---|---|---|---|---|---|---|---|---|---|
| 1 | 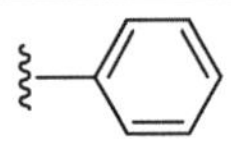 | **15a** | 12 | 72 | 4 | 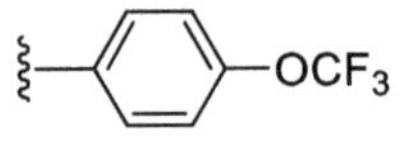 | **15d** | 13 | 77 |
| 2 | 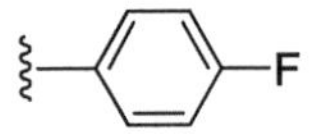 | **15b** | 12 | 79 | 5 | 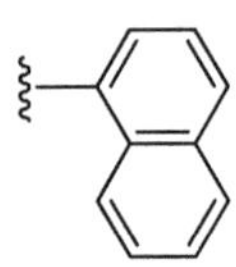 | **15e** | 14 | 84 |
| 3 | 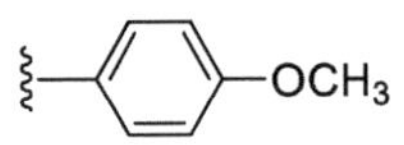 | **15c** | 11 | 76 | | | | | |

Considering the atomic size, fluorine is known to mimic hydrogen without much distortion of the geometry, but their electronegativity differs very much. So, replacement of hydrogen atoms or functional groups in nucleoside molecules by fluorine atom may cause a dramatic change in physical properties and biological activities of molecules [18, 19]. Here we have discussed the synthetic protocol of morpholino-modified 7′-*O*-*tert*-butyldiphenylsilyl-*N*-trityl-8-(4-fluorophenyl) adenosine (**15b**)

4. The solid residue was purified by flash column chromatography on silica gel (230–400 mesh, 5–6% acetone in $CH_2Cl_2$) to give compound **15b** (404 mg, 79%) as a colorless foamy solid. TLC ($CH_2Cl_2$/MeOH 95:5): $R_f$ 0.36.
5. $^1$H NMR (500 MHz, $CDCl_3$): $\delta$ 8.38 (s, 1H), 7.60–7.14 (comp, 27H), 6.88 (app t, $J$ = 8.5 Hz, 2H), 6.30 (dd, $J$ = 10.2, 2.2 Hz, 1H), 6.07 (br. s, 2H), 4.32–4.28 (m, 1H), 3.79 (dd, $J$ = 10.5, 4.5 Hz, 1H), 3.57 (dd, $J$ = 10.8, 6.2 Hz, 1 H), 3.22 (d, $J$ = 11.5 Hz, 1H), 2.89 (d, $J$ = 11.5 Hz, 1H), 2.25 (app t, $J$ = 11.0 Hz, 1H), 1.36 (app t, $J$ = 11.0 Hz, 1H), 0.97 (s, 9H) ppm. $^{13}$C NMR (125 MHz, $CDCl_3$): $\delta$ 163.9 ($^1J_{C,F}$ = 251.5 Hz), 155.1, 152.6, 151.3, 150.9, 135.6, 133.3, 133.2, 132.1 ($^3J_{C,F}$ = 7.5 Hz), 129.9, 129.8, 129.2, 127.9, 127.8, 126.6 ($^4J_{C,F}$ = 3.4 Hz), 126.4, 118.9, 115.6 ($^2J_{C,F}$ = 21.4 Hz), 80.9, 77.3, 77.0, 64.7, 49.7, 49.5, 26.9,19.3 ppm. IR (neat): $\nu$ 3440, 3019, 2931, 1636, 1215, 758 cm$^{-1}$. HRMS (ESI$^+$): $m/z$ calcd. for $C_{51}H_{49}N_6O_2FSi$ [M+Na] 847.3568; found 847.3566.

### 3.6 Synthesis of C-8-Substituted Guanosine Morpholino Monomers

Morpholino-modified guanosine monomer **17** was brominated using the same bromination method as that of adenosine monomer to obtain morpholino-modified $N^2$-isobutyryl-*N*-trityl-8-

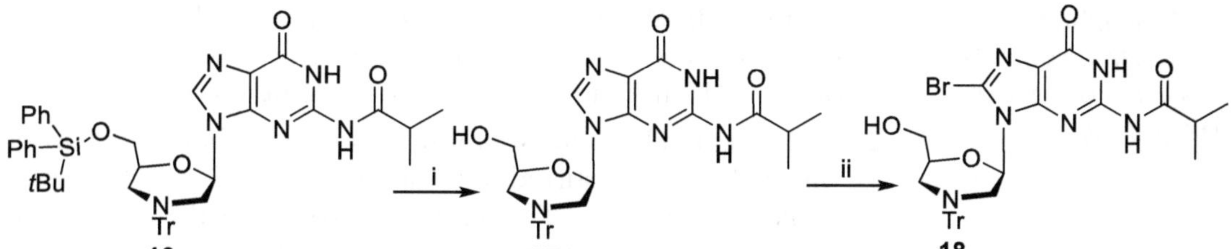

16 17 18

**Fig. 5** Synthesis of 8-bromoguanosine morpholino monomer **18** [ref. 10]. Reagents and conditions: (i) *n*$Bu_4NF$ (1 M in THF, 1.6 equiv.), $Et_3N$ (1.5 equiv.), THF, 15 °C to rt, 3 h, 91%; (ii) $Br_2$ in $Na_2HPO_4$ buffer (10% w/v, pH 7), dioxane, 15 °C to rt 6 h, conv. 71%, yield 82% (based on recovered starting material). Abbreviation: *Tr* trityl

bromoguanosine, **18**. Compound **17** was obtained after *O*-silyl group deprotection of compound **16** (*see* Fig. 5). Unlike adenosine monomer, here hydrophobic silyl group deprotection of **16** is necessary for bromination reaction. Because of more nonpolar nature, compound **16** was less soluble in water/dioxane mixture and gave poor yield (24%) in bromination conditions.

*3.6.1 Morpholino-Modified* N$^2$*-Isobutyryl-*N*-trityl-guanosine (**17**) [Ref. 10]*

1. In an oven-dried round-bottomed flask, compound **16** (2.88 g, 3.52 mmol) was taken in dry THF (30 mL) under Ar atmosphere.
2. To this stirred solution, *n*-$Bu_4NF$ (5.6 mL, 5.6 mmol, 1 M in THF) was added dropwise at 15 °C. The resulting mixture was stirred for 3 h at room temperature.
3. The reaction mixture was concentrated in vacuo, and the residue was dissolved in $CHCl_3$ (200 mL). The organic layer was washed with water and brine, respectively, and dried over anhydrous $Na_2SO_4$.
4. The solvent was removed under reduced pressure, and the solid residue was purified by flash column chromatography on silica gel (230–400 mesh, 2.5–4% MeOH in $CH_2Cl_2$) to give the product **17** (1.854 g, 91%) as a colorless solid. TLC ($CH_2Cl_2$/MeOH 95:5): $R_f$ 0.35.
5. $^1$H NMR (500 MHz, $CDCl_3$): $\delta$ 11.96 (br s, 1 H), 9.29 (br s, 1 H), 7.52 (s, 1 H), 7.43 (br s, 6 H), 7.27–7.24 (comp, 6 H), 7.15 (m, 3 H), 5.97 (dd, $J$ = 10.0, 2.0 Hz, 1 H), 4.26–4.23 (m, 1 H), 3.56 (br s, 3 H), 3.35 (d, $J$ = 11.5 Hz, 1 H), 3.09 (d, $J$ = 12.0 Hz, 1 H), 2.69 (heptet, $J$ = 6.8 Hz, 1 H), 1.79 (app t, $J$ = 10.5 Hz, 1 H), 1.51 (app t, $J$ = 11.2 Hz, 1 H), 1.24 (d, $J$ = 7.0 Hz, 3 H), 1.22 (d, $J$ = 6.5 Hz, 3 H) ppm. $^{13}$C NMR (125 MHz, $CDCl_3$): $\delta$ 179.3, 155.8, 148.0, 147.9, 136.6, 129.3, 128.0, 126.6, 120.5, 80.5, 77.8, 77.0, 63.5, 52.5, 49.0, 36.3, 19.2, 18.8 ppm. IR (KBr): $\nu$ 3408, 3186, 2972, 1682, 1607, 1406, 1024 cm$^{-1}$. HRMS (ESI+): *m/z* calcd. for $C_{33}H_{34}N_6O_4$ [M+Na] 601.2539, found 601.2535.

#### 3.6.2 Morpholino-Modified $N^2$-Isobutyryl-N-trityl-8-bromoguanosine (18) [Ref. 10]

1. Bromine (0.84 mL) was added to an aqueous $Na_2HPO_4$ solution (10% w/v, 100 mL), and the mixture was stirred vigorously for 15 min until most of the bromine was dissolved.
2. The decanted bromine solution (40 mL) was then added dropwise at 15 °C to a stirred solution of **17** (1.643 g, 2.84 mmol) in dioxane (50 mL). The reaction mixture was then stirred for 6 h at room temperature until became colorless.
3. The reaction mixture was extracted with $CHCl_3$ (2 × 125 mL). The combined organic layers were washed with brine (100 mL) and dried over anhydrous $Na_2SO_4$. The solvent was removed under reduced pressure.
4. The solid residue was purified by flash column chromatography on silica gel (230–400 mesh, 2–3% MeOH in $CH_2Cl_2$) to give the product **16** (1.326 g, conv. 71%, yield 82% based on recovered starting material) as a colorless solid together with unreacted **15** (230 mg). TLC ($CH_2Cl_2$/MeOH 95:5): $R_f$ 0.41.
5. $^1$H NMR (500 MHz, DMSO-$d_6$): $\delta$ 12.12 (br. s, 1H), 11.62 (br. s, 1 H), 7.44–7.32 (comp, 12H), 7.20 (br. s, 3H), 6.26 (d, $J$ = 10.0 Hz, 1H), 4.83 (t, $J$ = 5.2 Hz, 1H), 4.17 (br. s, 1 H), 3.46–3.44 (m, 1H), 3.31 (dd, $J$ = 11.0, 5.5 Hz, 1H), 3.16 (d, $J$ = 11.5 Hz, 1H), 3.10 (d, $J$ = 11.0 Hz, 1H), 2.84 (sept, $J$ = 6.8 Hz, 1H), 2.44 (app t, $J$ = 10.5 Hz, 1H), 1.42 (app t, $J$ = 11.0 Hz, 1 H), 1.14 (d, $J$ = 6.5 Hz, 6H) ppm. $^{13}$C NMR (125 MHz, DMSO-$d_6$): $\delta$ 180.2, 153.5, 149.6, 148.3, 128.8, 127.9, 126.4, 121.1, 120.3, 81.2, 77.5, 76.4, 61.8, 49.7, 49.4, 34.7, 18.83, 18.77 ppm. IR (KBr): $\nu$ 3376, 3160, 2930, 1639, 1607, 1559, 1449 cm$^{-1}$. HRMS (ESI$^+$): $m/z$ calcd. for $C_{33}H_{33}N_6O_4Br$ [M+Na] 681.1624; found 681.1631 [mass peak was simulated with Br 81 isotope].

### 3.7 Pd-Catalyzed Cross-Coupling Reaction of $N^2$-Isobutyryl-Protected 8-Bromoguanosine Morpholino Monomer 18

C-8 phenyl alkyne-substituted morpholino-modified guanosine derivative was obtained from the bromo compound **18** using Sonogashira coupling reaction with phenyl acetylene as coupling partner (*see* Fig. 6). Here similar reaction conditions were employed which were optimized for adenosine nucleobase.

#### 3.7.1 Sonogashira Coupling on 8-Bromoguanosine Morpholino Monomer 18

##### Morpholino-Modified $N^2$-Isobutyryl-*N*-trityl-8-(phenylethynyl)guanosine (19) [Ref. 10]

1. In an oven-dried round-bottomed flask, compound **18** (526 mg, 0.80 mmol) was taken in dry DMF (7.5 mL). The solution was purged with argon gas for 5–10 min.

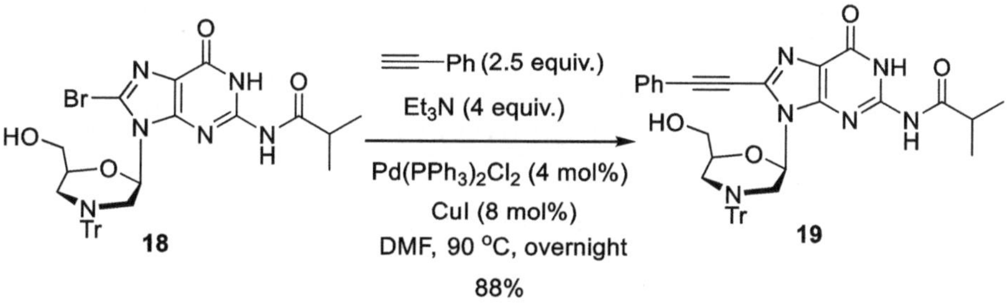


**Fig. 6** Sonogashira coupling of the morpholino-modified 8-bromoguanosine monomer **18**

2. To this stirred solution of **18**, triethylamine (445 μL, 3.2 mmol), phenylacetylene (220 μL, 2.0 mmol), Pd $(PPh_3)_2Cl_2$ (22 mg, 0.032 mmol), and CuI (12 mg, 0.064 mmol) were added, respectively, under Ar atmosphere. The resulting mixture was heated with stirring at 90 °C for overnight.
3. The reaction mixture was concentrated in vacuo, and the residue was dissolved in $CHCl_3$ (150 mL). The organic layers were washed with water and brine and dried with anhydrous $Na_2SO_4$. The solvent was removed under reduced pressure.
4. The solid residue was purified by flash column chromatography on silica gel (230–400 mesh, 2–3% MeOH in $CH_2Cl_2$) to give the compound **19** (478 mg, 88%) as a colorless solid. TLC ($CH_2Cl_2$/MeOH 95:5): $R_f$ 0.39 (*see* **Note 9**).
5. $^1$H NMR (500 MHz, DMSO-$d_6$): $\delta$ 12.16 (s, 1H), 11.80 (s, 1H), 7.49–7.25 (comp, 15H), 7.10–7.08 (m, 5H), 6.26 (d, $J$ = 10.0 Hz, 1H), 4.82 (t, $J$ = 5.5 Hz, 1H), 4.18–4.16 (m, 1H), 3.41–3.38 (m, 1H), 3.24–3.13 (m, 3H), 2.85 (sept, $J$ = 6.8 Hz, 1H), 2.57 (app t, $J$ = 10.8 Hz, 1H), 1.48 (app t, $J$ = 11.2 Hz, 1H), 1.15 (d, $J$ = 6.5 Hz, 6 H) ppm. $^{13}$C NMR (125 MHz, DMSO-$d_6$): $\delta$ 180.8, 154.6, 149.6, 148.6, 131.7, 130.6, 130.3, 129.2, 128.3, 126.9, 120.6, 120.3, 93.5, 80.2, 80.1, 77.6, 76.9, 62.5, 50.7, 50.2, 35.3, 19.4, 19.3 ppm. IR (KBr): $\nu$ 3434, 3061, 2997, 2876, 2218, 1713, 1682, 1599, 1543, 1190 cm$^{-1}$. HRMS (ESI$^+$): *m/z* calcd. for $C_{41}H_{38}N_6O_4$ [M+Na] 701.2852; found 701.2850.

#### 3.7.2 Suzuki Coupling on 8-Bromoguanosine Morpholino Monomer 18

C-8 phenyl-functionalized morpholino-modified guanosine derivative **20** was obtained from the bromo compound **18** using Suzuki coupling conditions already optimized for adenosine nucleobase (*see* Fig. 7).

##### Morpholino-Modified $N^2$-isobutyryl-$N$-trityl-8-phenylguanosine (20) [Ref. 10]

1. In an oven-dried round-bottomed flask, compound **18** (526 mg, 0.80 mmol) was taken in dry dioxane (8 mL) under Ar atmosphere.

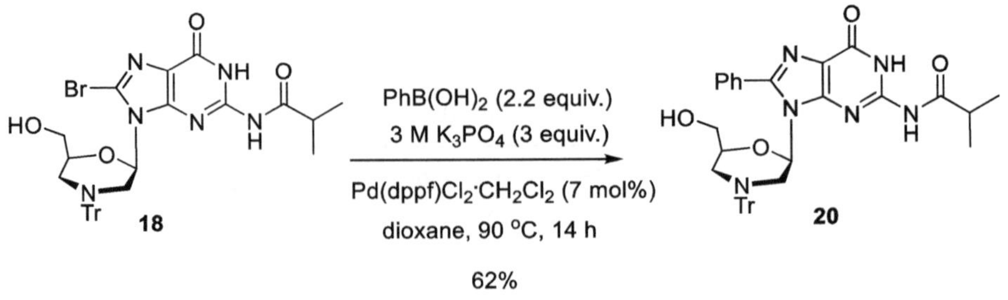


**Fig. 7** Suzuki couplings of the morpholino-modified 8-bromoguanosine monomer **18**

2. To a stirred solution of **18**, $K_3PO_4$ (3 M in $H_2O$, 800 μL, 2.4 mmol), phenylboronic acid (215 mg, 1.76 mmol), and Pd (dppf)$Cl_2$•$CH_2Cl_2$(45 mg, 0.056 mmol) were added, respectively. The resulting mixture was heated with stirring at 90 °C for 14 h.

3. The reaction mixture was concentrated in vacuo, and the residue was dissolved in $CHCl_3$ (150 mL). The organic phase was washed with water and brine and dried over anhydrous $Na_2SO_4$. The solvent was removed under reduced pressure.

4. The solid residue was purified by flash column chromatography on silica gel (230–400 mesh, 2–3% MeOH in $CH_2Cl_2$) to give compound **18** (325 mg, 62%) as a colorless solid. TLC ($CH_2Cl_2$/MeOH 95:5): $R_f$ 0.36.

5. $^1H$ NMR (500 MHz, $CDCl_3$): $\delta$ 11.87 (s, 1H), 9.14 (s, 1H), 7.61–7.59 (m, 2H), 7.55–7.42 (comp, 9H), 7.26–7.23 (m, 6H), 7.18–7.15 (m, 3H), 5.95 (dd, $J$ = 10.2, 2.8 Hz, 1H), 4.12–4.09 (m, 1H), 3.94 (br. s, 1H), 3.59 (br. s, 2H), 3.09 (d, $J$ = 11.0 Hz, 1H), 2.98 (d, $J$ = 12.0 Hz, 1H), 2.64 (app t, $J$ = 10.8 Hz,1H), 2.58 (sept, $J$ = 7.0 Hz, 1H), 1.68 (app t, $J$ = 11.2 Hz, 1H),1.22 (d, $J$ = 7.0 Hz, 3H), 1.18 (d, $J$ = 6.5 Hz, 3H) ppm. (*see* **Note 10**). $^{13}C$ NMR (125 MHz, $CDCl_3$): $\delta$ 178.8, 155.5, 149.4, 149.2, 146.8, 130.3, 129.7, 129.6, 129.4, 128.7, 127.9, 126.6, 121.2, 82.2, 77.0, 76.7, 63.3, 49.7, 48.4, 36.3, 19.2, 18.7 ppm. IR (KBr): $\nu$ 3426, 3196, 2972, 1684, 1607, 1557, 700 $cm^{-1}$. HRMS ($ESI^+$): *m/z* calcd. for $C_{39}H_{38}N_6O_4$ [M+Na] 677.2852; found 677.2852.

## 4 Notes

1. During washing of organic layer with $NaHCO_3$ solution, shaking of the separating funnel was very gentle because of vigorous effervescence of $CO_2$. The neutralization of acids was confirmed by checking the pH (~7) of aqueous layer using pH paper.

2. Compound **2** is moderately soluble in DCM, so during workup organic layer formed a white colloidal type solution. During washing of organic layer with water, shaking of the separating funnel should be gentle, and it takes more time for layer separation.
3. During washing of $Na_2SO_4$, ~10% MeOH in DCM was used to avoid material loss.
4. If compound **4** contains trace amount of water/MeOH, in the next *N*-tritylation step, the poor conversion of starting material to product was observed.
5. *N*-tritylation of compound **4** was carried out in dry DCM to get the selectively morpholine "*N*"-monoprotected product; otherwise the same reaction yielded a mixture of desired product **5** and exocyclic amine ($N^4$)-tritylated product when dry DMF was used as the solvent.
6. Half-saturated brine solution was used during extraction of the compounds **9** when $CH_2Cl_2$ was used in order to avoid the formation of emulsion. Shaking of the separating funnel was very gentle.
7. Bromine was used carefully inside the hood. The bromine solution in buffer was decanted carefully and used in the reaction for bromination of compounds **12** and **17**.
8. Sometimes, Pd-catalyst was contaminated with the cross-coupled product and was removed by second time column purification using solvent system like EtOAc/petroleum ether (bp 60–80 °C) or acetone/DCM depending on $R_f$ value.
9. $R_f$ of product **19** was very close to the staring material **18**, but **19** was short-wavelength UV active spot visualized in TLC. Reaction was stopped after the completion of the reaction which simplified the purification of **19**.
10. In $^1$H NMR, few protons have been represented as "app t" (apparent triplet) which are actually doublet of doublet, but overlapped due to same *J*-value. Complex protons (6–15H together with mixed *J*-value) are represented as "comp."

## Acknowledgment

This work was supported by the Council of Scientific and Industrial Research (CSIR), New Delhi, by a grant no. 02 (0204)/14/EMR-II.

## References


1. Lundin KE, Gissberg O, Smith CIE (2015) Oligonucleotide therapies: the past and the present. Hum Gene Ther 26:475–485
2. Corey DR, Abrams JM (2001) Morpholino antisense oligonucleotides: tools for investigating vertebrate development. Genome Biol 2: reviews1015.1–reviews1015.3
3. Summerton JE (2007) Morpholino, siRNA, and S-DNA compared: impact of structure and mechanism of action on off-target effects and sequence specificity. Curr Top Med Chem 7:651–660
4. Summerton J, Weller D (2009) Morpholino antisense oligomers: design, preparation, and properties. Antisense Nucleic Acid Drug Dev 7:187–195
5. Wagner RW, Matteucci MD, Lewis JG, Gutierrez AJ, Moulds C, Froehler BC (1993) Antisense gene inhibition by oligonucleotides containing C-5 propyne pyrimidines. Science 260:1510–1513
6. He J, Seela F (2002) Propynyl groups in duplex DNA: stability of base pairs incorporating 7-substituted 8-aza-7-deazapurines or 5-substituted pyrimidines. Nucleic Acids Res 30:5485–5496
7. Ihara T, Nakayama M, Murata M, Nakano K, Maeda M (1997) Gene sensor using ferrocenyl oligonucleotide. Chem Commun:1609–1610
8. Brázdilová P, Vrábel M, Pohl R, Pivoňková H, Havran L, Hocek M, Fojta M (2007) Ferrocenylethynyl derivatives of nucleoside triphosphates: synthesis, incorporation, electrochemistry, and bioanalytical applications. Chem Eur J 13:9527–9533
9. Paul S, Nandi B, Pattanayak S, Sinha S (2012) Synthesis of 5-alkynylated uracil–morpholino monomers using Sonogashira coupling. Tet Lett 53:4179–4183
10. Nandi B, Pattanayak S, Paul S, Sinha S (2013) Synthesis of nucleobase-functionalized morpholino-modified nucleoside monomers through palladium-catalyzed cross-coupling reactions. Eur J Org Chem:1271–1286
11. Paul S, Pattanayak S, Sinha S (2014) Synthesis and cell transfection properties of cationic uracil-morpholino tetramer. Tet Lett 55:1072–1076
12. Nandi B, Khatra H, Khan PP, Bhadra J, Pattanayak S, Sinha S (2017) Cationic cytosine morpholino-based transporters: synthesis and regulation of intracellular localization. ChemistrySelect 2:5059–5067
13. Graham D, Parkinson JA, Brown T (1998) DNA duplexes stabilized by modified monomer residues: synthesis and stability. J Chem Soc Perkins Trans 1:1131–1138
14. Ahmadian M, Zhang P, Bergstrom DE (1998) A comparative study of the thermal stability of oligodeoxyribonucleotides containing 5-substituted 2′-deoxyuridines. Nucleic Acids Res 26:3127–3135
15. Bhadra J, Pattanayak S, Sinha S (2015) Synthesis of morpholino monomers, chlorophosphoramidate monomers, and solid-phase synthesis of short morpholino oligomers. Curr Protoc Nucleic Acid Chem 62:4.65.1–4.65.26
16. Orain D, Mattes H (2005) Synthesis of two new azabicyclophosphinic acids as constrained analogues of TPMPA. Synlett (19):3008–3010
17. El-Saghee AH, Brown T (2010) Click chemistry with DNA. Chem Soc Rev 39:1388–1405
18. Liu P, Sharon A, Chu CK (2008) Fluorinated nucleosides: synthesis and biological implication. J Fluor Chem 129:743–766
19. Seela F, Xu K, Chittepu P, Ming X (2007) Fluorinated 7-deazapurine 2′-deoxyribonucleosides: modification at the nucleobase and the sugar moiety. Nucleosides Nucleotides Nucleic Acids 26:607–610

# Chapter 8

## Synthesis and Application of $^{L}K\gamma T$ Peptide Nucleic Acids

**Nathaniel Shank, Kara M. George Rosenker, Ethan A. Englund, Andrew V. Dix, Elizabeth E. Rastede, and Daniel H. Appella**

### Abstract

Displaying ligands in a succinct and predictable manner is essential for elucidating multivalent molecular-level binding events. Organizing ligands with high precision and accuracy provides a distinct advantage over other ligand-display systems, such as polymers, because the number and position of the ligand(s) can be accurately and fully characterized. Here we describe the synthesis of peptide nucleic acids (PNAs), which are oligonucleotide mimics with a pseudopeptide backbone that can hybridize to oligonucleotides through Watson-Crick base pair to form highly predictable and organized scaffold for organizing a ligand. The ligand(s) are covalently attached to the PNA through a squarate coupling reaction that occurs between a free amine on the ligand and a free amine appended to the pseudopeptide backbone of the PNA. In this chapter we describe the synthesis of a $^{L}K\gamma T$ monomer, which ultimately yields the free amine off the backbone of the PNA, incorporation of the monomer in a PNA oligomer, and the sequential squarate coupling to conjugate the ligand.

**Key words** Nonnatural nucleic acid, Peptide nucleic acid, Molecular scaffold, Ligand organization, Multivalent interactions

## 1 Introduction

Concerted multivalent interactions of ligand-protein cognates are critical to major roles in cell adhesion, growth, and intracellular signaling pathways [1, 2]. These highly synchronized interactions between multiple ligands and protein receptors act to increase effective concentration, selectivity, and specificity. To dissect the contributions of individual components and assess synergistic effects of these complex systems, researchers require the appropriate set of tools. Recently, our lab has used peptide nucleic acid (PNA) as a nano-scaffold for organizing multiple ligands in a controlled and predictable manner to better understand these interactions [3–5].

PNAs are synthetic mimics of DNA/RNA capable of forming hetero (PNA/DNA) or homo (PNA/PNA) duplexes through

Nathaniel Shank (ed.), *Non-Natural Nucleic Acids: Methods and Protocols*, Methods in Molecular Biology, vol. 1973,
https://doi.org/10.1007/978-1-4939-9216-4_8, © Springer Science+Business Media, LLC, part of Springer Nature 2019

Watson-Crick base pairing [6, 7]. Significantly, the sugar-phosphate backbone is replaced with a pseudopeptide backbone composed of 2-aminoethyl-glycine units linked by amide bonds. This structural difference makes PNA inherently resistant to nucleases or proteases and provides unique positions for modifying the oligomer (such as attaching ligands and incorporating chirality). Although PNAs were originally developed as candidates for antigen and antisense reagents [7], they have found applications as fluorescent probes [8–17], in diagnostic assays [18–20] and as scaffolds for organizing small molecules [3–5, 21–24].

PNAs are attractive scaffolds for organizing small molecules because of their ease of synthesis, programmability, and structural predictability. Synthetically, PNAs are accessible through standard Fmoc- or Boc-based solid-phase peptide synthesis (SPPS) techniques and equipment. Ligands can be attached on-resin or once the PNA has been cleaved and purified. The former approach, in conjunction with appropriate orthogonal protecting groups, can readily yield hetero-functionalized PNAs, while the latter is more amenable to creating homo-functionalized PNAs. Ligands spacing can be varied by attaching them at each terminus (and then adjusting the length of the oligomer) or, through chemistries developed in our lab as well as others [9, 21], they can be appended at discrete points along the pseudopeptide backbone. Rationally designed complementary template strands allow for the creation of unique arrays of homo- and/or hetero-ligand systems by simply changing the nucleobase sequence to match the desired PNA strand(s). In this way, from just a small set of PNAs and a handful of complementary strands, researchers can produce a diverse set of duplexes capable of probing multivalent interactions.

This chapter details the efficient incorporation of multiple ligands onto a strand of PNA after it has been cleaved from the resin. For post-cleavage modifications of PNAs, our group has found great success in the use of squaric acid, specifically, diethyl squarate. Squarate esters react readily with primary amines to yield amides, and when used in large excess the reaction is nearly quantitative with minimal intramolecular reactions. The low solubility of PNA in diethyl ether allows for the easy removal of excess squarate and further removes the necessity to purify the intermediate by HPLC, thus saving time and resources. The reaction of the PNA-squarate intermediate with a slight excess of ligand yields the final conjugate that can be precipitated again using diethyl ether and then purified by HPLC. Here we describe the protocol to attach three ligands, but we have successfully appended a 12-nucleotide PNA with five ligands (three off the backbone and one at each terminus). It should also be noted that while we only describe the synthesis of homogeneous ligand display, the creation of heterogeneous PNA-ligand systems can easily be created through proper orthogonal protecting groups of the lysine side chains.

## 2 Materials

All commercially available products and solvents were purchased and used without further purification. All solvents are purchased anhydrous or prepared that way using standard procedures. All reactions were carried out under nitrogen in oven-dried glassware unless otherwise specified. $^1$H NMR were collected on a 500 mHz NMR, and chemical shifts are reported in ppm relative to tetramethylsilane.

### 2.1 Reagents for γT Monomer Synthesis

1. Boc-Lysine(Fmoc)-OH.
2. *N*,*O*-dimethylhydroxylamine.
3. *N*-(3-Dimethylaminopropyl)-*N*′-ethylcarbodiimide hydrochloride (EDC).
4. *N*, *N*′-Diisopropylethylamine (DIEA).
5. Dichloromethane ($CH_2Cl_2$).
6. Dimethylhydroxylamine hydrochloride.
7. Ethyl acetate (EtOAc).
8. 1 M hydrochloric acid (1 M HCl).
9. Saturated $NaHCO_3$.
10. Saturated sodium chloride (NaCl) solution.
11. Saturated sodium bicarbonate ($NaHCO_3$) solution.
12. Anhydrous sodium sulfate ($Na_2SO_4$).
13. Sodium carbonate ($Na_2CO_3$).
14. Saturated sodium hydrogen sulfate ($NaHSO_4$) solution.
15. Lithium aluminum hydride ($LiAlH_4$).
16. Tetrahydrofuran (THF).
17. Benzyl glycinate hydrochloride.
18. Sodium triacetoxyborohydride ($NaHB(OAc)_3$).
19. Deionized water.
20. Silica gel 60 (230–400 mesh).
21. Thymine acetic acid.
22. 1-Hydroxybenzotriazole hydrate (HOBt).
23. *N*, *N*′-Dimethylformamide (DMF).
24. Methanol (MeOH).
25. Hydrogen gas ($H_{2(g)}$).
26. Palladium on carbon, 10 wt % (Pd/C).
27. Precoated silica plates (250 μm layer thickness).
28. UV lamp (for visualizing TLC plates).

### 2.2 Reagents for Synthesis of Peptide Nucleic Acid Oligomer Containing γT Monomer

1. Boc-8-amino-3,6-dioxaoctanoic acid • dicyclohexylamine (Boc-mPeg).
2. 4-Methylbenzhydrylamine hydrochloride, polymer-bound 100–200 mesh (MBHA resin).
3. *N*, *N*′-Dimethylformamide (DMF).
4. Dichloromethane (DCM).
5. 1-Methyl-2-pyrrolidone (NMP).
6. *Boc-deprotection solution*: 5% *m*-cresol in trifluoroacetic acid (TFA).
7. *Fmoc-deprotection solution:* 20% piperidine (Pip) in DMF.
8. *Capping solution:* acetic anhydride ($Ac_2O$), pyridine (Pyr), and NMP (1:25:25).
9. *Activator solution:* 0.19 M *O*-(Benzotriazol-1-yl)-*N,N,N*′,*N*′-tetramethyluronium hexafluorophosphate (HBTU) in DMF (2.16 g HBTU in 30 mL).
10. *Base solution:* 0.8 M *N*, *N*′-Diisopropylethylamine (DIEA) in NMP (4.18 mL DIEA in 25.8 mL with NMP).
11. *N*-(*N*-Boc-2-aminoethyl)-*N*-[(1-thyminylacetyl]-glycine (T monomer).
12. *N*-(*N*-Boc-2-aminoethyl)-*N*-[(*N*-4-Z-1-cytosyl)acetyl]-glycine (C monomer).
13. *N*-(*N*-Boc-2-aminoethyl)-*N*-[(*N*-6-Z-9-adenyl)acetyl]-glycine (A monomer).
14. *N*-(*N*-Boc-2-aminoethyl)-*N*-[(*N*-6-Z-9-guanyl)acetyl]-glycine (G monomer).
15. L-lysine γ-substituted T monomer ($^{L}K\gamma T$) (*see* Subheading 3.1).

### 2.3 Reagents for Cleavage and Purification of PNA Oligomer

1. Trifluoroacetic acid (TFA).
2. *m*-cresol.
3. Thioanisole.
4. Trifluoromethanesulfonic acid (TFMSA).
5. Cleavage cocktail solution: 150 μL *m*-cresol, 150 μL thioanisole, 300 μL TFMSA, and 900 μL TFA.
6. Diethyl ether ($Et_2O$).
7. Mobile phase A: 0.1% TFA in HPLC grade water.
8. Mobile phase B: 9:1 acetonitrile/water.
9. C18 semi-prep HPLC column, 5 μM, 10 × 250 mm (*see* **Note 1**).

#### 2.4 Reagents for Ligand-Squarate PNA Conjugation

1. Anhydrous ethanol (EtOH).
2. Anhydrous dimethylsulfoxide (DMSO).
3. Triethylamine (TEA).
4. 3,4-Diethoxy-3-cyclobutene-1,2-dione (diethyl squarate).
5. Diethyl ether.
6. Cyclo[Arg- Gly-Asp-D-Phe-Lys]-mPeg-mPeg (cRGD).

#### 2.5 Reagents for Annealing Ligand-Squarate PNA/DNA Duplex

1. Complementary DNA.
   (a) Multi-1: 5′-CAG TGA TCT ACT-3′.
   (b) Multi-2: 5′-CAG TGA TCT ACT CAG TGA TCT ACT-3′.
2. PBS: Phosphate buffer pH = 7.4.

## 3 Procedures

#### 3.1 Synthesis of $^{L}K\gamma T$ Monomer (Fig. 1)

##### 3.1.1 Boc-Lys(Fmoc)-N(Me)OMe (2)

1. Dissolve Boc-Lys(Fmoc)-OH (5.33 g, 11.4 mmol) in DCM (66 mL) in a 100 mL round-bottom flask.
2. Add EDC (2.72 g, 14.2 mmol) and HOBT (1.92 g, 14.2 mmol) to the stirring solution.
3. Cool the reaction to 0 °C with an ice bath, and, dropwise, add DIEA (2.5 mL, 14.3 mmol) to the stirring solution.
4. Stir for 5 min (*see* **Note 2**).
5. Add dimethylhydroxylamine hydrochloride (1.38 g, 14.2 mmol) and DIEA (2.5 mL, 14.3 mmol). Allow the ice bath/$H_2O$ to gradually warm up to room temperature, and stir for 12 h.

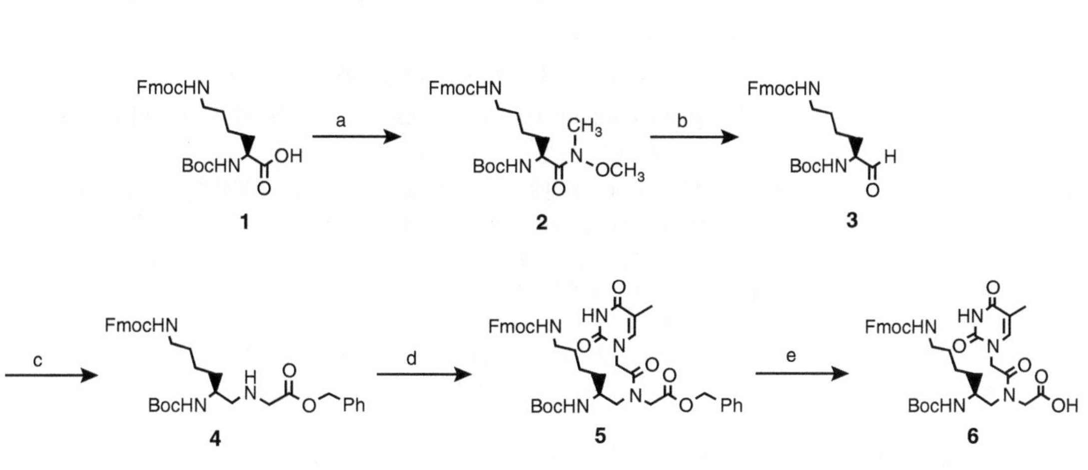


**Fig. 1** Synthesis of $^{L}K\gamma T$ monomer. Reagents and conditions: (**a**) *N,O*-dimethylhydroxylamine, EDC, DIEA, $CH_2Cl_2$, 85%; (**b**) $LiAlH_4$, THF, 78%; (**c**) benzyl glycinate, DIEA, $NaHB(OAc)_3$, $CH_2Cl_2$, 30%; (**d**) thymine acetic acid, EDC, HOBt, DMF, 82%; (e) $H_{2(g)}$, 10% Pd/C, 79%

6. Transfer the reaction mixture to a separatory funnel, and wash with 1 M HCl (2 × 30 mL), saturated $NaHCO_3$ (2 × 30 mL), and saturated NaCl (2 × 30 mL).
7. Dry the organic layer over anhydrous $Na_2SO_4$.
8. Evaporate under reduced pressure to afford **2** as a colorless, viscous oil in 99% yield (5.74 g) (*see* **Note 3**).
9. $^1$H NMR ($CDCl_3$): $\delta$ 7.77 (d, 2 H, Fmoc-CH), 7.60 (d, 2 H, Fmoc-CH), 7.40 (t, 2 H, Fmoc-CH), 7.32 (t, 2 H, Fmoc-CH), 5.22 (d, 1 H, NH), 4.88 (s, 1 H, NH), 4.68 (s, 1 H, HN–CH($CH_2$)–CO), 4.39 (d, 2 H, $OCH_2$–CH($CH)_2$), 4.21 (t, 1 H, O–$CH_2$–CH($CH)_2$), 3.77 (s, 3 H, $CH_3$), 3.20 (s, 3 H, $NCH_3$), 3.20 (m, 2 H, NH–$CH_2$–$CH_2$), 1.73–1.33 (m, 6 H, CH–$CH_2$–$CH_2$–$CH_2$–$CH_2$), 1.43 (s, 9 H, tert-butyl-$CH_3$).

*3.1.2 Boc-Lys(Fmoc)-CHO (**3**)*

1. In a 100 mL round-bottom flask, dissolve compound **2** (2.33 g, 5.5 mmol) in a minimal amount of DCM to form a gray film (approximately 5 mL) (*see* **Note 4**).
2. Add dry THF (60 mL) and cool the stirring solution to 0 °C with an ice bath.
3. Gradually, over three portions, add lithium aluminum hydride (365 mg, 9.60 mmol). The reaction mixture is stirred under nitrogen atmosphere at 0 °C for 45 min.
4. After 45 min, quench the reaction mixture with a saturated solution of $NaHSO_4$ (20 mL), and stir for an additional 10 min.
5. Transfer the quenched reaction solution to a separatory funnel and extract with EtOAc (3 × 55 mL).
6. Wash the combined organic layers with 1 M HCl (2 × 50 mL), saturated $NaHCO_3$ (2 × 50 mL), water (2 × 50 mL), and saturated NaCl (2 × 50 mL).
7. Dry the organic layer over anhydrous $Na_2SO_4$.
8. Evaporate under reduced pressure to afford **3** as a white solid in 85% yield (1.70 g).
9. $^1$H NMR (CDC13): $\delta$ 9.58 (s, 1 H, CH–COH), –7.76 (d, 2 H, Fmoc-CH), 7.59 (d, 2 H, Fmoc-CH), 7.40 (t, 2 H, Fmoc-CH), 7.31 (t, 2 H, Fmoc-CH), 5.20 (s, 1 H, NH), 4.89 (s, 1 H, NH), 4.41 (d, 2 H, O–$CH_2$–CH($CH)_2$), 4.21 (m, 1 H, O–$CH_2$–CH($CH)_2$), 4.21 (m, 1 H, HN–CH($CH_2$)–CO), 3.19 (d, 2 H, NH–$CH_2$–$CH_2$), 1.73–1.33 (m, 6 H, $CHCH_2$–$CH_2$–$CH_2$–$CH_2$), 1.44 (s, 9 H, tert-butyl-$CH_3$).

*3.1.3 Boc-Lys(Fmoc)-PNA Backbone (**4**)*

1. Combine **3** (2.45 g, 5.4 mmol) and benzyl glycinate hydrochloride (2.01 g, 5.96 mmol) in a 100 mL round-bottom flask. Dissolve in DCM (30 mL).

2. Slowly, add DIEA (1.04 mL, 5.9 mmol) to the stirring solution.
3. After 10 min, add sodium triacetoxyborohydride (1.6 g, 7.56 mmol) (*see* **Note 5**).

   Stir at room temperature for 2 h.
4. Quench the reaction with a saturated solution of $Na_2CO_3$ (15 mL), and stir for an additional 10 min.
5. Transfer the quenched reaction to a separatory funnel and extract with 3 × 30 mL of DCM.
6. Dry the combined organic layers over anhydrous $Na_2SO_4$.
7. Evaporate under reduced pressure to afford a clear, colorless oil.
8. Further purify the oil by flash column chromatography using ethyl acetate as the eluent ($R_f$ 0.5, TLC) to afford **4** as a colorless oil in 60% yield (1.96 g, 3.23 mmol).
9. $^1$H NMR ($CDCl_3$): $\delta$ 7.74 (d, 2 H, Fmoc-CH) 7.58 (d, 2 H, Fmoc-CH) 7.33 (m, 9H, phenyl, Fmoc-CH), 5.13 (s, 2 H, $CH_2$), 5.11 (s, 1 H, NH), 4.82 (d, 1 H, NH), 4.38 (m, 2 H, O–$CH_2$–CH), 4.19 (t, 1 H, $CH_2$–CH–$(CH)_2$), 3.62 (s, 1 H, NH–CH–$(CH_2)_2$), 3.41(dd, 2 H, NH–$CH_2$–CO), 3.16 (d, 2 H, NHFmoc–$CH_2$–$CH_2$), 2.61 (m, 2 H, NH–$CH_2$–CH (NHBoc)–$CH_2$), 1.39 (m, 6 H, CH–$CH_2$–$CH_2$–$CH_2$–$CH_2$), 1.43 (s, 9 H, tert-butyl-$CH_3$).

#### 3.1.4 γ-Boc-Lys-(Fmoc) Thymine Monomer Ester (5)

1. Combine **4** (570 mg, 0.94 mmol), thymine acetic acid (226 mg, 1.23 mmol), and HOBT (7 mg, 0.05 mmol) in a 50 mL round-bottom flask. Dissolve in DMF (20 mL).
2. Add EDC hydrochloride (272 mg, 1.42 mmol) to the stirring solution.
3. After 36 h, transfer reaction solution to a separatory funnel and add water (40 mL). A white precipitate will form.
4. Extract with EtOAc (3 × 40 mL).
5. Wash the combined organic layers with 1 M HCl (2 × 40 mL), saturated $NaHCO_3$ (2 × 40 mL), water (2 × 40 mL), and saturated NaCl (2 × 40 mL).
6. Dry the organic layer over $Na_2SO_4$.
7. Evaporate under reduced pressure to afford **5** as a white solid in 82% yield (593 mg).
8. $^1$H NMR (DMSO) major rotamer only (*see* **Note 6**) $\delta$ 11.31 (s, 1 H, Thymine-CH) $\delta$ 7.88 (d, 2 H, Fmoc-CH) 7.68 (d, 2 H, Fmoc-CH) 7.36 (m, 9 H, phenyl, Fmoc-CH), 7.20 (s, 1 H, Thymine-NH) 5.12 (s, 2 H, O–CH2–phenyl), 4.70 (dd, 2 H, OC–CH2–Thymine), 4.28 (s, 2 H, O–$CH_2$–CH), 4.20 (s, 1 H, $CH_2$–CH–$(CH)_2$), 4.09 (m, 2 H, N–$CH_2$–CO), 3.66

(s, 1 H, NH–CH–$(CH_2)_2$), 3.32 (m, 2 H, N–$CH_2$–CH (NHBoc)–$CH_2$), 2.97 (m, 2 H, NHFmoc–$CH_2$–$CH_2$), 1.74 (s, 3 H, Thymine–$CH_3$) 1.35 (s, 9 H, tert-butyl-$CH_3$), 1.35 (m, 6 H, CH–$CH_2$–$CH_2$–$CH_2$–$CH_2$).

#### 3.1.5 γ-Boc-Lys-(Fmoc) Thymine Monomer (6)

1. Dissolve **5** (126 mg, 0.16 mmol) into MeOH (15 mL) and sparge with an inert gas for 10 min.
2. Pd/C (10%, 30 mg) was slowly added to the stirring solution over 2 min (*see* **Note 7**).
3. Place solution under an atmosphere of $H_2$ (balloon pressure) for 6 h or until completion by TLC ($R_f = 0$, EtOAc).
4. Filter off Pd/C with celite and wash the pad with additional MeOH.
5. Evaporate under reduced pressure to afford **6** as an off-white solid in 79% yield (86 mg) (*see* **Note 8**).
6. $^1$H NMR (DMSO) major rotamer only: $\delta$ 11.29 (s, 1 H, Thymine-CH) $\delta$ 7.89 (d, 2 H, Fmoc-CH) 7.68 (d, 2H, Fmoc-CH) 7.33 (m, 5 H, Thymine-NH, Fmoc-CH), 6.81 (d, 1 H, NH), 6.70 (d, 1 H, NH), 4.66 (dd, 2 H, $CH_2$), 4.28 (s, 2 H, O–$CH_2$–CH), 4.20 (s, 1 H, $CH_2$–CH–$(CH)_2$), 3.96 (m, 2 H, N–$CH_2$–CO), 3.66 (s, 1 H, NH–CH–$(CH_2)_2$), 3.32 (m, 2 H, NHFmoc$CH_2$–$CH_2$), 2.96 (m, 2 H, N–$CH_2$–CH(NHBoc)–$CH_2$), 1.74 (s, 3 H, Thymine–$CH_3$) 1.37 (s, 9 H, tert-butyl-$CH_3$), 1.37 (m, 6 H, CH–$CH_2$–$CH_2$–$CH_2$–$CH_2$); HRMS (ESI-MS $m/z$) mass calcd for $C_{35}H_{43}N_5O_9$ [M+H]+, 678.3094, found 678.3140.

### 3.2 Solid-Phase Peptide Synthesis of γT Containing Peptide Nucleic Acid Oligomer (Fig. 2)

All $^L$KγT containing PNA oligomers were synthesized on an automated peptide synthesizer using the general sequence GTC-ACT-AGA-TGA (C → N terminal) on MBHA resin downloaded with Boc-mPeg (5 μmol scale, 50 mg; *see* **Note 9**). All deprotections, reactions, and washings were performed with agitation.

1. Monomers are weighed (*see* **Note 10**) and then dissolved in 175 μL of NMP to yield a 0.15 M solution for each coupling. Transfer this solution to the cartridges and load them on the synthesizer in the appropriate order.
2. Set the instrument to swell the resin for 104 min.
3. For the first coupling reaction:
   (a) Deprotect the terminal Boc group twice with the Boc-deprotection solution for 3 min each. *Subsequent couplings are only treated once with Boc-deprotection solution.*
   (b) Wash the resin with DCM 6× and NMP 6× to remove all of the TFA.

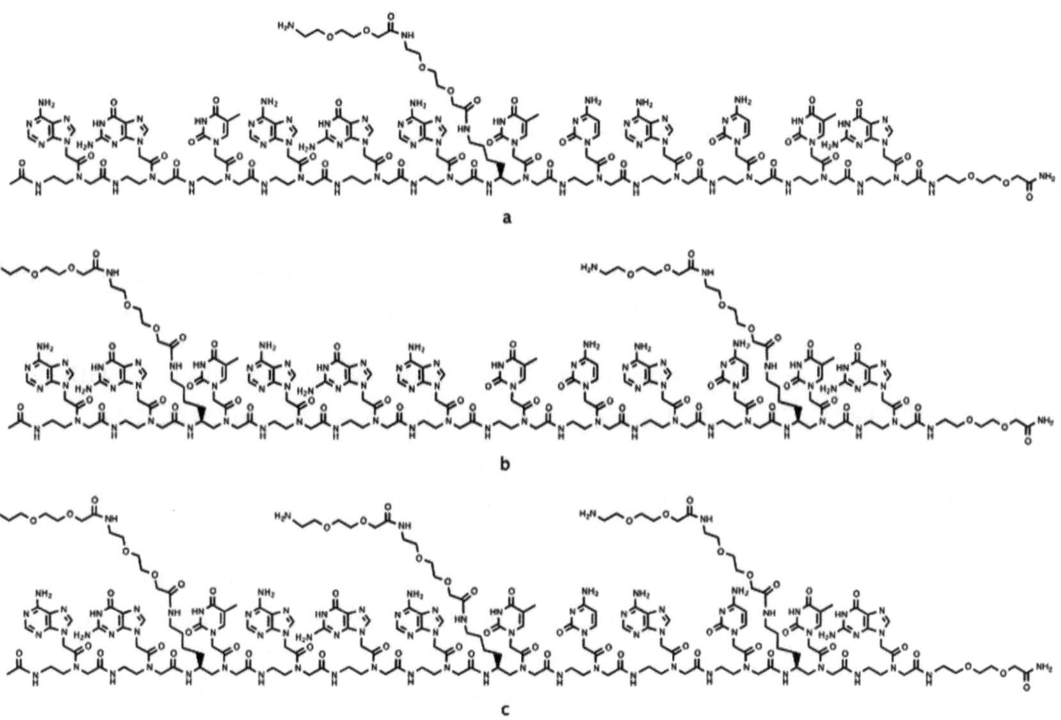

**Fig. 2** Structures of PNA oligomers containing $^{L}K\gamma T$ monomer. (**a**) 1γPNA, (**b**) 2γPNA, and (**c**) 3γPNA. Oligomers are drawn in the N to C direction

(c) Activate the monomer by adding 125 μL of the activator solution and 125 μL of the base solution to the monomer cartridge.

(d) Let the solution stand for 1 min to activate and then transfer to the reaction vessel and then couple for 90 min.

(e) At the end of the coupling reaction, wash the resin with NMP 6×, and then treat with the capping solution for 1 min.

4. Repeat **steps a–e** with the following modifications (*see* **Note 11**):
   - When coupling $^{L}K\gamma T$ monomers, use extended monomer-coupling time (90 min).
   - Monomers following $^{L}K\gamma T$ residues were coupled with 2 × Boc-deprotection solution and extended reaction times (120 min).
   - Standard coupling time is 30 min.
5. To acylate the N-terminus, after the final monomer:

   (a) Deprotect the terminal Boc group with the Boc-deprotection solution for 3 min.

   (b) Treat with the capping solution.

6. To attach mPeg spacer on the γT lysine side chains:
   (a) Deprotect the Fmoc protecting groups with 3 × 20% Pip in DMF deprotection for 3 min.
   (b) Couple the Boc-mPeg linker using **step 3b–e** with the following modifications: extended coupling time 120 min, double couple (repeat **steps c** and **d** twice).
   (c) Couple the second Boc-mPeg linker using **steps 3a–e** with the following modifications: extended coupling time 120 min, double couple (repeat **steps c** and **d** twice).

### 3.3 Isolation of the PNA Oligomer

#### 3.3.1 Cleavage of the PNA from MHBA Resin

1. Transfer the resin to a glass reaction vial and rinse with DCM.
2. Wash resin twice with TFA for 4 min (*see* **Note 12**).
3. Remove TFA from reaction vessel.
4. Slowly add one aliquot of the cleavage cocktail solution to the resin, and let it stand for 1 h.
5. Elute the solution into a glass 20 mL scintillation vial.
6. Repeat **steps 4** and **5** for a total of three cleavages, combining the newly eluted solution with the previous.
7. Using inert gas (argon or nitrogen), evaporate the volatiles to reduce the overall volume (about 30–90 min).
8. Disperse up to 200 μL of elution solution into four 2 mL microfuge tubes. Add $Et_2O$ to fill each vial and mix thoroughly (*see* **Note 13**).
9. Place the tubes on dry ice for at least 10 min (*see* **Note 14**).
10. Centrifuge to pellet the precipitated PNA.
11. Carefully, decant the supernatant and then add fresh diethyl ether and resuspend pellet.
12. Repeat **steps 9–11** three more times.
13. Allow remaining ether to evaporate (standing in fume hood for about 15–30 min).
14. Store in the freezer if the oligomer is not purified immediately.

#### 3.3.2 Purification of PNA Oligomers

1. Dissolve residual pellet in water or water/acetonitrile mixture (2:1 recommended).
2. Purify by reversed-phase liquid chromatography using the following gradient at 4 mL/min:
   - 0–2 min hold at 0% mobile phase B.
   - 2–5 min ramp to 10% mobile phase B.
   - 5–20 min ramp to 20% mobile phase B (12.5—elution time).

   Monitor at 220, 260, and 315 nm.

3. Pool fractions identified to contain the product (mass spec) and concentrated under reduced pressure (lyophilize) to afford 7 (TFA adduct) as a fluffy white solid (*see* **Note 15**).

### 3.4 Synthesis of Ligand-Squarate PNA Conjugate (Fig. 3)

#### 3.4.1 3γPNA-Squarate (8)

1. In a 2 mL centrifuge tube, add DMSO (280 μL) to the lyophilized PNA-TFA salt (5.8 mg, 1.27 μmol).
2. Add EtOH (140 μL), followed by TEA (31.8 μL, 228 μmol) and diethyl squarate (16.9 μL, 114 μmol) (*see* **Note 16**).
3. Place under argon/nitrogen, seal, and agitate for 3 h.
4. Concentrate under reduced pressure to afford viscous oil.

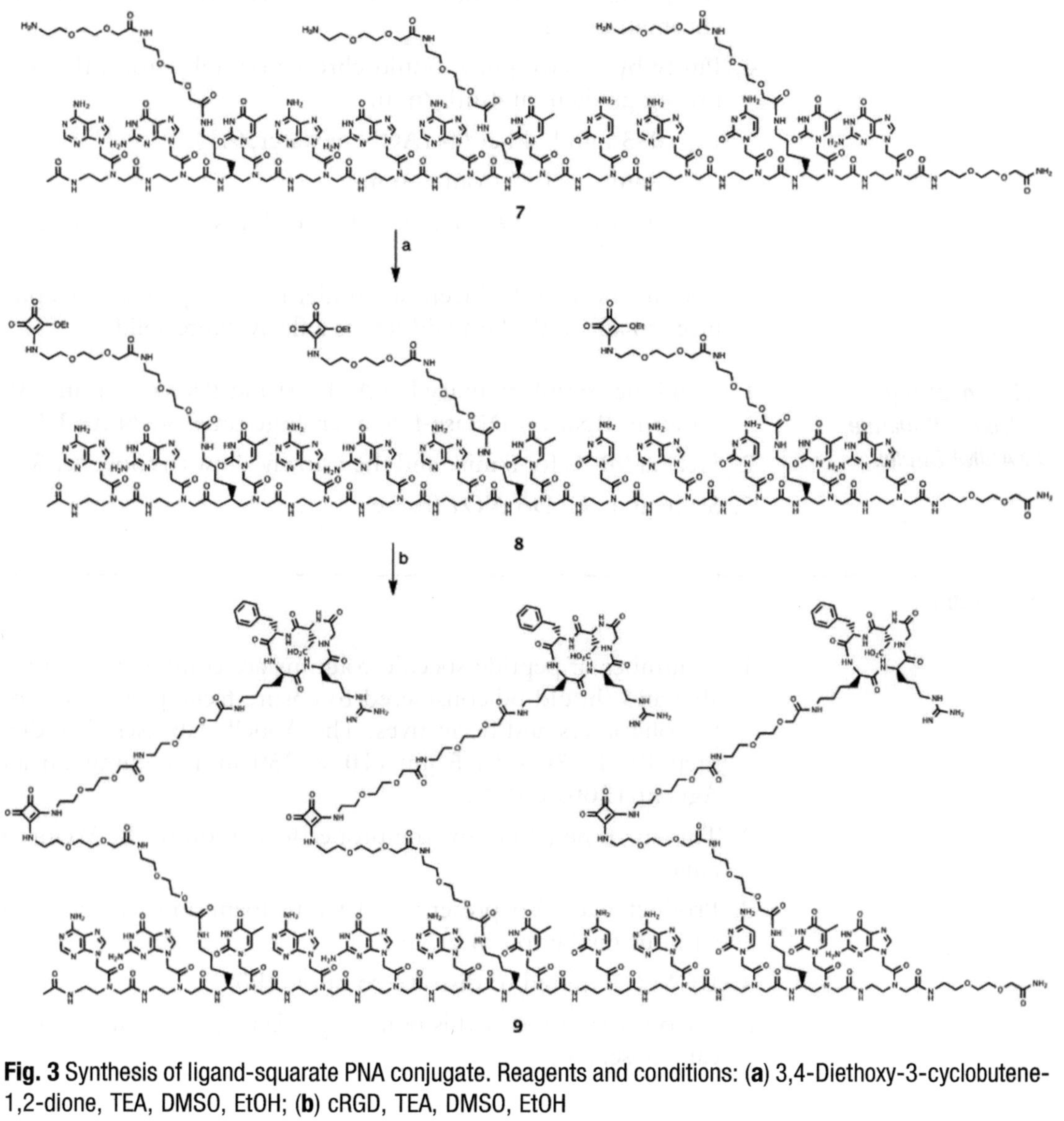

**Fig. 3** Synthesis of ligand-squarate PNA conjugate. Reagents and conditions: (**a**) 3,4-Diethoxy-3-cyclobutene-1,2-dione, TEA, DMSO, EtOH; (**b**) cRGD, TEA, DMSO, EtOH

5. Add 2 mL of $Et_2O$, agitate to mix thoroughly, and place in dry ice for 10 min.
6. Centrifuge and decant supernatant (*see* **Note 17**).
7. Repeat **steps 5** and **6** for a total of three ether washes.
8. Dry PNA-Sq under vacuum for 30 min to afford **8** as an off-white flaky solid.

#### 3.4.2 3γPNA-cRGD (9)

1. Dissolve **8** in DMSO (420 μL), followed by addition of EtOH (210 μL) and then TEA (15.9 μL, 114 μmol).
2. Add cRGD (68 mg, 76.1 μmol) (*see* **Note 18**).
3. Place under inert atmosphere and agitate for 72 h or until completion. Monitor by reversed-phase liquid chromatography.
4. Purify by reverse-phase liquid chromatography using the following gradient at 4 mL/min:
   (a) 0–3 min hold at 0% (ACN in water, 9:1).
   (b) Ramp to 10% from 3–6 min.
   (c) Increase to 29% over 6–30 min (3γPNA-cRGD elutes at 28.05 min).
5. Pool fractions and concentrate under reduced pressure (lyophilize) to afford **9** (TFA adduct) as a fluffy white solid.

### 3.5 Annealing of Ligand-Squarate PNA/DNA Duplex

1. Combine complementary DNA (1 eq) and PNA (1 eq) in PBS buffer in RNase-/DNase-free microfuge tube (*see* **Note 19**).
2. Heat at 90 °C for 5 min, and then slowly allow to cool over 3 h.
3. Store at 4 °C. Do *NOT* freeze.

## 4 Notes

1. A number of peptide specific columns are commercially available and should be considered to ensure facile purification of the oligomers and derivatives. The Appella lab uses X-bridge prep BEH 130 C18 5 μm (10 × 250 mm) column on an Agilent 1200s HPLC.
2. This stir time allows for the proper formation of the Weinreb amide.
3. Product may also present as a white foam and can be used without further purification.
4. DCM is required to dissolve compound 3.
5. A cloudy solution at this point is perfectly normal and is typically observed.

6. Rotamers are present in most PNA monomers. Typically the $^{1}H$ signal of the less stable conformation is in close proximity to the major peak.
7. It is suggested to use a slightly longer needle to carry out the sparging and then only remove the septum enough to allow access to the flask while adding the Pd/C. This technique will help maintain the inert atmosphere and reduce the potential of the Pd/C from igniting.
8. This reaction proceeds cleanly and does not require a purification step. However, if the means are available, it is suggested that the final monomer is purified using a reverse-phase flash column, eluting with water and an increasing concentration of ethanol. UV active fractions are collected and lyophilized.
9. The Appella lab follows the standard practice of incorporating Boc-mPeg. Other variations of the multivalent ligand display have used *N*,*N*-dimethyl lysine to maintain aqueous solubility of PNA/PNA duplexes at longer PNA lengths.
10. Quantities for each monomer per coupling (5 equivalents) are as follows: Boc-T monomer = 10.1 mg; Boc-C monomer = 13.2 mg; Boc-A monomer = 13.9 mg; Boc-G monomer = 14.2 mg; and $^{L}K\gamma T$ = 17.8 mg. G monomers dissolve slowly, but the solution can be heated in a heat block (80 °C) or sonicated at room temperature.
11. The extended coupling time for the first monomer helps increase the final yield of the oligomer.
12. Do not let the resin dry out. Be sure to force the resin to the bottom of the reaction vessel so that it will be fully covered with the cleavage solution.
13. The brown color of the cleavage solution should disappear, and ideally a white cloudy precipitate will readily form.
14. This step is unnecessary if the PNA precipitates at room temperature.
15. Any time water is removed from PNA, we recommend freezing and lyophilizing.
16. For the synthesis of 1γ or 2γ PNA oligomers, use 60 equivalents TEA and 30 equivalents diethyl squarate per γ incorporation. The remaining conditions are constant.
17. The PNA-squarate is very flaky at this point and often does not pellet well. Caution should be taken when decanting the ether washes, or use a pipette to remove the liquid.
18. For the synthesis of 1γ or 2γ cRGD-PNA oligomers, use 30 equivalents TEA and 20 equivalents cRGD per γ incorporation. The remaining conditions are constant.

19. For each repeating 12-mer sequence on the complementary single-strand DNA, use one equivalent of PNA. For example, for a Multi-1 sequence, DNA/PNA used 1:1; for a Multi-2 system, DNA/PNA used 1:2. A slight 1% excess is advisable to ensure full complementarity of higher-length duplexes.

## References


1. Fasting C, Schalley CA, Weber M, Seitz O, Hecht S, Koksch B, Dernedde J, Graf C, Knapp EW, Haag R (2012) Multivalency as a chemical organization and action principle. Angew Chem Int Ed Engl 51:10472–10498
2. Mammen M, Choi S-K, Whitesides GM (1998) Polyvalent Interactions in Biological Systems: Implications for Design and Use of Multivalent Ligands and Inhibitors. Angew Chem Int Ed 37:2754–2794
3. Dix AV, Moss SM, Phan K, Hoppe T, Paoletta S, Kozma E, Gao ZG, Durell SR, Jacobson KA, Appella DH (2014) Programmable nanoscaffolds that control ligand display to a G-protein-coupled receptor in membranes to allow dissection of multivalent effects. J Am Chem Soc 136:12296–12303
4. Dix AV, Conroy JL, George Rosenker KM, Sibley DR, Appella DH (2015) PNA-based multivalent scaffolds activate the dopamine D2 receptor. ACS Med Chem Lett 6:425–429
5. Englund EA, Wang D, Fujigaki H, Sakai H, Micklitsch CM, Ghirlando R, Martin-Manso G, Pendrak ML, Roberts DD, Durell SR, Appella DH (2012) Programmable multivalent display of receptor ligands using peptide nucleic acid nanoscaffolds. Nat Commun 3:614
6. Nielsen PE, Egholm M, Berg RH, Buchardt O (1991) Sequence-selective recognition of DNA by strand displacement with a thymine-substituted polyamide. Science 254:1497–1500
7. Egholm M, Buchardt O, Christensen L, Behrens C, Freier SM, Driver DA, Berg RH, Kim SK, Norden B, Nielsen PE (1993) PNA hybridizes to complementary oligonucleotides obeying the Watson–Crick hydrogen-bonding rules. Nature 365:566–568
8. Datta B, Schmitt C, Armitage BA (2003) Formation of a PNA2–DNA2 Hybrid Quadruplex. J Am Chem Soc 125:4111–4118
9. Englund EA, Appella DH (2005) Synthesis of γ-substituted peptide nucleic acids: a new place to attach fluorophores without affecting DNA binding. Org Lett 7:3465–3467
10. Kohler O, Jarikote DV, Seitz O (2005) Forced intercalation probes (FIT Probes): thiazole orange as a fluorescent base in peptide nucleic acids for homogeneous single-nucleotide-polymorphism detection. Chembiochem 6:69–77
11. Kuhn H, Demidov VV, Coull JM, Fiandaca MJ, Gildea BD, Frank-Kamenetskii MD (2002) Hybridization of DNA and PNA molecular beacons to single-stranded and double-stranded DNA targets. J Am Chem Soc 124:1097–1103
12. Kohhler O, Jarikote DV, Singh I, Parmar VS, Weinhold E, Seitz O (2005) Forced intercalation as a tool in gene diagnostics and in studying DNA–protein interactions. Pure Appl Chem 77:327–339
13. Moustafa ME, Hudson RH (2011) An azo-based PNA monomer: synthesis and spectroscopic study. Nucleosides Nucleotides Nucleic Acids 30:740–751
14. Ortiz E, Estrada G, Lizardi PM (1998) PNA molecular beacons for rapid detection of PCR amplicons. Mol Cell Probes 12:219–226
15. Robertson KL, Yu L, Armitage BA, Lopez AJ, Peteanu LA (2006) Fluorescent PNA probes as hybridization labels for biological RNA. Biochemistry 45:6066–6074
16. Roy S, Tanious FA, Wilson WD, Ly DH, Armitage BA (2007) High-affinity homologous peptide nucleic acid probes for targeting a Quadruplex-forming sequence from a MYC promoter element. Biochemistry 46:10433–10443
17. Xi C, Balberg M, Boppart SA, Raskin L (2003) Use of DNA and peptide nucleic acid molecular beacons for detection and quantification of rRNA in solution and in whole cells. Appl Environ Microbiol 69:5673–5678
18. Nielsen PE, Appella DH (2014) Peptide nucleic acids: methods and protocols. Humana Press, New York
19. Ray A, Nordén B (2000) Peptide nucleic acid (PNA): its medical and biotechnical applications and promise for the future. FASEB J 14:1041–1060
20. Corradini R, Sforza S, Tedeschi T, Totsingan F, Manicardi A, Marchelli R (2011) Peptide nucleic acids with a structurally biased backbone. Updated review and emerging challenges. Curr Top Med Chem 11:1535–1554

21. Englund EA, Appella DH (2007) Gamma-substituted peptide nucleic acids constructed from L-lysine are a versatile scaffold for multifunctional display. Angew Chem Int Ed Engl 46:1414–1418
22. Manicardi A, Guidi L, Ghidini A, Corradini R (2014) Pyrene-modified PNAs: Stacking interactions and selective excimer emission in PNA2DNA triplexes. Beilstein J Org Chem 10:1495–1503
23. Scheibe C, Wedepohl S, Riese SB, Dernedde J, Seitz O (2013) Carbohydrate-PNA and aptamer-PNA conjugates for the spatial screening of lectins and lectin assemblies. Chembiochem 14:236–250
24. Winssinger N (2012) DNA display of PNA-tagged ligands: a versatile strategy to screen libraries and control geometry of multidentate ligands. Artif DNA PNA XNA 3:105–108

# Chapter 9

# Aminoglycoside Functionalization as a Tool for Targeting Nucleic Acids

**Derrick Watkins, Krishnagopal Maiti, and Dev P. Arya**

## Abstract

Aminoglycoside functionalization as a tool for targeting natural and unnatural nucleic acids holds great promise in their development as diagnostic probes and medicinally relevant compounds. Simple synthetic procedures designed to easily and quickly manipulate amino sugar (neomycin, kanamycin) to more powerful and selective ligands are presented in this chapter. We describe representative procedures for (a) aminoglycoside conjugation and (b) preliminary screening for their nucleic acid binding and selectivity.

**Key words** Aminoglycoside, Nucleic acid binding, Neomycin, RNA, DNA, DNA–RNA, miRNA, Kanamycin

## 1 Introduction

Aminoglycosides have been used as antibacterial agents for decades [1]. Well known for their bactericidal activity [2] and binding to ribosomal RNA [3, 4], aminoglycosides have been shown as useful scaffolds for targeting a variety of nucleic acid structures [5, 6]. Amino sugar binding to RNA structures such as hairpin ribozymes [7], group I introns [8], hammerhead ribozyme [9], hepatitis delta virus ribozyme [10], TAR [11–14], and RRE [15], and DNA structures such as duplexes [16–24], triplexes [24–32], quadruplexes [33–35], and DNA–RNA hybrid helixes [36, 37] has been reported. A "shape readout" survey [6] for aminoglycoside recognition of nucleic acid structure revealed that neomycin displays the highest affinity for A-form structures, whereas the dimer binds B-form DNA [38]. Neomycin's affinity correlated directly to the width of the nucleic acid structure's major groove with higher binding affinities being found for structures having narrower and deeper major grooves, as shown in Fig. 1. A strategy that couples the aminoglycoside's major groove affinity with a second nucleic acid binding mode (such as intercalation, backbone, or minor groove) can be used to obtain designer ligands selective for nucleic

Nathaniel Shank (ed.), *Non-Natural Nucleic Acids: Methods and Protocols*, Methods in Molecular Biology, vol. 1973, https://doi.org/10.1007/978-1-4939-9216-4_9, © Springer Science+Business Media, LLC, part of Springer Nature 2019

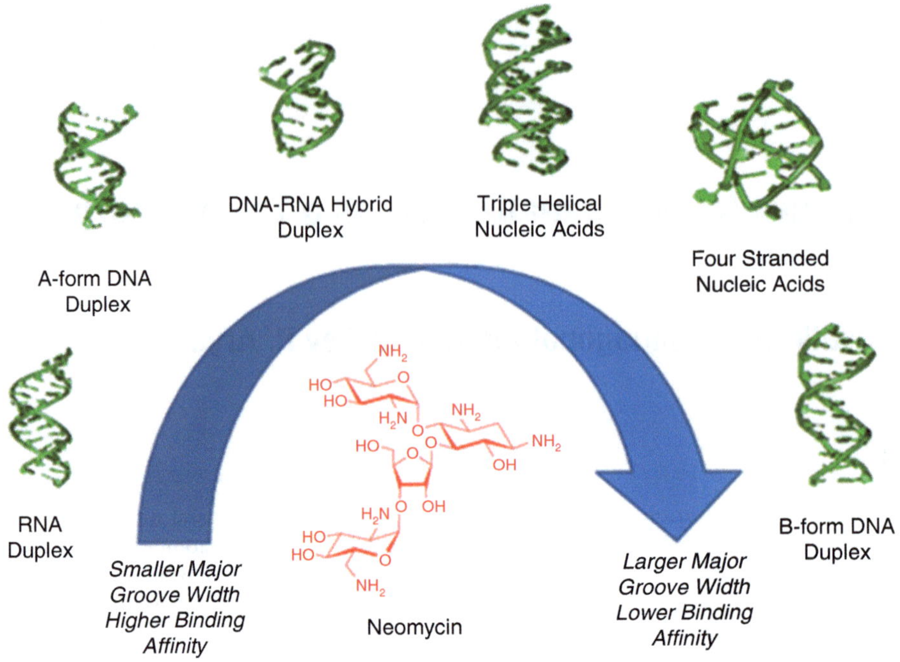


**Fig. 1** Graphic depicting neomycin's binding affinity for nucleic acids as they transit from A-form RNA to B-form DNA nucleic acid structure. Reprinted and adapted with permission from ref. 5. Copyright © 2011 American Chemical Society

acid conformations and structures. In this chapter, we present selective examples of nucleic acid targeted ligand design using neomycin as the groove binding amino sugar (*see* **Notes 1** and **2**).

## *1.1 Aminoglycoside Conjugates*

### *1.1.1 Functionalization of Neomycin*

Preparation of the aminoglycoside for conjugation requires the selective functionalization at a site not required for nucleic acid binding, as shown in Fig. 2. The synthetic strategy utilizes the primary 5″ hydroxyl on neomycin's ring III in **1a** as the site of modification [39–42]. All three modes of conjugation begin by protection of the amino groups with di-tert-butyl dicarbonate. From this universal intermediate, functionalization necessary for synthesizing ligands (**1b, 2a, 2b, 3–5**) for the three binding modes ($NH_2$-nucleophilic, NCS-electrophilic, and $N_3$-cycloaddition) is easily achieved (Fig. 2).

### *1.1.2 Conjugation of Neomycin*

Three approaches for aminoglycoside coupling are presented; formation of amides, thioureas, and triazoles. In addition to small organic molecules, polymers such as deoxyoligonucleotides and peptide nucleic acids can be conjugated. The intermediates **2–5** can be conjugated to reagents of choice in solution or solid phase. Figure 3 represents a small population of the total number of potential aminoglycoside conjugates that can be synthesized.

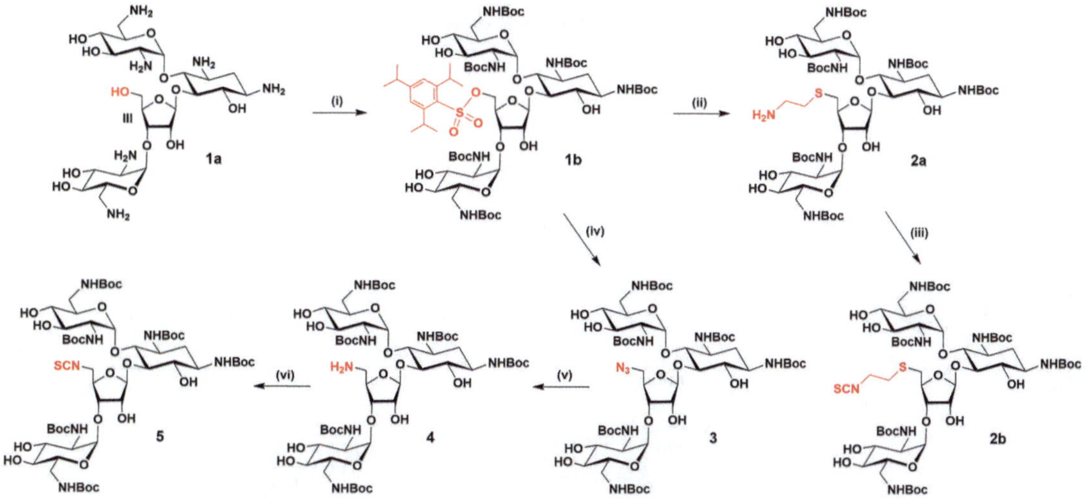


**Fig. 2** The chemical transformation of 5″-OH group in neomycin's ring III: (i) a. $(Boc)_2O$, DMF, $H_2O$, $Et_3N$, 60 °C, 5 h, 60%; b. 2,4,6-triisopropylbenzenesulfonyl chloride, pyridine, rt., 40 h, 50%. (ii) HCl•$H_2NCH_2CH_2SH$, NaOEt–EtOH, rt., 18 h, 50%. (iii, vi) 1,1′-thiocarbonyldi-2(1*H*)-pyridone, $CH_2Cl_2$, room temperature, overnight, 85–90%. (iv) $NaN_3$, DMF–$H_2O$ (10:1), 12 h, 90 °C, 90%. (v) $H_2$, Pd/C, MeOH, 12 h, 75%

### *1.1.3 Examples of Neomycin Conjugation and Nucleic Acid Stabilization/Selectivity*

#### Neomycin–Methidium Conjugate (Amide linkage, Target: DNA–RNA Hybrids) [36]

A substituted methidium chloride derivative was coupled to neomycin amine, as shown in Fig. 4. Studied with poly(dA)•poly(rU) duplex, the neomycin–methidium conjugate (**NM**) showed a remarkable affinity with $K_a = 4.8 \times 10^{10}$ $M^{-1}$ [36]. In order to ascertain the conjugate's selectivity, a mixed melting thermal denaturation assay was conducted (Fig. 5). An analysis of the assay revealed that **NM** preferentially stabilized the poly(dA)•poly(rU) duplex over all others, a clear demonstration of the conjugate's increased selectivity, making it the most potent and selective DNA–RNA hybrid reported to date.

#### Neomycin–Hoechst 33258 Conjugate (Thiourea Linkage, Target: DNA Duplex and RNA Duplex)

A neomycin–Hoechst 33258 conjugate was prepared via the formation of a thiourea linkage, as shown in Fig. 6 [20]. UV-thermal denaturation, CD spectroscopy, differential scanning spectroscopy (DSC), and fluorescence data all showed enhanced conjugate stabilization of A-T-rich sequences over Hoechst 33258 alone, and recently a triple recognition ligand containing neomycin, Hoechst, and pyrene has been reported [18]. Hoechst–neomycin conjugates have been shown to bind other nucleic acid targets such as quadruplexes and RNA structures [43, 44].

#### Neomycin Dimers (Triazole Linkage, Target: TAR RNA, Bacterial rRNA) [11, 13, 21, 45, 46]

A triazole linkage was used to synthesize a series of neomycin dimers, one of which is represented in Fig. 7. The library was created to target the trans activation response (TAR) region of the HIV virus. UV thermal denaturation studies showed a 3–10 °C $T_m$ increase with the neomycin dimers versus the monomer. Displacement titrations using ethidium bromide revealed

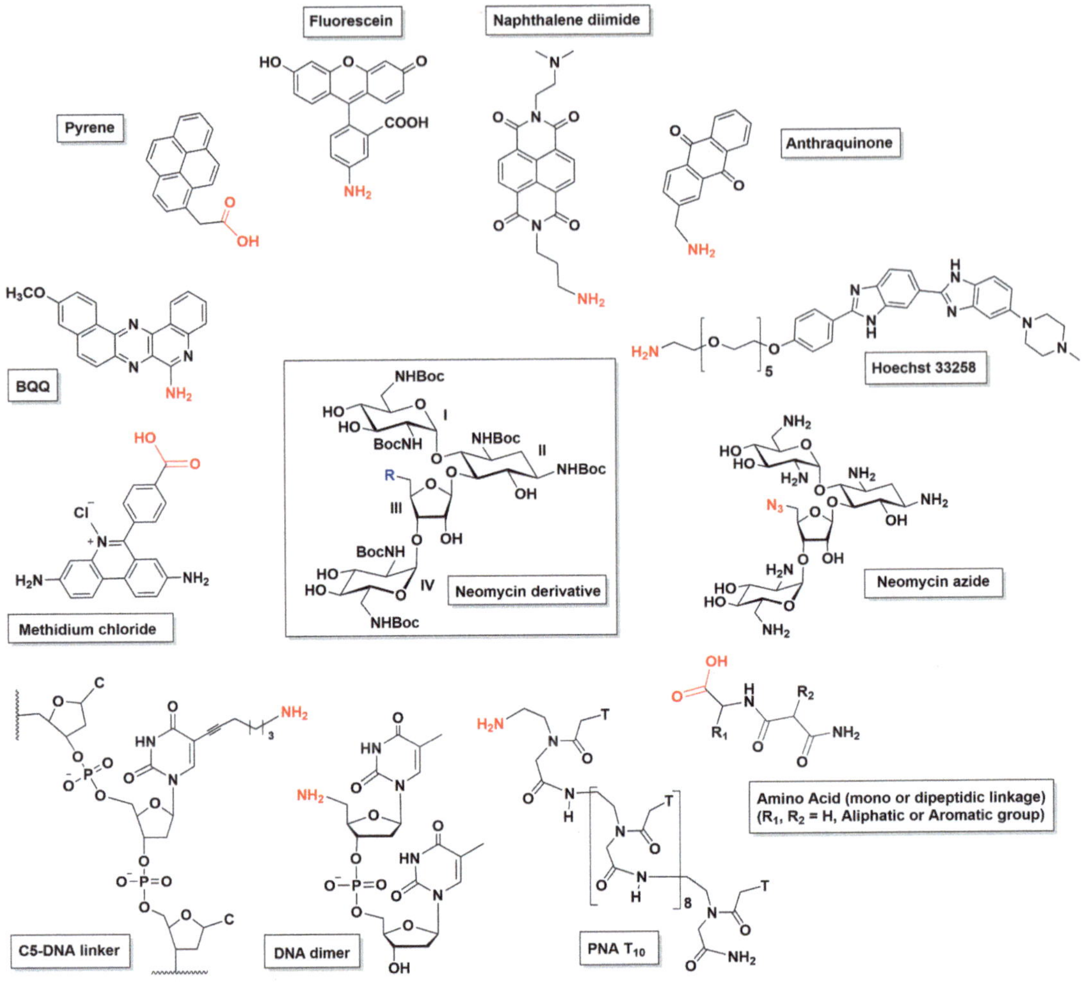


**Fig. 3** A graphic depicting the potential diversity for neomycin's conjugation with other ligands. The **R** group on neomycin represents an amine, azide, or isothiocyanate as described in Fig. 2 (*2a–5*). Conjugation can therefore be achieved by amide bond, thiourea linkage, or triazole formation using the complimentary functional groups (red) on the ligands shown

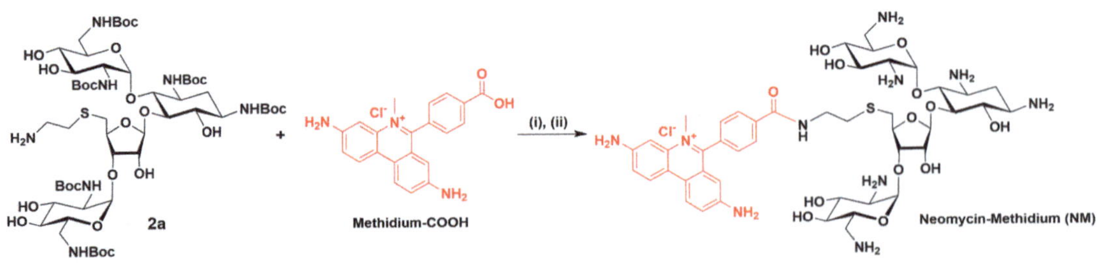


**Fig. 4** Reagents and conditions: (i) *N,N*-dicyclohexylcarbodiimide (DCC), DMAP, DMF, rt., 28 h, 80%; (ii) TFA/ $CH_2Cl_2$, $SHCH_2CH_2SH$, rt., 30 min, 90%. Reprinted and adapted with permission from ref. 36. Copyright © 2008 Elsevier

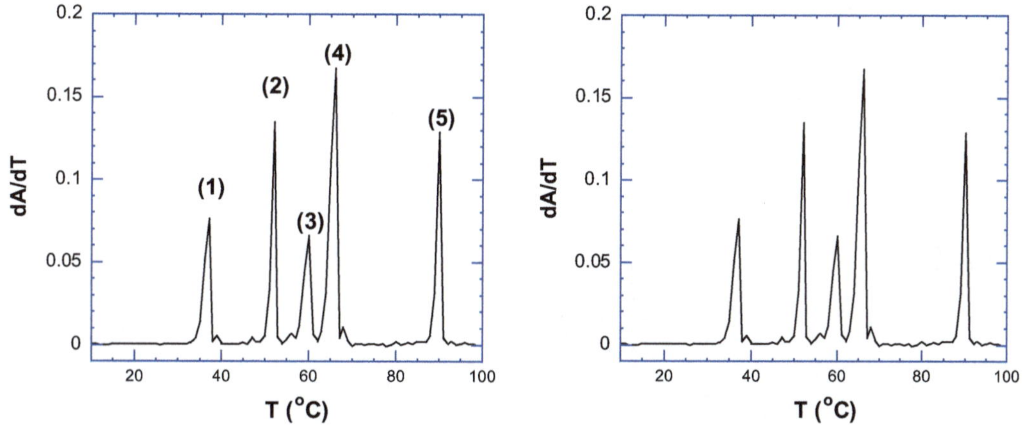


**Fig. 5** Mixed melting profile of various nucleic acids. Each panel shows the melting mixture of poly(dG):poly (dC); peak 5, poly(dA):poly(dT); peak 4, poly(rA):poly(dT); peak 3, poly(rA):poly(rU); peak 2, poly(dA):poly(rU); peak 1. In each panel, the solid line reflects the melting in the absence of ligand, while the dashed line represents the mixed melting in the presence of *NM*. Reprinted and adapted with permission from ref. 36. Copyright © 2008 Elsevier

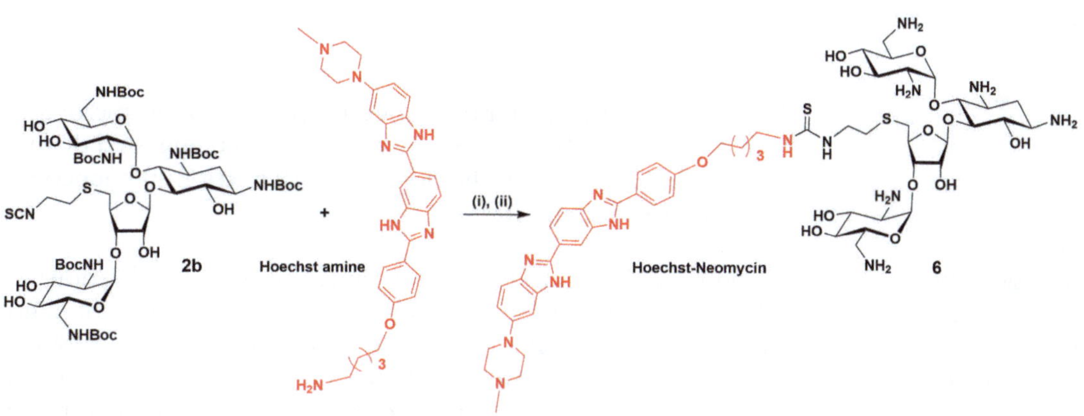


**Fig. 6** Conjugation of *2b* and Hoechst amine via a thiourea linkage. (i) Pyridine, DMAP, 52%; (ii) 1:1 TFA/$CH_2Cl_2$, 94% [20]

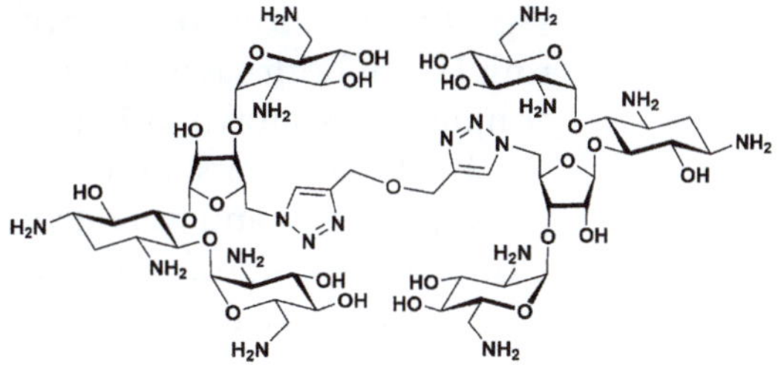

**Fig. 7** Structure of a triazole linked neomycin dimer [11]

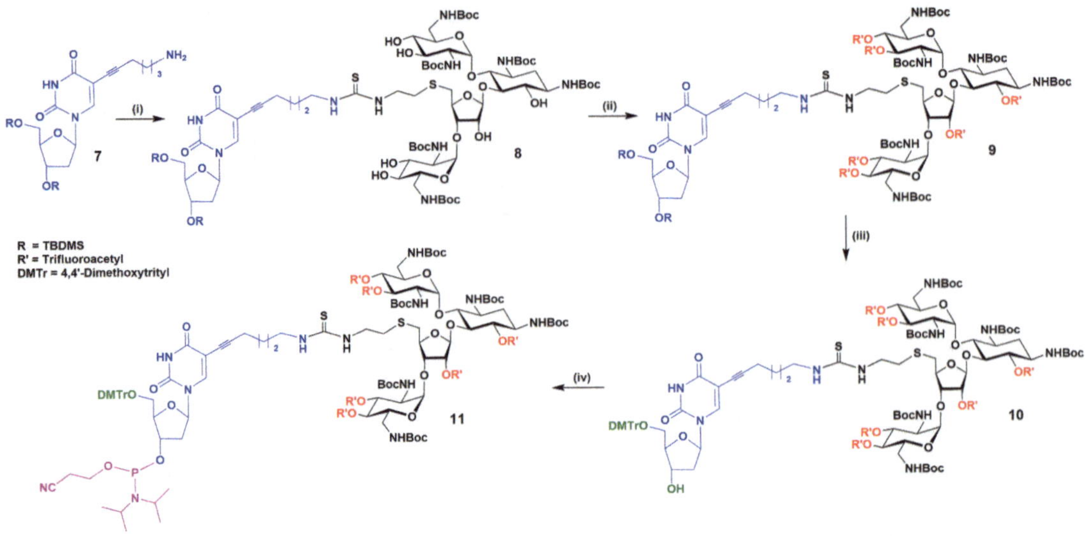


**Fig. 8** *Reagents and Conditions:* (i) *2b*, DMAP, pyridine (76%); (ii) $(CF_3CO_2)O$, *i*-$Pr_2EtN$, $CH_2Cl_2$ (75%); (iii) TBAF, DMF; and then DMTrCl, pyridine, DMAP (78%); (iv) $NCCH_2CH_2OP[N(iPr)_2]_2$, bis(diisopropylammonium) tetrazolide and $CH_2Cl_2$ (89%). Reprinted and adapted with permission from ref. 42. Copyright © 2007 American Chemical Society

dimer $IC_{50's}$ in the nanomolar range as opposed to the micromolar binding produced by the monomer. Both results suggested that the dimer was able to exploit the close proximal relationship of neomycin to TAR's binding sites.

#### Aminoglycoside–Oligonucleotide Conjugates (Thiourea Linkage, Target: Ribosomal RNA, miRNA) [41, 42]

A possible application for aminoglycoside stabilization of DNA–RNA antisense structures involves the conjugation of oligonucleotides with an amino sugar. To test this hypothesis, DNA–aminoglycoside conjugates were synthesized, as shown in Figs. 8 and 9. The synthesis of these conjugates was achieved through two similar but separate strategies. In Fig. 8, thiourea coupling was used to link **2b** to C5-position of 2′-deoxyuridine that had been functionalized with an amine and the phosphoramidite used to synthesize the polymer on solid phase. Kanamycin was used in Fig. 9 in addition to neomycin by functionalizing the 6″-OH on ring III. In both cases, thiourea coupling was used to link 2b to the conjugate, as shown in Fig. 9.

PNA oligomers were synthesized using standard PNA solid-phase protocols and were deprotected from solid support (Fig. 10). Covalent linkage of neomycin (**2b**) was achieved via formation of thiourea linkage using a catalytic amount of DMAP in pyridine for 12 h.

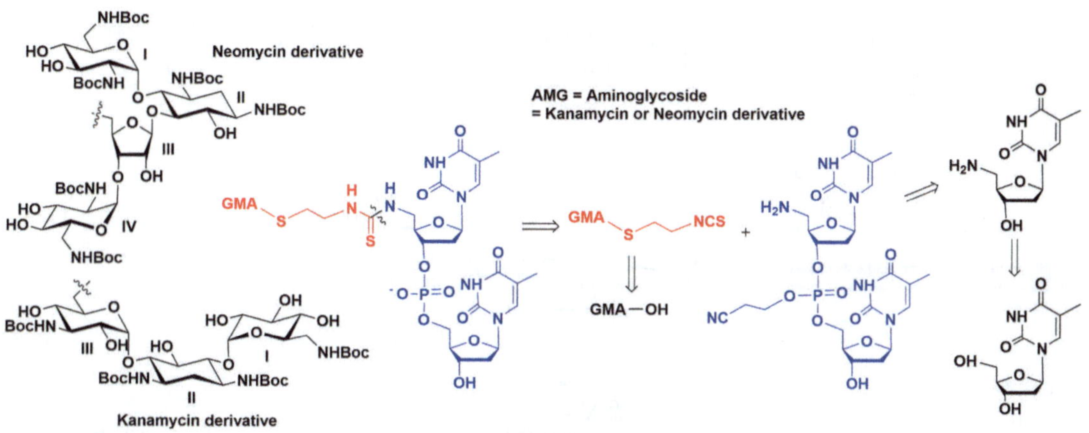


**Fig. 9** Retrosynthetic analysis of covalently linked aminoglycoside to DNA dimer [40]

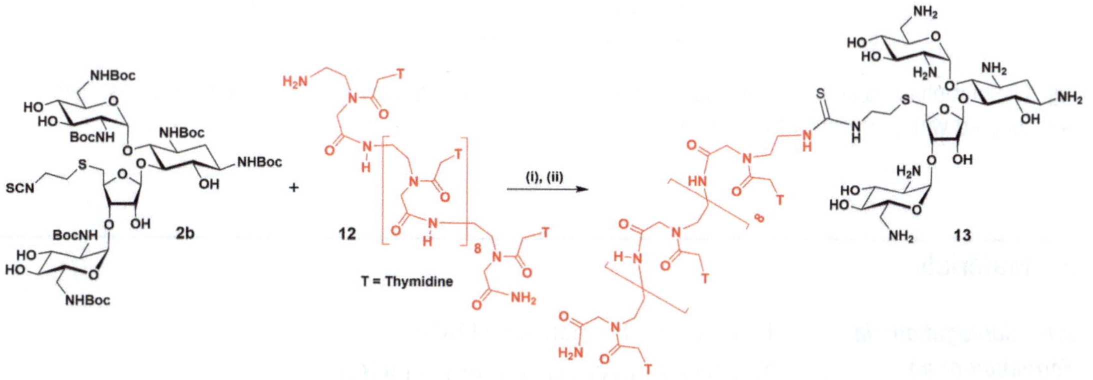


**Fig. 10** Synthesis of PNA–Neomycin Conjugates. (i) DMAP, Pyridine, 12 h; (ii) 1 (M) HCl–dioxane, 1,2-ethanedithiol, 5 min

Aminoglycoside–Amino Acid Conjugates (Amide Linkage, Target: Ribosomal RNA, miRNA, ncRNA, DNA) [39]

Aminoglycoside (neomycin and kanamycin) was also conjugated with amino acids (mono and di) via amide linkage through solid-phase technique (Fig. 11). Antibacterial studies with several bacterial strains and the binding studies with bacterial 16S ribosomal RNA were carried out with neomycin peptidic library. A rapid synthesis of an aminoglycoside library through a solid-phase technique and screening out of compounds against the bacterial activity provide an optimal binding environment which helps to recognize structure and activity relationship. In the similar fashion, monopeptidic and dipeptidic kanamycin conjugates were synthesized through solid-phase technique and their antibacterial activities were evaluated with several bacterial strains [47, 48].

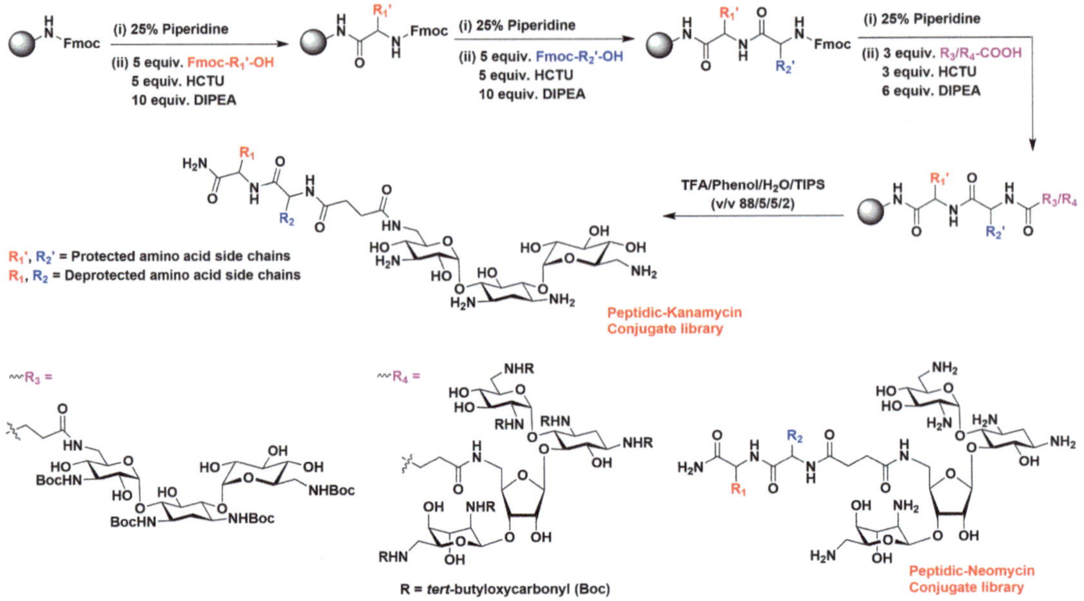


**Fig. 11** Solid-phase scheme for synthesis of amino acid peptidic kanamycin–neomycin conjugates. Reprinted and adapted with permission from ref. 47

## 2 Materials

***2.1 Conjugation via Formation of an Amide Bond***

1. Dimethylformamide (DMF).
2. Dicyclohexylcarbodiimide (DCC).
3. 4-dimethylaminopyridine (DMAP)
4. Methylene chloride ($CH_2Cl_2$).
5. Methanol (MeOH).
6. Trifluoroacetic acid (TFA).
7. 1,2-ethanedithiol.
8. Diethylether ($Et_2O$).

***2.2 Conjugation via Formation of a Thiourea Linkage***

1. Pyridine.
2. 4-dimethylaminopyridine (DMAP).
3. Methanol (MeOH).
4. Methylene chloride ($CH_2Cl_2$).
5. Trifluoroacetic acid (TFA).

***2.3 Conjugation via Formation of a Triazole Linkage***

1. Toluene.
2. Copper iodide (CuI).
3. Diisopropylethylamine (DIPEA).

4. Ethanol (EtOH).
5. Methylene chloride ($CH_2Cl_2$).
6. Dioxane.
7. Hydrochloric acid (HCl).
8. Diethylether ($Et_2O$).
9. Hexane.

#### 2.4 Conjugation via Formation of an Amide Linkage Using Solid Phase

1. Rink amide MBHA resin.
2. Dimethylformamide (DMF).
3. 2-(6-Chloro-1-*H*-benzotriazole-1-yl)-1,1,3,3-tetramethylaminium hexafluorophosphate (HCTU).
4. Diisopropylethylamine (DIPEA).
5. Piperidine.
6. *N*-methyl-2-pyrrolidone (NMP).
7. Methylene chloride ($CH_2Cl_2$).
8. Trifluoroacetic acid (TFA).
9. Phenol.
10. Water.
11. Triisopropylsilane (TIPS).
12. Ethyl acetate (EtOAc).
13. Neomycin or kanamycin acid monomer.

#### 2.5 A Rapid Assay to Monitor Nucleic Acid Binding and Selectivity: Fluorescence Intercalator Displacement (FID) Assay

1. Thiazole orange (TO).
2. 96 well plate.
3. Plate reader.
4. Buffer: 100 mM NaCl, and 0.5 mM EDTA, 10 mM sodium cacodylate, pH 6.5.

#### 2.6 A Rapid Assay for the Detection of Ribosomal Binding Drugs: Fluorescein–Neomycin (F-neo) Displacement (FND) Assay [49]

1. RNA (5′-GGCGUCACACCUUCGGGUAAGUCGCC-3′).
2. Diethylpyrocarbonate (DEPC)-treated water.
3. 96-well plate.
4. Plate reader.
5. Fluorescein–neomycin (F-neo), a proprietary probe available through NUBAD LLC.
6. Buffer 2×: 100 mM NaCl, and 1.0 mM EDTA, 20 mM MOPSO (7.0).

## 3 Methods

### *3.1 Conjugation via Formation of an Amide Bond (R′-COOH + 2a or 4)*

1. Mix 0.02–0.1 mmol of carboxylate of choice (R′-COOH) dissolved in 3.0–10.0 mL of dry DMF with one equivalent of dicyclohexylcarbodiimide and 1.0 mg of 4-dimethylaminopyridine in a nitrogen-purged round-bottom flask with a stir bar. Stir the solution under an atmosphere of nitrogen or argon for 3 h.
2. Add one equivalent of *2a* or *4* dissolved in 3 mL of dry DMF to the reaction flask. Be sure to keep the reaction conditions dry. Stir the reaction mixture at room temperature for a period of 12–28 h and follow the reaction by TLC.
3. Remove the solvent using a rotary evaporator. Wash the solid with an organic solvent ($CH_2Cl_2$) and dry it.
4. Purify the solid with flash chromatography using a solvent gradient from 0 to 25% MeOH in $CH_2Cl_2$ to yield the Boc-protected conjugate (60–90%) as a purple solid.
5. Deprotect the Boc-protected conjugate by adding 3.0–10 mL dichloromethane with an equal volume of trifluoroacetic acid (TFA) and 100 μL of 1,2-ethanedithiol (if using *2a*) in a round-bottom flask. Stir for 30 min.
6. Remove the solvents under vacuum and wash the resulting oil or solid with diethylether ($Et_2O$) to afford a solid product.

### *3.2 Conjugation via Formation of a Thiourea Linkage (R′-NH$_2$ + 2b or 5)*

1. Mix 0.015–0.1 mmol of an amine of choice (R′-NH$_2$) with a solution of one equivalent of neomycin isothiocyanate (2b or 5) dissolved in 2.0–10 mL of dry pyridine and a catalytic amount of DMAP (1 crystal) in a nitrogen-purged round-bottom flask. Stir the reaction mixture with a stir bar at room temperature overnight.
2. Remove the solvent using a rotary evaporator.
3. Purify the solid with flash chromatography utilizing a solvent gradient from 0 to 6% MeOH in $CH_2Cl_2$ and evaporate the solvents to yield the desired product as a solid (70–90%).
4. Deprotect the Boc-protected conjugate by adding 2.0–10 mL dichloromethane with an equal volume of trifluoroacetic acid in a round-bottom flask and stir at room temperature for 3 h.
5. Remove the solvent under vacuum and dissolve the solid in 5 mL of deionized water. Wash the aqueous layer with 3 × 5 mL aliquots of ether. Lyophilize the aqueous solution to yield the conjugate as a yellow solid (90–100%).

### *3.3 Conjugation via Formation of a Triazole Linkage (R′-alkyne + 3)*

1. Mix 0.025–0.1 mmol of an alkyne (R′-alkyne) and one equivalent of neomycin azide (3) dissolved in 5.0 mL of dry toluene in an argon-purged round-bottom flask with a stir bar. Add one equivalent CuI and 2 equiv. DIPEA to the flask and stir at room temperature for 18 h. Be sure to maintain a dry environment.

2. Remove the solvent using a rotary evaporator.
3. Purify the solid with flash chromatography utilizing a solvent gradient from 0 to 10% EtOH in $CH_2Cl_2$ to yield the desired product as a solid (80–90%).
4. Deprotect the Boc-protected conjugate by adding 3.0–10 mL of dioxane and 1–3 mL of a 4 M HCl–dioxane mixture to the conjugate in a small round-bottom flask.
5. Stir the mixture at room temperature until a precipitate (generally white) forms (approximately 15 min).
6. Centrifuge for 10 min at 5000 rpm, decant the solvent and collect the solid.
7. Wash the solid with a (3 × 5 mL each, v/v) diethyl ether–hexane solution.
8. Dissolve the solid in water and lyophilize to afford a solid (90–99%).

#### 3.4 Conjugation via Formation of an Amide Linkage Using Solid Phase (See Notes 3 and 4)

1. *Coupling of N-protected amino acid with resin:* Rink amide MBHA resin (20 mg) with 0.52 mmol/g loading was swollen in DMF (2 mL, overnight) with stirring. The resin was deprotected by stirring with 20% piperidine in DMF (3 mL, 2 × 15 min). An N-Fmoc-protected amino acid (Fmoc-$R_1'$-OH, 5 equiv.) was then coupled to resin in DMF in the presence of HCTU (5 equiv.) and DIPEA (10 equiv.) for 1 h. The resin was washed extensively with and DMF (4 × 1 mL) between reactions. After the completion of coupling of first amino acid, the resin was deprotected with 20% piperidine in DMF (2 mL, 2 × 15 min) and then washed extensively with DMF (4 × 1 mL). Another *N*-Fmoc protected amino acids (Fmoc-$R_2'$-OH, 5 equiv.) was then coupled to the resin in the similar fashion.
2. *Coupling of neomycin/kanamycin acid monomer:* After the completion of coupling of amino acids, the resin was deprotected with 20% piperidine in DMF (2 mL, 2 × 15 min) and then washed extensively with DMF (4 × 1 mL). Resin was mixed with neomycin/kanamycin acid monomer (2.5 equiv.) in NMP and DMF (1:1) in the presence of HCTU (2.5 equiv.) and DIPEA (5 equiv.) for 24 h. The same step was repeated twice. After coupling, the resin was washed with DMF (4 × 1 mL) and $CH_2Cl_2$ (4 × 1 mL), and then dried under room atmosphere for 1 h.
3. *Side chain deprotection and cleavage of peptides from resin*: The peptides were cleaved from the resin and all protecting groups were removed by a 2 h treatment with a solution containing 88:5:5:2 TFA–phenol–$H_2O$–TIPS (v/v; 400 μL). After the treatment, TFA was evaporated under nitrogen pressure. To the resulting resin, 1.5 mL of deionized water was then added

deprotection buffer removed by evaporation using a SpeedVac. Resuspend the RNA in DEPC treated water to the desired stock concentration. Determine the concentration of the RNA by the absorbance at 260 nm and 10 mm path length using the extinction coefficient provided by the manufacturer.

3. The F-neo fluorescent probe should be present to an equivalent final concentration of that of the model A-site, 0.1 μM, in each well (*see* **Note 6**).
4. The compound tested should be added to a final concentration of 0.3 μM, three times the molar concentration of the F-neo probe and the model A-site.
5. The final concentration for the reaction buffer is 10 mM MOPSO (7.0), 50 mM NaCl, and 0.5 mM EDTA (*see* **Note 7**).
6. For high-throughput assays, the F-neo is premixed with the model ribosomal A-site at 0.2 μM, twice the final concentration, in a 20 mM MOPSO (7.0), 100 mM NaCl, and 0.5 mM EDTA buffer.
7. Each compound of interest should be diluted to 0.6 μM, twice the final reaction concentration, in water.
8. In the 96-well plate experiments 100 μL of the A-site/F-neo solution is added to each well, and 100 μL of each compound of interest is added to a designated well to give a total final volume of 200 μL.
9. Every 96-well plate should contain a negative control of 100 μL of water added to the 100 μL of the 0.2 μM A-site/F-neo solution and a positive control of 100 μL of a 0.6 μM neomycin added to the 100 μL of the 0.2 μM A-site/F-neo solution.
10. All compounds and controls should be run in duplicate and the difference in the duplicate measurements should be less than 10%.
11. The instrument should be set to excite at 485 nm and record the emission at 535 nm, the excitation and emission of fluorescein.
12. The data is processed by subtracting the fluorescence of well containing the drug of interest ($F$) from the fluorescence of the negative control ($F_0$) to give the change in the fluorescence due to the displacement of F-neo (ΔF) by the compound.
13. In order to normalize the affinity of the compound for the A-site, the displacement is calculated as the percent displacement of F-neo by the compound of interest as compared to the displacement of F-neo by neomycin to give a relative percent binding using the Eq. 1 (*see* **Notes 6** and **8**).

$$\%\text{binding} = \left(\text{DF}_{\text{drug}}/\text{DF}_{\text{neomycin}}\right) \times 100\% \qquad (1)$$

## 4 Notes

1. Small molecules that regulate nucleic acid function are useful as tools (probes) as well as therapeutics. This report describes select examples of rapid chemistries and screening methodologies for targeting any given nucleic acid (rRNA, miRNA, DNA, DNA–RNA hybrids). While aminoglycoside functionalization is not a trivial task, this protocol attempts to make some of these assays and modifications accessible to the general practitioner.
2. A number of these probes and screening libraries will be available from NUBAD LLC (www.nubadllc.com).
3. Solution chemistries (Subheadings 3.1–3.3) tend to be more labor and purification intensive; hence, the recent development of solid-phase approaches (Subheading 3.4) is a much needed tool to make these libraries accessible. For example, the usage of amino acid–amino sugar conjugates opens up the selective targeting of functional RNAs (rRNA, miRNA) using small molecules.
4. In general, the purity of the compounds is greater than 90% and does not require any purification. For a few compounds per library, lower purity and yields are seen. These should be carefully repeated with pure starting materials and reagents. In rare cases, if impure compound is obtained, one should look for possible side reactions.
5. Probes such as TO (Subheading 3.5) have been used routinely for nucleic acid screening, but these intercalating probes bind to numerous sites on a nucleic acid, altering its solution structure.
6. The advent of groove binding probes (F-neo) offers a useful alternative as F-neo does not alter the nucleic acid structure by binding in a much lower stoichiometry than TO [16]. Given these synthesis (Subheadings 3.1–3.4) and screening approaches (Subheadings 3.5–3.6), small molecule assays and leads can in principle be generated for any given nucleic acid target of interest.
7. We have found that replacing the MOPSO with HEPES or the NaCl with KCl has no effect on the results.
8. In cases where F-neo cannot be replaced by the displacing ligand, alternate aminoglycoside or linker may have to be used for optimal reversible binding.

## Acknowledgments

We thank the National Institutes of Health (R42GM097917, AI114114, AI120303) for financial support.

### References


1. Arya DP (2007) Aminoglycoside antibiotics: from chemical biology to drug discovery. Wiley-Interscience, Hoboken, N.J, p 319
2. Willis B, Arya DP (2006) An expanding view of aminoglycoside–nucleic acid recognition. Adv Carbohydr Chem Biochem 60:251–302
3. Lynch SR, Puglisi JD (2001) Structural origins of aminoglycoside specificity for prokaryotic ribosomes. J Mol Biol 306:1037–1058
4. Davies JE (1964) Studies on the ribosomes of streptomycin-sensitive and resistant strains of *Escherichia Coli*. Proc Natl Acad Sci U S A 51:659–664
5. Arya DP, Xue L, Willis B (2003) Aminoglycoside (neomycin) preference is for a-form nucleic acids, not just RNA: results from a competition dialysis study. J Am Chem Soc 125:10148–10149
6. Xi H, Davis E, Ranjan N, Xue L, Hyde-Volpe-D, Arya DP (2011) Thermodynamics of nucleic acid "shape readout" by an aminosugar. Biochemistry 50:9088–9113
7. Earnshaw DJ, Gait MJ (1998) Hairpin ribozyme cleavage catalyzed by aminoglycoside antibiotics and the polyamine spermine in the absence of metal ions. Nucleic Acids Res 26:5551–5561
8. Hoch I, Berens C, Westhof E, Schroeder R (1998) Antibiotic inhibition of RNA catalysis : neomycin B binds to the catalytic core of the td group I intron displacing essential metal ions. J Mol Biol 282:557–569
9. Clouet-d'Orval B, Stage TK, Uhlenbeck OC (1995) Neomycin inhibition of the hammerhead ribozyme involves ionic interactions. Biochemistry 34:11186–11190
10. Chia JS, Wu HL, Wang HW, Chen DS, Chen PJ (1997) Inhibition of hepatitis delta virus genomic ribozyme self-cleavage by aminoglycosides. J Bio Med Sci 4:208–216
11. Kumar S, Kellish P, Robinson WE, Wang D, Appella DH, Arya DP (2012) Click dimers to target HIV TAR RNA conformation. Biochemistry 51:2331–2347
12. Kumar S, Arya DP (2011) Recognition of HIV TAR RNA by triazole linked neomycin dimers. Bioorg Med Chem Lett 21:4788–4792
13. Kumar S, Ranjan N, Kellish P, Gong C, Watkins D, Arya DP (2016) Multivalency in the recognition and antagonism of a HIV TAR RNA–TAT assembly using an aminoglycoside benzimidazole scaffold. Org Biomol Chem 14:2052–2056
14. Ranjan N, Kumar S, Watkins D, Wang D, Appella DH, Arya DP (2013) Recognition of HIV-TAR RNA using neomycin–benzimidazole conjugates. Bioorg Med Chem Lett 23:5689–5693
15. Tok JB, Dunn LJ, Des Jean RC (2001) Binding of dimeric aminoglycosides to the HIV-1 rev responsive element (RRE) RNA construct. Bioorg Med Chem Lett 11:1127–1131
16. Watkins D, Gong C, Kellish P, Arya DP (2017) Probing A-form DNA: a fluorescent aminosugar probe and dual recognition by anthraquinone-neomycin conjugates. Bioorg Med Chem 25:1309–ch1319. S0968-0896(16)30730-1 [pii]
17. Willis B, Arya DP (2006) Major groove recognition of DNA by carbohydrates. Curr Org Chem 10:663–673
18. Willis B, Arya DP (2009) Triple recognition of B-DNA. Bioorg Med Chem Lett 19: 4974–4979
19. Willis B, Arya D (2009) Triple recognition of B-DNA by a neomycin−Hoechst 33258−pyrene conjugate. Biochemistry 49: 452–469
20. Willis B, Arya DP (2006) Recognition of B-DNA by neomycin−Hoechst 33258 conjugates. Biochemistry 45:10217–10232
21. Kumar S, Xue L, Arya DP (2011) Neomycin-neomycin dimer: an all-carbohydrate scaffold with high affinity for AT-rich DNA duplexes. J Am Chem Soc 133:7361–7375
22. Kumar S, Spano MN, Arya DP (2015) Influence of linker length in shape recognition of B* DNA by dimeric aminoglycosides. Bioorg Med Chem 23:3105–3109. https://doi.org/10.1016/j.bmc.2015.04.082
23. Arya DP, Coffee RL, Xue L (2004) From triplex to B-form duplex stabilization: reversal of target selectivity by aminoglycoside dimers. Bioorg Med Chem Lett 14:4643–4646

24. Arya DP, Willis B (2003) Reaching into the major groove of B-DNA: synthesis and nucleic acid binding of a neomycin-Hoechst 33258 conjugate. J Am Chem Soc 125:12398–12399
25. Arya DP (2005) Aminoglycoside-nucleic acid interactions: the case for neomycin. Top Curr Chem 253:149–178
26. Arya DP, Coffee RL Jr, Charles I (2001) Neomycin-induced hybrid triplex formation. J Am Chem Soc 123:11093–11094
27. Arya DP, Coffee RL Jr (2000) DNA triple helix stabilization by aminoglycoside antibiotics. Bioorg Med Chem Lett 10:1897–1899
28. Arya DP (2010) New approaches toward recognition of nucleic acid triple helices. Acc Chem Res 44:134–146
29. Arya DP, Coffee RL, Willis B, Abramovitch AI (2001) Aminoglycoside–nucleic acid interactions: remarkable stabilization of DNA and RNA triple helices by neomycin. J Am Chem Soc 123:5385–5395
30. Arya DP, Xue L, Tennant P (2003) Combining the best in triplex recognition: synthesis and nucleic acid binding of a BQQ–neomycin conjugate. J Am Chem Soc 125:8070–8071
31. Xue L, Ranjan N, Arya DP (2011) Synthesis and spectroscopic studies of the (neomycin)-perylene conjugate binding to human telomeric DNA. Biochemistry 50:2838–2849
32. Xue L, Charles I, Arya DP (2002) Pyrene–neomycin conjugate: dual recognition of a DNA triple helix. Chem Commun 1:70–71
33. Ranjan N, Davis E, Xue L, Arya DP (2013) Dual recognition of the human telomeric G-quadruplex by a neomycin–anthraquinone conjugate. Chem Commun 49:5796–5798
34. Ranjan N, Arya DP (2013) Targeting C-myc G-quadruplex: dual recognition by aminosugar-bisbenzimidazoles with varying linker lengths. Molecules 18:14228–14240
35. Watkins D, Ranjan N, Kumar S, Gong C, Arya DP (2013) An assay for human telomeric G-quadruplex DNA binding drugs. Bioorg Med Chem Lett 23:6695–6699
36. Shaw NN, Xi H, Arya DP (2008) Molecular recognition of a DNA:RNA hybrid: sub-nanomolar binding by a neomycin–methidium conjugate. Bioorg Med Chem Lett 18:4142–4145
37. Shaw NN, Arya DP (2008) Recognition of the unique structure of DNA:RNA hybrids. Biochimie 90:1026–1039
38. Hamilton PL, Arya DP (2012) Natural product DNA major groove binders. Nat Prod Rep 29:134–143
39. Jiang L, Watkins D, Jin Y, Gong C, King A, Washington AZ, Green KD, Garneau-Tsodikova S, Oyelere AK, Arya DP (2015) Rapid synthesis, RNA binding, and antibacterial screening of a peptidic-aminosugar (PA) library. ACS Chem Biol 10:1278–1289
40. Charles I, Xue L, Arya DP (2002) Synthesis of aminoglycoside–DNA conjugates. Bioorg Med Chem Lett 12:1259–1262
41. Charles I, Arya DP (2005) Synthesis of neomycin-DNA/peptide nucleic acid conjugates. J Carbohydr Chem 24:145–160
42. Charles I, Xi H, Arya DP (2007) Sequence-specific targeting of RNA with an oligonucleotide-neomycin conjugate. Bioconjug Chem 18:160–169
43. Ranjan N, Arya DP (2016) Linker dependent intercalation of bisbenzimidazole-aminosugars in an RNA duplex; selectivity in RNA vs. DNA binding. Bioorg Med Chem Lett 26:5989–5994
44. Willis B, Arya DP (2014) Recognition of RNA duplex by a neomycin–Hoechst 33258 conjugate. Bioorg Med Chem 22:2327–2332
45. King A, Watkins D, Kumar S, Ranjan N, Gong C, Whitlock J, Arya DP (2013) Characterization of ribosomal binding and antibacterial activities using two orthogonal high-throughput screens. Antimicrob Agents Chemother 57:4717–4726. https://doi.org/10.1128/AAC.00671-13
46. Jin Y, Watkins D, Degtyareva NN, Green KD, Spano MN, Garneau-Tsodikova S, Arya DP (2016) Arginine-linked neomycin B dimers: synthesis, rRNA binding, and resistance enzyme activity. Medchemcomm 7:164–169. https://doi.org/10.1039/C5MD00427F
47. Kukielski C, Maiti K, Bhaduri S, Story S, Arya DP (2018) Rapid solid-phase syntheses of a peptidic-aminoglycoside library. Tetrahedron 74:4418–4428
48. Ghosh A, Degyatoreva N, Kukielski C, Story S, Bhaduri S, Maiti K, Nahar S, Ray A, Arya DP, Maiti S (2018) Targeting miRNA by tunable small molecule binders: peptidic aminosugar mediated interference in miR-21 biogenesis reverts epithelial to mesenchymal transition. Med Chem Commun 9:1147–1154
49. Watkins D, Norris FA, Kumar S, Arya DP (2013) A fluorescence-based screen for ribosome binding antibiotics. Anal Biochem 434:300–307. https://doi.org/10.1016/j.ab.2012.12.003

# Chapter 10

# Preparation and Purification of Oligodeoxynucleotide Duplexes Containing a Site-Specific, Reduced, Chemically Stable Covalent Interstrand Cross-Link Between a Guanine Residue and an Abasic Site

**Maryam Imani Nejad, Xu Guo, Kurt Housh, Christopher Nel, Zhiyu Yang, Nathan E. Price, Yinsheng Wang, and Kent S. Gates**

## Abstract

Methods for the preparation of DNA duplexes containing interstrand covalent cross-links may facilitate research in the fields of biochemistry, molecular biology, nanotechnology, and materials science. Here we report methods for the synthesis and isolation of DNA duplexes containing a site-specific, chemically stable, reduced covalent interstrand cross-link between a guanine residue and an abasic site. The method uses experimental techniques and equipment that are common in most biochemical laboratories and inexpensive, commercially available oligonucleotides and reagents.

**Key words** Interstrand DNA cross-link, Cross-link synthesis, Abasic site, Gel electrophoresis, Reductive amination, Post-synthetic oligonucleotide modification

## 1 Introduction

Covalent interstrand cross-links in duplex DNA have significant biological effects because they prevent the strand separation that is required for readout of the genetic information stored in the nucleotide sequence of the double helix [1, 2]. Interstrand cross-links can arise from environmental chemicals, anticancer drugs, and unavoidable endogenous processes [1, 3–9]. The repair of interstrand cross-links is important in human health and disease, but the complex processes required for cellular recognition and repair of cross-links are not yet well understood [2, 10–13]. In the fields of

Nathaniel Shank (ed.), *Non-Natural Nucleic Acids: Methods and Protocols*, Methods in Molecular Biology, vol. 1973, https://doi.org/10.1007/978-1-4939-9216-4_10, © Springer Science+Business Media, LLC, part of Springer Nature 2019

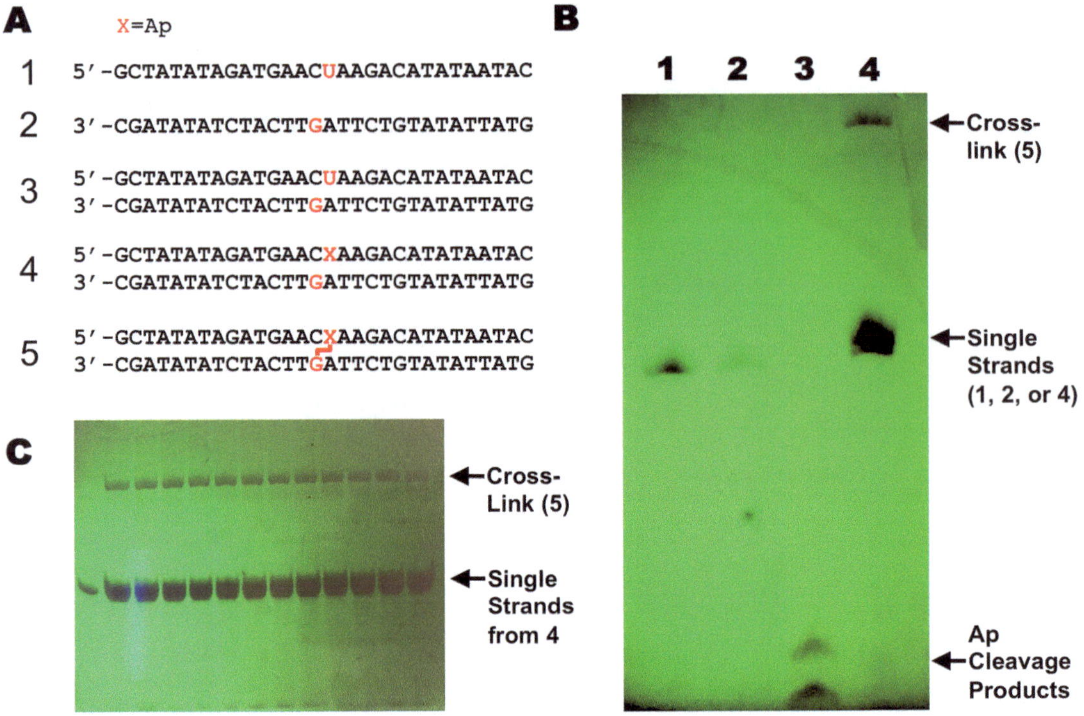


**Fig. 3** Generation of a covalent cross-link between a guanine residue and a reduced abasic site in duplex DNA. Panel (**a**) DNA sequences used in this study (X = Ap site). Panel (**b**) Photograph of UV-shadowed preparative denaturing polyacrylamide gel. Lane 1, single strand **1**; lane 2, single strand **2**; lane 3, single strand **2** treated with piperidine (0.1 M, 90 °C, 30 min) to induce strand cleavage; lane 4, cross-linking reaction mixture containing a mixture of duplexes **4** and **5**. Panel (**c**) Photograph of UV-shadowed preparative gel from which the cross-linked DNA was isolated

## 2 Materials

### 2.1 Analytical Detection of Cross-Link Formation in $^{32}$P-Labeled 2′-Deoxy-oligonucleotide Duplexes

### 2.2 Materials for Preparative Cross-Linking Reaction

1. 1 mM oligodeoxynucleotide **1** dissolved in HPLC grade water (*See* **Note 1**).
2. 1 mM oligodeoxynucleotide **2** in water.
3. Uracil-DNA glycosylase "10×" reaction buffer (UDG buffer; New England Biolabs).
4. Uracil-DNA glycosylase enzyme (UDG, New England Biolabs).
5. 25:24:1 phenol-chloroform-isoamyl alcohol mixture.

6. 500 mM 3-(*N*-morpholino)propanesulfonic acid, pH 7 (MOPS).
7. 1 M sodium chloride.
8. 3 M sodium acetate, pH 5.2.
9. 2.5 M sodium cyanoborohydride, freshly prepared.
10. 1.5 mL microcentrifuge tubes.
11. Vortex mixer.
12. Thermostat-controlled oven-incubator set at 37 °C.
13. Benchtop centrifuge.
14. Benchtop centrifuge in 4 °C cold room.
15. Speed-vac concentrator.
16. Thermostat-controlled heating block.

#### 2.3 Materials for Purification and Isolation of a Cross-Linked Duplex Using Denaturing Polyacrylamide Gel Electrophoresis

1. 20% polyacrylamide solution (19:1 acrylamide/bis-acrylamide, 8 M urea).
2. *N,N,N′,N′*-tetramethylethylenediamine (TEMED).
3. 10% (w/v) aqueous ammonium persulfate (for molecular biology, for electrophoresis).
4. Tris-borate-EDTA buffer (TBE, 89 mM tris-borate, 2 mM EDTA, pH 8.3) prepared by dilution of a 10× stock solution.
5. Formamide loading buffer composed of 17.75 M formamide (deionized), 0.01 M EDTA, and bromophenol blue dye (ACS reagent) in water [30].
6. Elution buffer (0.2 M NaCl, 0.001 M EDTA, pH 8).
7. Crushed dry ice.
8. Water (HPLC grade).
9. Glass gel plates (16 × 19.7 cm) with 2-mm-thick spacers and 12-well comb.
10. 1.5 and 2 mL microcentrifuge tubes.
11. 100 mL beaker and magnetic stir bar.
12. 20 mL disposable syringe and 18 gauge, 1.5 in needle.
13. Electrophoresis power source.
14. Vortex mixer.
15. All-purpose laboratory wrap (clear, Saran-type wrap).
16. UV lamp and silica gel TLC plate impregnated with UV-254 fluorophore.
17. Disposable, single-edge razor blade.
18. Glass rod (round tip, 5 mm × 15 cm).
19. Poly-Prep chromatography column (spin column, Bio-Rad Laboratories).
20. Clinical centrifuge (swinging bucket type).

21. C-18 Sep-Pak cartridges (1 mL, 100 mg, Waters, cat. no. WAT023590).
22. Speed-vac concentrator.
23. Syringe-tip filter, Titan3 PES (polyethersulfone, Thermo Fisher Scientific).
24. Vertical slab gel electrophoresis stand.
25. 3 M sodium acetate, pH 5.2.
26. Ethanol, 200 proof.
27. Micro-90 concentrated cleaning solution.
28. Precision wipes.
29. Medium size binder clips (heavy-duty office-type paper clips).

## 3 Methods

All manipulations were carried out at room temperature unless otherwise noted.

### 3.1 Preparation of the dU-Containing Oligonucleotide Duplex 3

1. Concentrations of single-stranded oligonucleotides were determined using UV-vis absorbance and the extinction coefficient provided by the manufacturer (or calculated using the IDT oligo analyzer web app).
2. To a 1.5 mL microcentrifuge tube, add 5 μL of oligonucleotide **1**, 5 μL of oligonucleotide **2**, 2.5 μL of 500 mM pH 7.0 MOPS buffer, 5 μL of 1 M NaCl, and 32.5 μL water (final concentrations of 25 mM MOPS and 100 mM NaCl and 5 nmols of DNA). In the method described here, 12 reactions each containing 5 nmols of DNA were run in parallel.
3. Mix for 5 s using a vortex mixer.
4. Incubate in a thermostat-controlled aluminum heating block for 5 min at 95 °C.
5. Remove the aluminum block containing the reaction tube from the heating source, and cool to room temperature overnight to allow annealing of the duplex.

### 3.2 Preparation of the Ap-Containing Oligonucleotide Duplex 4

1. Mix the solution containing the annealed dU-duplex for 30 s using a vortex mixer.
2. Add 10 μL "10×" UDG buffer, 33 μL water, and 7 μL UDG enzyme (35 U).
3. Mix for 10 s using a vortex mixer.
4. Incubate for 1 h at 37 °C.
5. Add 100 μL of phenol-chloroform-isoamyl alcohol, and vortex for a few seconds until the solution gets cloudy.

6. Centrifuge for 3 min at 15,000 rpm (approx. 17,000 × *g*) using a benchtop microcentrifuge at 24 °C.
7. Remove the top aqueous layer containing the DNA using a micropipetter set to 100 μL.
8. Precipitate the DNA by addition of 10 μL of 3 M sodium acetate, pH 5.2 to the aqueous solution, followed by vortex mixing for 10 s.
9. Add 550 μL of cold absolute ethanol (−20 °C) and mix the solution for 10 s.
10. Place the sample on crushed dry ice for 1 h.
11. Centrifuge the sample for 1 h at 13,300 rpm (approx. 14,000 × *g*) at 4 °C.
12. Remove the supernatant, being careful not to disturb the DNA pellet at the bottom of the tube.
13. Wash the pellet by addition of 90 μL of 8:2 ethanol/water, and centrifuge for 20 min at 13,300 rpm at 4 °C.
14. Remove the supernatant, being careful not to disturb the DNA pellet at the bottom of the tube.
15. Dry the Ap-containing oligonucleotide duplex **4** under vacuum in a speed-vac concentrator for 2 to 3 min at room temperature (24 °C, *see* **Notes 2** and **3**).

### 3.3 Preparation of the Cross-Linked Duplex 5

1. Add 33.5 μL water to the freshly prepared Ap-containing duplex **4**.
2. Add to the microcentrifuge tube containing the Ap-duplex **4**, 12.5 μL of 3 M sodium acetate, pH 5.2, and 5 μL of 2.5 M $NaCNBH_3$ in water, and vortex for 10 s (final concentrations: $NaCNBH_3$ 250 mM and sodium acetate 750 mM) (*see* **Notes 4** and **5**).
3. Incubate the sample for 24 h at 37 °C.
4. Ethanol precipitate the DNA as described above in **steps 8–15**, and store the sample dry at −20 °C until gel purification.

### 3.4 Gel Electrophoresis Setup

CAUTION: Gel electrophoresis involves the use of a high-voltage power source and the toxic chemicals, acrylamide and bis-acrylamide. Use a gel apparatus with proper safety designs after receiving proper safety training, and take appropriate precautions when dispensing and working with acrylamide and bis-acrylamide.

1. Wash the glass gel plates and spacers with Micro-90 concentrated cleaning solution, and rinse them well with DI water. Hold the plates on the edges and rinse them with absolute ethanol.
2. Dry the plates with precision wipes. Assemble the plates separated by spacers (a small amount of grease on the top and bottom of the spacers will keep them from moving during

plate assembly), clamp the assembly together with binder clips, and add some grease to the corners to prevent leaking when the gel is poured.

3. In a 100 mL beaker equipped with a magnetic stir bar, pour 40 mL of 20% polyacrylamide solution containing 8 M urea (*see* **Note 6**).
4. Add TEMED (neat, 20 to 25 μL) to the stirred 100 mL beaker from **step 3**.
5. Add 200 μL of aqueous ammonium persulfate to the stirred 100 mL beaker from **step 4**, and stir for 1–2 min.
6. Holding the gel assembly at an incline, with the top of the assembly approximately 5–8 cm above the top, pour the mixture into the gel plate assembly.
7. Insert a 12-well comb between the plates assembly. Well size 5 mm wide × 9 mm high × 2 mm thickness.
8. Allow gel polymerization to occur (1.5 to 2 h at room temperature).
9. After polymerization is complete, remove the comb and bottom spacer slowly, and flush the wells with 10 mL of 1 × TBE in a 20 mL disposable syringe equipped with a disposable needle.
10. Mount the gel plate assembly onto the gel stand according to manufacturer's instructions (*see* **Note 7**). Fill the top and bottom buffer trays of the gel stand with 1× TBE.
11. Before loading the samples on the gel, electrophorese the gel for 30 min at 300 V.

### 3.5 Separation of Cross-Linked Duplex 5 from Uncross-Linked DNA Using Denaturing Gel Electrophoresis

This protocol describes the purification and isolation of approximately 5 nmol of cross-linked duplex **5** via denaturing polyacrylamide gel electrophoresis.

1. Add 20 μL formamide loading buffer to the dry cross-linked duplex **5** sample and vortex mix for 5 s.
2. Spin the sample to the bottom of the tube using a brief (3 s) spin in a benchtop microcentrifuge.
3. Before loading on the gel, heat the samples at 95 °C for 2 min to denature uncross-linked material.
4. Turn off the power supply, and completely disconnect the gel apparatus from the power supply. Load the first sample into a well of the gel. Each reaction is loaded into a separate lane of the gel.
5. Flush the wells of the gel once again with 10–15 mL 1 × TBE dispensed from a 20 mL disposable syringe equipped with a 18 gauge needle. Flushing the wells removes urea that diffuses from the gel. If the wells are not flushed, the samples in loading

buffer will not sink properly to the bottom of the wells when loaded. Samples should be loaded soon after flushing the wells to ensure proper layering of the sample on the bottom of the well.

6. Load the sample into a well of the gel.
7. Rinse the sample tube with 10 μL formamide loading buffer, and spin the sample to the bottom of the tube using a brief (3 s) spin in a benchtop microcentrifuge, and load the rinse into the same lane as in **step 5**.
8. Repeat **steps 6** and 7 to load each reaction into a separate well of the gel.
9. Carefully reconnect the gel apparatus to the power supply. Turn the power supply on and electrophorese the gel at 300 V.
10. Turn off the power supply, and disconnect the gel from the power supply once the dye has traveled approximately 12–14 cm from the well (approximately 4 h).
11. Remove the plate assembly from the stand, and separate the plates carefully, and remove the gel from both plates.

### 3.6 Isolation of Cross-Linked Duplex 5 from the Gel

1. Wrap the gel with plastic wrap, and place on top of a large (20 × 20 cm) silica gel TLC plate containing UV-254 fluorophore.
2. Illuminate the gel from above with a handheld UV-254 lamp to "UV-shadow" the DNA bands in the gel [30]. The DNA will appear as purple bands against the light green fluorescence background of the TLC plate. The cross-linked DNA will be located approximately 5–7 cm from the wells for the oligonucleotides used here (Fig. 3c, *see* **Notes 8** and **9**). Minimize the amount of time the DNA is exposed to the handheld UV light. UV light can cause DNA damage. Using a permanent marker, outline the location of the band on the plastic wrap covering the gel.
3. Cut the band from the gel with a disposable razor blade. Peel the plastic wrap from the gel slice, and place the gel slice into a 2 mL microcentrifuge tube.
4. Crush the gel slice with the tip of a glass rod and add 0.3 mL elution buffer.
5. Agitate the sample utilizing a vortex mixer for at least 1 h.
6. Spin the sample through a Poly-Prep column according to manufacturer's instructions.
7. Ethanol precipitate the cross-linked duplex **5** as described above.
8. Resuspend the sample in 300 μL water, and filter it through a Titan3 PES syringe filter to remove residual polyacrylamide.

9. Rinse the filter with 300 μL water.
10. Freeze the solution on crushed dry ice for 2–3 min.
11. Lyophilize the sample using a Speed-vac concentrator at room temperature (approximately 3 h).
12. Estimate the concentration of the cross-linked duplex when dissolved in aqueous solution using the extinction coefficient of the uncross-linked duplex **3** calculated by literature methods [31]. The procedures described here, involving the isolation of DNA from 12 separate 5 nmol reactions, purified in 12 separate lanes of a preparative gel, yield approximately 9 nmols of cross-linked duplex (15% overall yield) (*see* **Notes 10–13**).

## 4 Notes

1. The sequence of purchased oligonucleotides can be confirmed using ESI-TOF-MS [32] and Maxam-Gilbert sequencing reactions [33].
2. Following ethanol precipitation, care should be taken to dry oligonucleotides for only 2–3 min under vacuum at room temperature in the Speed-vac concentrator. Excessive drying can dehydrate the DNA making it difficult (or impossible) to resuspend.
3. Heating the Ap-containing duplex during Speed-vac evaporation causes unwanted strand cleavage via β-elimination at the Ap site [34, 35].
4. Read the material safety data sheet (MSDS) for the reagent $NaCNBH_3$, and take appropriate precautions. For example, avoid mixing $NaCNBH_3$ with strong acid.
5. Cross-links can form at adenine residues in the absence of $NaCNBH_3$ in the sequence 5′-ApT/AA [24]. Our unpublished data indicates that incubation of duplexes containing the sequence 5′CApT/AAG, in the presence of $NaCNBH_3$, generates a mixture of cross-linked duplexes (separable on a denaturing polyacrylamide sequencing gel) presumably including both the dA-Ap and dG-$Ap_{red}$ cross-links. Such sequences should be avoided if generation of a single cross-linked species is desired.
6. Polyacrylamide is neurotoxic. Read the MSDS sheet and use appropriate safety precautions.
7. Gel electrophoresis involves the use of a high-voltage power source. Use a gel apparatus with proper safety designs after receiving proper safety training.
8. It may be useful to load a sample of single-stranded oligonucleotide in one lane of the gel to provide a size marker (Fig. 3).

9. It may be useful to carry out a piperidine work-up on a small sample of the single-stranded Ap-oligonucleotide treated with UDG to ensure that the enzyme is active in the conversion of dU to Ap (Fig. 3b, lane 3). The cleavage products will migrate faster than full-length oligonucleotides.

10. Differences in flanking sequence can affect cross-link yields. Analytical gel measurements using $^{32}$P-labeled oligonucleotides indicate that the cross-link yields for duplexes that contain various bases flanking the 5′-CAp/AG cross-link site produce cross-linked DNA in yields ranging from 10 to 40%. The sequence employed here typically gives approximately 20% yield of cross-linked DNA using $^{32}$P-labeled oligonucleotides using $NaCNBH_3$ under the conditions described here [22].

11. The dG-$Ap_{red}$ cross-linkage can be generated in duplexes of about 20 base pairs and longer. Shorter Ap-containing duplexes may not be fully hybridized at room temperature and therefore may provide poor cross-link yields.

12. The cross-linked duplex obtained by this protocol displays temperature-dependent hyperchromicity consistent with melting of the helical regions flanking the cross-link. When the cross-linked duplex is annealed with two slow heat-cool (95 °C to room temp) cycles, the melting curve such as that shown in Fig. 4 will be obtained. As expected [23, 36], the

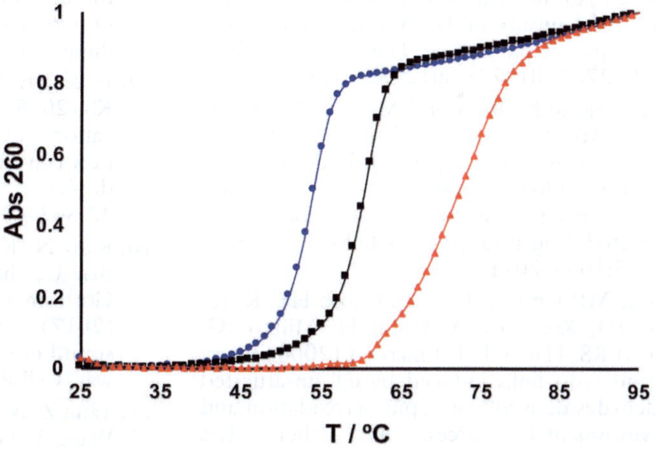


**Fig. 4** The cross-linked duplex **5** displays temperature-dependent hyperchromicity and melts at a substantially higher temperature than uncross-linked duplexes **3** and **4**. Duplexes (1.4 μM) were dissolved in 50 mM HEPES (pH 7) containing 100 mM NaCl, and the absorbance was measured while heating at 0.5 °C/min. The melting curves were normalized to set the starting absorbance to 0 and the final absorbance to 1. The total change in absorbance upon melting was similar for duplexes **3**–**5**, at approximately 0.2 units. From left to right, the curves show melting of the Ap-duplex (blue, **4**), the dU-duplex (black, **3**), and the cross-linked duplex (red, **5**)

melting temperature of the cross-linked duplex is higher than the corresponding uncross-linked dU-duplex **3** or the Ap-duplex **4**. Finally, consistent with the presence of double-helical character in the DNA flanking the cross-link, we have shown that a 21-base pair duplex insert containing a centrally located dG-$Ap_{red}$ cross-link can be enzymatically ligated into a linearized plasmid [25].

13. Vendors are identified for products where we believe that there could be significant differences between suppliers or proprietary technology makes it difficult to identify comparable products.

## Acknowledgment

We are grateful to the National Institutes of Health for supporting this work (ES021007).

### References

1. Schärer OD (2005) DNA interstrand cross-links: natural and drug-induced DNA adducts that induce unique cellular responses. Chembiochem 6:27–32
2. Clauson C, Schärer OD, Niedernhofer LJ (2013) Advances in understanding the complex mechanisms of DNA interstrand cross-link repair. Cold Spring Harb Perspect Biol 5. a012732/012731-a012732/012725
3. Imani-Nejad M, Johnson KM, Price NE, Gates KS (2016) A new cross-link for an old cross-linking drug: the nitrogen mustard anticancer agent mechlorethamine generates cross-links derived from abasic sites in addition to the expected drug-bridged cross-links. Biochemistry 55:7033–7041
4. Stone MP, Cho YJ, Huang H, Kim HY, Kozekov ID, Kozekova A, Wang H, Minko IG, Lloyd RS, Harris TM, Rizzo CJ (2008) Interstrand cross-links induced by α,β-unsaturated aldehydes derived from lipid peroxidation and environmental sources. Acc Chem Res 41:793–804
5. Greenberg MM (2014) Abasic and oxidized abasic site reactivity in DNA: enzyme inhibition, cross-linking, and nucleosome-catalyzed reactions. Acc Chem Res 47:646–655
6. Rajski SR, Williams RM (1998) DNA cross-linking agents as antitumor drugs. Chem Rev 98:2723–2795
7. Yang Z, Price NE, Johnson KM, Wang Y, Gates KS (2017) Interstrand cross-links arising from strand breaks at true abasic sites in duplex DNA. Nucleic Acids Res 45:6275–6283
8. Catalano MJ, Liu S, Andersen N, Yang Z, Johnson KM, Price NA, Wang Y, Gates KS (2015) Chemical structure and properties of the interstrand cross-link formed by the reaction of guanine residues with abasic sites in duplex DNA. J Am Chem Soc 137:3933–3945
9. Price NE, Catalano MJ, Liu S, Wang Y, Gates KS (2015) Chemical and structural characterization of interstrand cross-links formed between abasic sites and adenine residue in duplex DNA. Nucleic Acids Res 43:3434–3441
10. Kato N, Kawasoe Y, Williams HL, Coates E, Roy U, Shi Y, Beese LS, Schärer OD, Yan H, Gottesman ME, Takahashi TS, Gautier J (2017) Sensing and processing of DNA interstrand crosslinks by the mismatch repair pathway. Cell Rep 21:1375–1385
11. Yang Z, Nejad MI, Gamboa Varela J, Price NE, Wang Y, Gates KS (2017) A role for the base excision repair enzyme NEIL3 in replication-dependent repair of interstrand cross-links derived from psoralen and abasic sites. DNA Repair 52:1–11
12. Semlow DR, Zhang J, Budzowska M, Drohat AC, Walter JC (2016) Replication-dependent unhooking of DNA interstrand cross-links by the NEIL3 glycosylase. Cell 167:498–511
13. Huang J-C, Liu S, Bellani MA, Thazhathveetil AK, Ling C, de Winter JP, Wang Y, Wang W,

Seidman MM (2013) The DNA translocase FANCM/MHF promotes replication traverse of DNA interstrand cross-links. Mol Cell 52:434–446

14. Tomás-Gamasa M, Serdjukow S, Su M, Müller M, Carell T (2014) "Post-it" type connected DNA created with a reversible covalent cross-link. Angew Chem Int Ed Eng 53:796–800
15. Imani-Nejad M, Shi R, Zhang X, Gu L-Q, Gates KS (2017) Sequence-specific covalent capture coupled with high-contrast nanopore detection of a disease-derived nucleic acid sequence. Chembiochem 18:1383–1386
16. Vieregg JR, Nelson HM, Stoltz BM, Pierce NA (2013) Selective nucleic acid capture with shielded covalent probes. J Am Chem Soc 135:9691–9699
17. Peng X, Greenberg MM (2008) Facile SNP detection using bifunctional cross-linking oligonucleotide probes. Nucleic Acids Res 36:e31
18. Fujimoto K, Yamada A, Yoshimura Y, Tsukaguchi T, Sakamoto T (2013) Details of the ultrafast DNA photo-cross-linking reaction of 3-cyanovinylcarbazole nucleoside: cis-trans isomeric effect and the application for SNP-based genotyping. J Am Chem Soc 135:16161–16167
19. Rajendran A, Endo M, Katsuda Y, Hidaka K, Sugiyama H (2011) Photo-cross-linking-assisted thermal stability of DNA origami structures and its application for higher-temperature self-assembly. J Am Chem Soc 133:14488–14491
20. Chen W, Schuster GB (2013) Structural stabilization of DNA-templated nanostructures: cross-linking with 2,5-bis(2-thienyl)-pyrrole monomers. Org Biomol Chem 11:35–40
21. Dutta S, Chowdhury G, Gates KS (2007) Interstrand crosslinks generated by abasic sites in duplex DNA. J Am Chem Soc 129:1852–1853
22. Johnson KM, Price NE, Wang J, Fekry MI, Dutta S, Seiner DR, Wang Y, Gates KS (2013) On the formation and properties of interstrand DNA-DNA cross-links forged by reaction of an Abasic site with the opposing guanine residue of 5'-CAp sequences in duplex DNA. J Am Chem Soc 135:1015–1025
23. Gamboa Varela J, Gates KS (2015) A simple, high-yield synthesis of DNA duplexes containing a covalent, thermally-reversible interstrand cross-link at a defined location *Angew.* Chem Int Ed Eng 54:7666–7669
24. Price NE, Johnson KM, Wang J, Fekry MI, Wang Y, Gates KS (2014) Interstrand DNA–DNA cross-link formation between adenine residues and Abasic sites in duplex DNA. J Am Chem Soc 136:3483––3490
25. Price NE, Li L, Gates KS, Wang Y (2017) Replication and repair of a reduced 2-′-deoxyguanosine-abasic site cross-link in human cells. Nucleic Acids Res 45:6486–6493
26. Gamboa Varela J, Gates KS (2016) Simple, high-yield syntheses of DNA duplexes containing interstrand DNA-DNA cross-links between an $N^4$-aminocytidine residue and an abasic site. Curr Protoc Nucleic Acid Chem 65:5.16.11–15.16.15
27. Varshney U, van de Sande JH (1991) Specificities and kinetics of uracil excision from uracil-containing DNA oligomers by *Escherichia coli* uracil DNA glycosylase. Biochemistry 30:4055–4061
28. Borch RF, Hassid AI (1972) A new method for the methylation of amines. J Org Chem 37:1673–1674
29. Romero RM, Rojsittisak P, Haworth IS (2013) Electrophoretic mobility of duplex DNA cross-linked by mechlorethamine at a cytosine-cytosine mismatch pair. Electrophoresis 34:917–924
30. Sambrook J, Fritsch EF, Maniatis T (1989) Molecular cloning: a lab manual. Cold Spring Harbor Press, Cold Spring Harbor, NY
31. Tataurov AV, You Y, Owczarzy R (2008) Predicting ultraviolet spectrum of single stranded and double stranded deoxyribonucleic acids. Biophys Chem 133:66–70
32. Melton D, Lewis CD, Price NE, Gates KS (2014) Covalent adduct formation between the antihypertensive drug hydralazine and abasic sites in double- and single-stranded DNA. Chem Res Toxicol 27:2113–2118
33. Maxam AM, Gilbert W (1980) Sequencing end-labeled DNA with base-specific chemical cleavages. Methods Enzymol 65:499–560
34. Gates KS (2009) An overview of chemical processes that damage cellular DNA: spontaneous hydrolysis, alkylation, and reactions with radicals. Chem Res Toxicol 22:1747–1760
35. Gates KS, Nooner T, Dutta S (2004) Biologically relevant chemical reactions of N7-alkyl-2′-deoxyguanosine adducts in DNA. Chem Res Toxicol 17:839–856
36. Shi Y-B, Hearst JE (1986) Thermostability of double-stranded deoxyribonucleic acids: effects of covalent additions of a psoralen. Biochemistry 25:5895–5902

# Chapter 11

# Copper-Catalyzed Alkyne-Azide Cycloaddition on the Solid Phase for the Preparation of Fully Click-Modified Nucleic Acids

**Malte Rosenthal, Franziska Pfeiffer, and Günter Mayer**


## Abstract

Click chemistry has become a widely used method to insert modifications into DNA. Due to its commercial availability, 5′-ethynyl-deoxyuridine (EdU) is commonly incorporated into the DNA for subsequent modification by click reaction. However, it is partially oxidized during deprotection during solid-phase synthesis, resulting in a ketone that is no longer accessible for click modification. To enable the high-fidelity solid-phase synthesis of EdU-containing DNA, this protocol describes a procedure to perform the click reaction on the solid phase before deprotection. Afterwards, the DNA can be deprotected and purified according to standard procedures, and the full modification of EdU with the azide of choice can be analyzed by HPLC and HPLC/MS.


**Key words** CuAAC, Click chemistry, Aptamer, Solid-phase synthesis, Nucleobase modification

## 1 Introduction

The introduction of chemical building blocks into DNA with the help of copper-catalyzed alkyne-azide cycloaddition (CuAAC) allows the modification of its biophysical properties as well as the addition of functional groups for further reactions, e.g., biotinylation for subsequent recognition with streptavidin [1, 2]. This reaction can be used in an adapted SELEX approach, termed click-SELEX, to select clickmers (click-modified aptamers) with novel recognition properties [3].

As 5′-ethynyl-deoxyuridine (EdU) is commercially available as phosphoramidite and triphosphate, EdU is often used for incorporation into DNA molecules as alkyne and subsequently modified with one of the large number of commercially available or easily synthesized azides [4]. However, large amounts of DNA that contain "clickable" EdU modifications are hard to produce by PCR in an economic manner. Therefore, production of the desired

Nathaniel Shank (ed.), *Non-Natural Nucleic Acids: Methods and Protocols*, Methods in Molecular Biology, vol. 1973,
https://doi.org/10.1007/978-1-4939-9216-4_11, © Springer Science+Business Media, LLC, part of Springer Nature 2019

oligonucleotides by solid-phase synthesis is recommended. Conventional alkaline workup of solid-phase-synthesized EdU-bearing DNA leads to strands with partially oxidized ethynyl moieties, so-called ketone deoxyuridine (KdU) [5]. Since KdU residues can no longer be click-functionalized, homogenous products are not accessible by this procedure.

To overcome this limitation, an altered strategy for solid-phase synthesis and subsequent workup has been developed [6]. Here, the EdU-containing DNA is functionalized via click reaction while it is still attached to the solid support. Subsequent purification of the DNA leads to homogenous functionalized products, alleviating the KdU residues.

This chapter describes how the click reaction is performed in an aqueous suspension on the solid support in the presence of the desired azide and a Cu(I) chelating ligand that prevents oxidative damage to the DNA [7, 8]. The workup and purification of the click-functionalized DNA is then performed according to standard procedures. Concluding investigations of the produced DNA is done by HPLC and HPLC/MS to ensure successful conversion of every EdU building block into the corresponding triazole.

## 2 Materials

### 2.1 Click-Catalyst Solution

1. Sodium ascorbate.
2. Copper sulphate.
3. Tris(3-hydroxypropyltriazolylmethyl)amine (THPTA).

### 2.2 Click Reaction on Solid Support

1. Phosphate buffer (10×): 1.36 M NaCl, 27 mM KCl, 100 mM $Na_2HPO_4$, 18 mM $KH_2PO_4$ (pH 7.0).
2. Click-catalyst solution (*see* Subheading 2.1).
3. Azide of choice (1 M) in DMSO.

### 2.3 HPLC/MS of Clicked Oligonucleotides

1. Mobile phase (A1) for HPLC/MS of oligonucleotides: 10 mM triethylamine (TEA), 100 mM hexafluoroisopropanol (HFIP). All reagents have to be LC-MS grade.
2. Mobile phase (B): Acetonitrile (LC-MS grade).
3. HPLC column for oligonucleotides: (C18) 2.1 × 100 mm, 5 μm.

### 2.4 Digestion to Nucleosides

1. 5× S1 nuclease reaction buffer.
2. S1 nuclease (100 U/μl).
3. 10× alkaline phosphatase buffer.
4. Alkaline phosphatase (CIAP) (1 U/μl).
5. Snake venom phosphatase I (1 U/μl).

6. Benzonase nuclease (250 U/μl).
7. Ultracentrifugal filters with 3 K molecular weight cutoff (MWCO).

***2.5 HPLC/MS of Nucleosides***

1. Mobile phase (A2): 20 mM ammonium acetate, acidified with acetic acid (pH 5.4). Reagents have to be LC-MS grade.
2. Mobile phase (B): Acetonitrile (LC-MS grade).
3. HPLC column for nucleosides: (C18) 2.1 × 100 mm, 5 μm.

## 3 Methods

***3.1 Preparation of Click-Catalyst Solution***

1. 10 mg of water-free sodium ascorbate powder are dissolved in 500 μl $H_2O$ to obtain a 100 mM solution.
2. A 100 mM $CuSO_4$ solution in $H_2O$ must be prepared. This solution can be prepared in a large stock and can be stored for several months at room temperature.
3. The 100 mM THPTA solution is prepared by solvation of 43.4 mg THPTA in 1 ml $H_2O$. The solution can be used for several months, however should be stored at −20 °C.
4. The catalyst solution is prepared by mixing 70 μl $H_2O$ with 25 μl 100 mM sodium ascorbate, 1 μl 100 mM $CuSO_4$, and 4 μl 100 μM THPTA (*see* **Notes 1** and **2**).

***3.2 Click Reaction on Solid Support (See Fig. 1)***

1. 75 μl 1 M solution of the desired azide in DMSO is added to the solid support (*see* **Note 3**).
2. 165 μl $H_2O$, 30 μl catalyst solution, and 30 μl (10×) phosphate buffer (pH 7) are added, and the suspension is vortexed.
3. The suspension is incubated for 60 min at 37 °C and 1400 rpm on a thermocycler. To avoid sedimentation of the solid support, the suspension is additionally mixed by pipetting up and down every 10 min.
4. After 1 h, the coupling solution is removed from the beads. Therefore, the suspension is centrifuged, and then the liquid is carefully removed from the solid support. To wash away

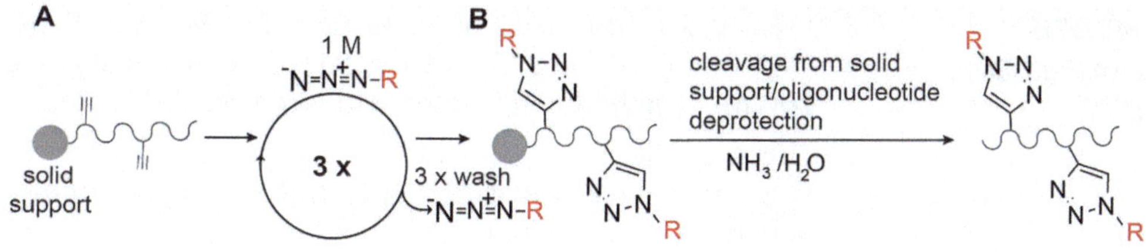


**Fig. 1** (**a**) EdU-containing DNA is synthesized on a solid support, whereon it is clicked with the desired azide. (**b**) The click-modified DNA is cleaved from the solid support and prepared for subsequent HPLC purification

excessive azide, the solid support is washed 3× with 500 μl acetonitrile and 500 μl water alternatingly.

5. Repeat **steps 1–4** twice.
6. DNA is cleaved of the solid support by treatment with ammonium hydroxide for 1 h. Removal of the protection groups from heterocyclic bases and phosphates is performed, and the aqueous ammonium hydroxide solution is evaporated. Oligonucleotides are purified over HPLC (*see* **Note 4**).

#### 3.3 HPLC/MS of Clicked Oligonucleotides (See Note 5)

1. The mobile phase (A1), a 10 mM trimethylamine (TEA) solution with 100 mM hexafluoroisopropanol (HFIP) in $H_2O$, is prepared.
2. A gradient of 0–30% acetonitrile (B) in 20 min with a flow of 0.5 ml/min is applied.
3. Measurement was done using ultrascan in the negative mode with the following settings: nebulizer, 50 psi; dry gas, 10 ml/min; dry temperature, 265 °C; SPS, 1000 *m/z*; ICC, 70,000; and scan, 500–1500 *m/z*.

#### 3.4 Enzymatic Digestion to Nucleotides

1. 300 pmol click-modified DNA are solubilized in 27 μl $H_2O$. 3 μl 5× S1 nuclease reaction buffer and 0.5 μl S1 nuclease (100 U/μl) are added to the solution.
2. Incubate at 800 rpm and 37 °C for 60 min on a thermocycler.
3. 3.5 μl phosphatase buffer as well as 0.5 μl alkaline phosphatase (CIAP) (1 U/μl), 0.5 μl snake venom phosphatase I (1 U/μl), and 0.5 μl Benzonase nuclease (250 U/μl) are added.
4. The solution is incubated for 120 min at 37 °C and 800 rpm on a thermocycler.
5. To stop the nuclease digestion, the sample is heated to 95 °C for 3 min.
6. Centrifuge for 3 min at 12,000 × *g* (*see* **Note 6**).
7. 30 μl of the sample are loaded onto an ultracentrifugal filter column with 3 k cutoff (*see* **Note 7**).
8. Centrifuge for 10 min at 15,000 × *g*.
9. Collect the flow-through (*see* **Note 8**).

#### 3.5 HPLC/MS of Nucleosides (See Note 9)

1. Prepare 20 mM ammonium acetate (LC-MS grade) pH 4.5 (A2) by solubilizing 154.16 mg $NH_4OAc$ in a final volume of 100 ml. Adjust the pH to 4.5 with acetic acid.
2. Use the following gradient: 0% acetonitrile (B) for 3 min, followed by a gradient from 0% to 20% B in 17 min, and a flow rate of 0.3 ml/min.
3. For mass spectrometry, the samples are measured with the standard enhanced mode (alternating) with the following

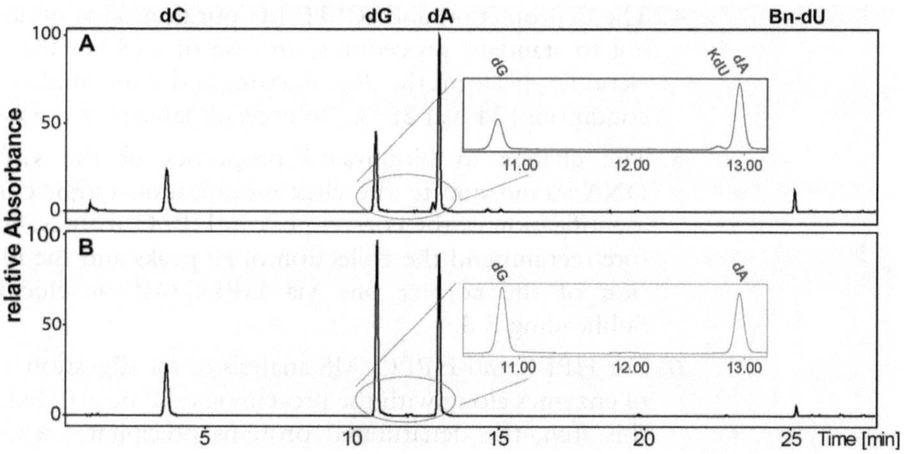


**Fig. 2** HPLC/MS-chromatogram of nucleoside digests. Oligonucleotides are clicked with benzyl azide. (**a**) Shown is the digestion of a randomized DNA pool that is clicked after the solid-phase synthesis and cleavage from the solid support. It can be seen that a fraction of EdU is partially oxidized to KdU and therefore is not accessible for the click reaction. (**b**) A determined oligonucleotide sequence is synthesized and clicked before it is cleaved from the solid support. The undesirable KdU cannot be observed for this procedure

settings: nebulizer, 40 psi; dry gas, 9 l/min; dry temperature, 365 °C; SPS, 400 *m/z*; $ICC_{neg}$, 50,000; $ICC_{pos}$, 100,000; and scan, 120–800 *m/z*.

4. Assign the detected masses to the four peaks in the UV trace. The chromatogram should only show four peaks (*see* Fig. 2) (*see* **Note 10**).

## 4 Notes

1. The ascorbate solution should be prepared freshly for each preparation of the click-catalyst solution. The resulting catalyst solution should be colorless. If the solution is of greenish/blueish color, some noncomplexed Cu(I) is still present, and the solution should not be used for click reaction.
2. In order to allow reduction of $Cu^{II}$ to $Cu^{I}$, let the catalyst solution rest for at least 10 min at RT before adding it to the solid support.
3. The reaction can be performed in a 1.5 ml reaction tube. Protective gas is not necessary for this step. Because the azide concentration for the click reaction on solid support is much higher than for click reactions in solution, it can happen that some azides will precipitate under the described conditions. This should not have a negative effect on the efficiency of the reaction if at least some azide is in solution.

4. The deprotection and RP-HPLC purification is done according to standard procedures. In case of Cy5-labeled oligonucleotides, perform the deprotection under the mildest possible conditions (24 h at 25 °C) to prevent damage of the Cy5 label.
5. The change in biophysical properties of the synthesized DNA strand due to the click modification might complicate identification of the correct peak in HPLC analysis. We therefore recommend the collection of all peaks and the identification of the correct one via HPLC/MS as elucidated in Subheading 3.3.
6. For HPLC and HPLC/MS analysis of the digestion, injection of enzymes along with the proteins should be avoided. During this step, the denaturated proteins precipitate, whereas the nucleosides stay in solution.
7. This step is to ensure further removal of leftover proteins. Please keep in mind that in contrast to normal uses of molecular weight cutoff columns, the flow-through is needed for the next steps, while the retained proteins are discarded.
8. The resulting volume of the nucleoside digestion procedure should be around 30 μl. If the volume is much higher than this after the Amicon purification, the volume should be reduced in a Speedvac before HPLC analysis is started (Subheading 3.5).
9. In theory, it is also possible to analyze the nucleoside digest by HPLC instead of HPLC/MS. In that case, a highly sensitive UV detector is needed as especially EdU and the click-modified nucleoside normally show a very low UV absorption [6]. Obviously, the digestion and then injection of higher amounts of DNA/nucleosides would also be an option to circumvent this.
10. The order of elution and the corresponding retention times should be dC (3.6 min), dG (10.7 min), and dA (14.9 min). Non-clicked EdU would appear at 11.9 min. The retention time of the click-modified nucleoside depends on the nature of the azide used for the click reaction.

## Acknowledgment

This work has been made possible by funds from the German Research Council (DFG) to G.M. (MA3442/4-2).

### References

1. El-Sagheer AH, Brown T (2010) Click chemistry with DNA. Chem Soc Rev 39:1388–1405
2. Gierlich J, Gutsmiedl K, Gramlich PM, Schmidt A, Burley GA, Carell T (2007) Synthesis of highly modified DNA by a combination of PCR with alkyne-bearing triphosphates and click chemistry. Chemistry 13:9486–9494

3. Tolle F, Brandle GM, Matzner D, Mayer G (2015) A versatile approach towards nucleobase-modified aptamers. Angew Chem Int Ed Engl 54:10971–10974
4. Suzuki T, Ota Y, Ri M, Bando M, Gotoh A, Itoh Y et al (2012) Rapid discovery of highly potent and selective inhibitors of histone deacetylase 8 using click chemistry to generate candidate libraries. J Med Chem 55:9562–9575
5. Ingale SA, Mei H, Leonard P, Seela F (2013) Ethynyl side chain hydration during synthesis and workup of "clickable" oligonucleotides: bypassing acetyl group formation by triisopropylsilyl protection. J Org Chem 78:11271–11282
6. Tolle F, Rosenthal M, Pfeiffer F, Mayer G (2016) Click reaction on solid phase enables high fidelity synthesis of nucleobase-modified DNA. Bioconjug Chem 27:500–503
7. Abel GR Jr, Calabrese ZA, Ayco J, Hein JE, Ye T (2016) Measuring and suppressing the oxidative damage to DNA during Cu(I)-catalyzed azide-alkyne cycloaddition. Bioconjug Chem 27:698–704
8. Besanceney-Webler C, Jiang H, Zheng T, Feng L, Soriano del Amo D, Wang W et al (2011) Increasing the efficacy of bioorthogonal click reactions for bioconjugation: a comparative study. Angew Chem Int Ed Engl 50:8051–8056

# Chapter 12

# Labeling Peptide Nucleic Acids with Indium-111

Igor G. Panyutin

## Abstract

Peptide nucleic acids (PNA) are widely used DNA mimics that bind sequence specifically to single- and double-stranded nucleic acids. Hence they are of interest in the design of gene-targeted radiotherapeutics that could deliver radiodamage to designated DNA and/or RNA sites. Here I describe a procedure for incorporation of gamma-emitting radionuclide $^{111}$In into PNA oligomers. Diethylenetriaminepentaacetic acid (DTPA) was conjugated to a lysine-containing mixed-base PNA. $^{111}$In-labeled PNAs were obtained by chelation of PNA-DTPA conjugates with $^{111}In^{3+}$ in an acidic aqueous solution.

**Key words** DTPA, $^{111}$In, Peptide nucleic acids, DNA damage, Gene targeting

## 1 Introduction

Various biophysical, biochemical, and biological studies utilizing synthetic analogues of nucleic acids would be impossible without conjugation to labels easily detected in various experimental settings. Introduction of radioactive isotopes into the arsenal of molecular biological research caused a revolution in cellular biology several decades ago elucidating such intimate processes inside cells such as replication and recombination. Currently, due to safety issues, fluorescent labels have been mostly replaced radiolabels in biomedical research. However, fluorescent labels are usually large moieties that can affect the outcome of experiments. Furthermore, detection of radioactive labeling is orders of magnitude more sensitive than fluorescent labeling. For this reason, radiolabeled oligonucleotides are widely applied and have become indispensable for in vivo imaging studies. Thus, radiolabeled oligonucleotides have been utilized in bio-distribution studies [1], gene and antisense radiotherapy [2–4], and for various imaging techniques [5]. Unfortunately, methods developed for radiolabeling natural nucleic acids using radioisotopes of their naturally occurring elements, and enzymes that work on natural nucleic acids are not always appropriate to synthetic oligonucleotides.

Nathaniel Shank (ed.), *Non-Natural Nucleic Acids: Methods and Protocols*, Methods in Molecular Biology, vol. 1973,
https://doi.org/10.1007/978-1-4939-9216-4_12, © Springer Science+Business Media, LLC, part of Springer Nature 2019

In this procedure, I describe a simple method to label non-natural oligonucleotides with $^{111}$In, a clinically used radiometal. The protocol has been optimized for PNAs, which are currently widely used in biomedical research and have become a popular alternatives to natural oligonucleotides [6]. The labeling is based on $^{111}$In chelation to DTPA conjugated to a PNA oligomer via a primary amino group. DTPA was conjugated to free amines of PNA oligomers at the carboxy (the side-chain lysine amino group) and/or amino (the end-located backbone amino group) termini via the acylation reaction [7, 8]. This method can be easily extended to other synthetic oligonucleotides that contain or can be conjugated to linkers containing primary amines. The described procedure utilizes HPLC and gel electrophoresis techniques that are commonly available in most laboratories.

## 2 Materials

### *2.1 Reagents for Radiolabeling*

1. In-111 solution with a specific activity of approximately 45,000 Ci/mmol and concentration of ~700 mCi/ml in 0.05 M HCl.
2. 1.25 M $NaHCO_3$ buffer (pH 8.5).
3. 0.2 M sodium acetate and 0.02 M sodium citrate (SAC) buffer (pH 4.5).
4. Chelex 100 ion-exchange resin (Bio-Rad, USA) (*see* **Note 1**).
5. Anhydrous DMSO.
6. Diethylenetriaminepentaacetic acid dianhydride (DTPAA) (Sigma, St. Louis, MO).
7. G-50 Sephadex gel-filtration mini-columns (MicroSpin G50, GE HealthCare, USA, or similar).

### *2.2 DNA Oligonucleotide Purification and Analysis*

1. DNA oligonucleotide (TTGAGATTACACAGATAGAGA-TAA-CTAGATACTTACGAC-fluorescein) was synthesized on an automated synthesizer (in this case Applied Biosystems, Foster City, CA) using 3′-fluorescein CPG. The oligomers were further purified by HPLC and polyacrylamide gel electrophoresis (PAGE) [9]. Alternatively, DNA oligonucleotides could be purchased from any commercial source.
2. Novex precast 20% TBE or 15% TBE-urea gels.
3. XCell SureLock Mini-Cell Gel Electrophoresis System.
4. 89 mM Tris, 89 mM boric acid, and 2 mM EDTA (1× TBE) buffer.
5. HP 1050 (Agilent, CA, USA) or similar HPLC system.
6. 0.1 M triethylamine acetate (TEAA, pH 7.5).

7. Anhydrous acetonitrile (ACN).
8. Reverse phase PRP-1 Hamilton C18 column (150 × 4.6 mm).

### 2.3 PNA Oligonucleotide

1. PNA (H-TAGTTATCTCTATCT-Lys-$NH_2$) was purchased from Applied Biosystems (Waltham, MA, USA).
2. 0.1% trifluoroacetic acid (TFA).
3. Low-adhesion microcentrifuge tubes 0.5 ml.
4. UV-Vis spectrophotometer.

## 3 Methods

### 3.1 PNA Purification

1. Dissolve PNA sample in 0.1% TFA by mixing for 30 min at 60 °C.
2. Estimate PNA concentration using a UV-Vis spectrophotometer and Oligonucleotide Properties Calculator (http://biotools.nubic.northwestern.edu/OligoCalc.html).
3. Lyophilize 1 nmol aliquots of PNA sample under vacuum in low-adhesion microcentrifuge tubes.

### 3.2 Conjugation PNA with DTPA

1. Mix 2 μl of 1.25 M $NaHCO_3$ buffer (pH 8.5) and 2 μl of a fresh solution of DTPAA in anhydrous DMSO (~400 nmol DTPAA), and add to 1 nmol dried PNA in a 0.5 ml low-adhesion polypropylene microcentrifuge tube (*see* **Note 2**).
2. Incubate the mixture at room temperature (~22 °C) overnight (*see* **Note 3**).
3. Purify PNA-DTPA conjugate from unconjugated DTPA by reverse phase HPLC. Use PRP-1 C18 column at flow rate of 0.7 ml/min with a linear gradient of solvent A (0.1% TFA) in 100% ACN) and solvent B (0.1%TFA in 1% ACN) from 1/99 to 100/0 (v/v) for 45 min at 55 °C. Collect the peak at approximately 7.5 min retention time, and lyophilize the sample under vacuum (*see* **Note 4**).
4. Prepare a stock solution of PNA-DTPA conjugates in SAC buffer (pH 4.5) that was pretreated with Chelex 100 resin to remove multivalent metal cations (*see* **Note 5**).
5. Mix an aliquot of PNA-DTPA conjugates with 10 times molar excess of the complementary DNA oligonucleotide sequence in SAC buffer in total volume 10 μl. Incubate at 37 °C for 30 min. Analyze samples with Novex precast 20% TBE gels at 100 v for 30 min using XCell SureLock Mini-Cell Gel Electrophoresis System (*see* **Note 5**).

**PNA** H-TAGTTATCTCTATCT-Lys-NH2

**DNA** TTGAGATTACAC**AGATAGAGATAACTA**GATACTTACGAC-FLUO

**Fig. 1** Sequences of PNA and DNA oligonucleotides. PNA sequence is written from N- to C-terminus, DNA sequence is written from the 5′- to the 3′-end, and the PNA binding site is shown in bold letters. *Lys* Lysine

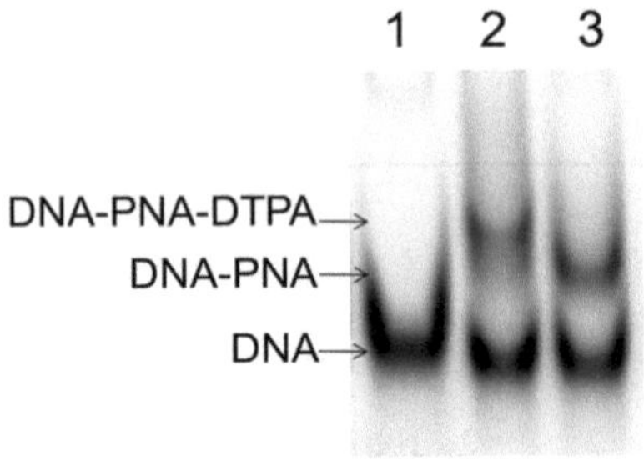


**Fig. 2** Analysis of PNA-DTPA conjugation by 20% non-denaturing PAGE at 4 °C. Lane 1, fluorescently labeled DNA oligonucleotide; lane 2, PNA-DTPA + DNA oligonucleotide; lane 3, PNA + DNA oligonucleotide

The extent of PNA conjugation with DTPA can be checked with gel shift analysis performed in 20% non-denaturing PAGE. Because PNA oligomers are usually weakly charged (PNA used here consists of a neutral backbone and carries two primary amines positively charged at physiological pH), it migrates very slowly during electrophoresis. Therefore, to detect a change in electrophoretic mobility caused by PNA conjugation with DTPA, PNA was annealed to a DNA oligonucleotide containing complementary sequence to increase the PNA gel mobility (Fig. 1).

Figure 2 shows the results of such a band-shift experiment. Binding PNA to DNA oligonucleotide resulted in the appearance of slow-migrating bands in lanes 2 and 3. The band corresponding to the duplex formed between DNA oligonucleotide and PNA-DTPA conjugate is shifted up in the gel (lane 2) relative to the band corresponding to the DNA/PNA duplex (lane 3). Conjugation yield was measured as a ratio of the intensity of the shifted band to the total intensity of the shifted and the original bands in lane 2, and was estimated to be >90%.

### 3.3 Labeling PNA-DTPA with $^{111}$In

1. Mix ~20 pmol (~1 mCi or 37 MBq) of fresh $^{111}$In in 0.05 N HCl (PerkinElmer, Boston, MA) and 10 pmol PNA-DTPA conjugate in total volume of 10 μl of SAC buffer (pH 4.5) that was demetalized by Chelex 100 resin (*see* **Note 6**).
2. Incubate the sample for 10 min at room temperature (further incubation did not result in a significant increase of $^{111}$In incorporation).

**Fig. 3** Analysis of PNA-DTPA labeling with $^{111}$In by 12% denaturing PAGE. The top band is PNA-DTPA-$^{111}$In conjugates, and the bottom band is DTPA-$^{111}$In

3. Quench the reaction with 1 μl 0.2 mM DTPA and bring the total volume to 20 μl with the SAC buffer.
4. Assess the extent of $^{111}$In incorporation into PNA-DTPA conjugate by 15% TBE-urea gels and XCell SureLock Mini-Cell Gel Electrophoresis System running at 150 V for 30 min.
5. Remove the remaining $^{111}$In-DTPA by column chromatography through the Sephadex G50 spin columns.

Binding of $^{111}$In to PNA-DTPA was monitored by denaturing PAGE (Fig. 3). The PNA-DTPA-[$^{111}$In] migrated in the gel very slow, giving rise to a smeared band on the top of the lane. This may reflect a smaller charge of the PNA molecule as compared to DNA molecule. The lower band corresponds to $^{111}$In bound to free DTPA that was added to quench the reaction. The initial molar ratio of $^{111}$In to PNA-DTPA was 2:1. Because approximately 40% of $^{111}$In was associated with the PNA-DTPA band, we estimated that ca. 80% of the PNA-DTPA conjugates were labeled with $^{111}$In.

The procedure described here was designed for induction of sequence-specific DNA breaks produced by decay of $^{111}$In and, therefore, was meant to obtain the highest specific activity of the PNA-DTPA-$^{111}$In conjugates. For this reason, a slight molar excess of $^{111}$In over the PNA-DTPA conjugates was used to ensure that all of the conjugates were labeled. However, for tracing and imaging purposes, an excess of PNA-DTPA conjugates can be used instead, thus insuring complete $^{111}$In incorporation into PNA-DTPA conjugates.

## 4 Notes

1. All radiolabeling procedures should be done according to radiation safety regulations. In-111 was purchased from PerkinElmer (USA) as a solution with a specific activity of approximately 45,000 Ci/mmol and concentration of ~700 mCi/ml in 0.05 M HCl. Thus, the concentration of $^{111}$In in the solution was ~15 μM (concentration and specific activity slightly vary from batch to batch and therefore are given as approximate). Multivalent metal ions should be removed from all buffers using Chelex 100 ion-exchange resin (Bio-Rad, USA).
2. DTPAA conjugation to PNA is performed at a DTPAA to PNA molar ratio of 400. Therefore, it is important to thoroughly purify the product from the excess of unreacted DTPAA, as otherwise it will interfere with $^{111}$In labeling. The concentration of PNA and the pH of the reaction were optimized to minimize the formation of the intermolecular cross-linking of two PNA molecules to one DTPAA. G-rich PNA may be hard to dissolve; one can use DMSO to dissolve PNA first and then adjust the volume with $NaHCO_3$ buffer.
3. Because of the very small volume of the sample, special care should be taken to prevent evaporation; a parafilm tape could be used to ensure the tight closure of the tube.
4. Alternatively, PNA-DTPA conjugate can be purified from excess DTPA by three consecutive separations on G-50 Sephadex gel-filtration mini-columns equilibrated with SAC buffer.
5. In our case, the target DNA was labeled with fluorescein, and the bands can be directly visualized and quantitated using a Fluorimager (model 595, Molecular Dynamics, Sunnyvale, CA, or similar). Alternatively, DNA oligonucleotide can be labeled with P-32 using a standard method (as, e.g., described in [10]), and then the bands can be visualized with autoradiography on an X-ray film or by a phosphorimaging device.
6. In-111 arrives in a very small volume, i.e., the whole required amount could be in a volume a little more than 1 μl. Therefore, PNA-DTPA conjugate can be added directly to the source vial of $^{111}$In. It is important to keep the volume of the mixture as low as possible to increase the rate and the extent of labeling.

## Acknowledgment

The study was supported by the Intramural Research Program of Clinical Center, NIH.

## References


1. Hnatowich DJ (1996) Pharmacokinetics of 99mTc-labeled oligonucleotides. Q J Nucl Med 40:202–208
2. Dewanjee MK, Haider N, Narula J (1999) Imaging with radiolabeled antisense oligonucleotides for the detection of intracellular messenger RNA and cardiovascular disease. J Nucl Cardiol 6:345–356
3. Iyer AK, He J (2011) Radiolabeled oligonucleotides for antisense imaging. Curr Org Synth 8:604–614
4. Panyutin IG, Neumann RD (2005) The potential for gene-targeted radiation therapy of cancers. Trends Biotechnol 23:492–496
5. Gambhir SS (2002) Molecular imaging of cancer with positron emission tomography. Nat Rev Cancer 2:683–693
6. Nielsen PE (2010) Sequence-selective targeting of duplex DNA by peptide nucleic acids. Curr Opin Mol Ther 12:184–191
7. Egholm M, Nielsen PE (1999) Peptide nucleic acids: protocols and applications. Horizon Scientific, Wymondham
8. Lewis MR, Jia F, Gallazzi F, Wang Y, Zhang J, Shenoy N, Lever SZ, Hannink M (2002) Radiometal-labeled peptide-PNA conjugates for targeting bcl-2 expression: preparation, characterization, and in vitro mRNA binding. Bioconjug Chem 13:1176–1180
9. Panyutin IG, Neumann RD (1994) Sequence-specific DNA double-strand breaks induced by triplex forming 125I labeled oligonucleotides. Nucleic Acids Res 22:4979–4982
10. He Y, Panyutin IG, Karavanov A, Demidov VV, Neumann RD (2004) Sequence-specific DNA strand cleavage by 111In-labeled peptide nucleic acids. Eur J Nucl Med Mol Imaging 31:837–845

# Chapter 13

# Site-Specific Labeling of DNA via PCR with an Expanded Genetic Alphabet

**Michael P. Ledbetter, Denis A. Malyshev, and Floyd E. Romesberg**

## Abstract

The polymerase chain reaction (PCR) is a universal and essential tool in molecular biology and biotechnology, but it is generally limited to the amplification of DNA with the four-letter genetic alphabet. Here, we describe PCR amplification with a six-letter alphabet that includes the two natural dA-dT and dG-dC base pairs and an unnatural base pair (UBP) formed between the synthetic nucleotides d**NaM** and d**5SICS** or d**TPT3** or analogs of these synthetic nucleotides modified with linkers that allow for the site-specific labeling of the amplified DNA with different functional groups. Under standard conditions, the six-letter DNA may be amplified with high efficiency and with greater than 99.9% fidelity. This allows for the efficient production of DNA site-specifically modified with different functionalities of interest for use in a wide range of applications.

**Key words** PCR, Unnatural base pair, Expanded genetic alphabet, Hydrophobic, Nucleic acid labeling

## 1 Introduction

The sequence-specific amplification of DNA via the polymerase chain reaction (PCR) has revolutionized virtually all of the biological sciences, enabling now routine and essential applications ranging from cloning to sequencing, as well as emerging applications in chemical biology, materials science, and diagnostics. However, many of these applications are limited by the fact that oligonucleotides may only be amplified if they are composed of the four natural deoxyribonucleotides (A, C, G, and T) or variants with a restricted range of modifications that do not perturb their recognition by DNA or RNA polymerases [1–14]. For example, while the evolution of oligonucleotides by in vitro directed evolution [15, 16] is extraordinarily powerful for the identification of oligonucleotides with interesting binding or catalytic properties, such properties are limited by the physicochemical diversity of the purines, pyrimidines, ribose, and phosphate moieties found in

Nathaniel Shank (ed.), *Non-Natural Nucleic Acids: Methods and Protocols*, Methods in Molecular Biology, vol. 1973,
https://doi.org/10.1007/978-1-4939-9216-4_13, © Springer Science+Business Media, LLC, part of Springer Nature 2019

natural DNA. Moreover, while DNA is receiving increased attention for nanomaterial fabrication [17–20], due to its sequence-controlled assembly, high monodispersity, and nm-scale length control, the applications of these potential materials are again limited by the physicochemical properties of DNA. Recent work has expanded the chemical space available to functional DNAs and RNAs by evolving polymerases to accept nucleotides with modified sugars [21] or nucleobases [22]; however, these methods are not site-specific. A more general approach that would allow for the site-specific inclusion of any functionality of interest would be made possible by the availability of an expanded genetic alphabet, in which synthetic nucleotides that form an unnatural base pair (UBP) could be incorporated at specific positions and also used to attach the functionality of interest.

Hydrogen-bonding (H-bonding) complementarity between the natural nucleobases, originally described by Watson and Crick [23] and Franklin [24], clearly plays a central role in the efficient and high-fidelity replication of natural DNA. Correspondingly, the first efforts to develop a replicable UBP, pioneered by Benner and coworkers, relied on the use of nucleotide analogs bearing nucleobases that pair via complementary but unique H-bonding patterns [25]. Early efforts along these lines were hindered by low-fidelity replication and chemical instability [26], but modifications have been found more recently that overcome these limitations, and DNA containing such unnatural pairs may be amplified by PCR [27]. However, H-bonds are not essential for nucleotide pairing [28], and we have worked toward developing UBPs whose selective pairing is driven by the hydrophobic effect and packing interactions, with the most notable being those formed by d**NaM** and d**5SICS** or d**TPT3** (Fig. 1) [29, 30]. This strategy has also been pursued by Hirao and coworkers who have developed shape complementary UBPs that are compatible with PCR amplification [31, 32]. Both the Benner and Hirao groups have reported the selection of aptamers containing UBPs [33–37].

DNA containing the d**NaM**-d**5SICS** UBP is both efficiently replicated [38] and transcribed [39] in vitro and, remarkably, despite its having no H-bonding or structural homology with a natural base pair, can be PCR amplified with efficiencies and fidelities approaching those of natural DNA and importantly with little to no sequence bias [40]. In 2014, we demonstrated that with import of the associated triphosphates, *E. coli* is able to replicate DNA containing the d**NaM**-d**5SICS** UBP, creating the first semi-synthetic organism (SSO) that stores increased genetic information. However, the d**NaM**-d**TPT3** UBP is replicated with higher fidelity, both in the SSO and in vitro, and we have also recently demonstrated that within the SSO, it may be transcribed into mRNAs and tRNAs with cognate unnatural codon-anticodon pairs that can efficiently direct the incorporation of noncanonical

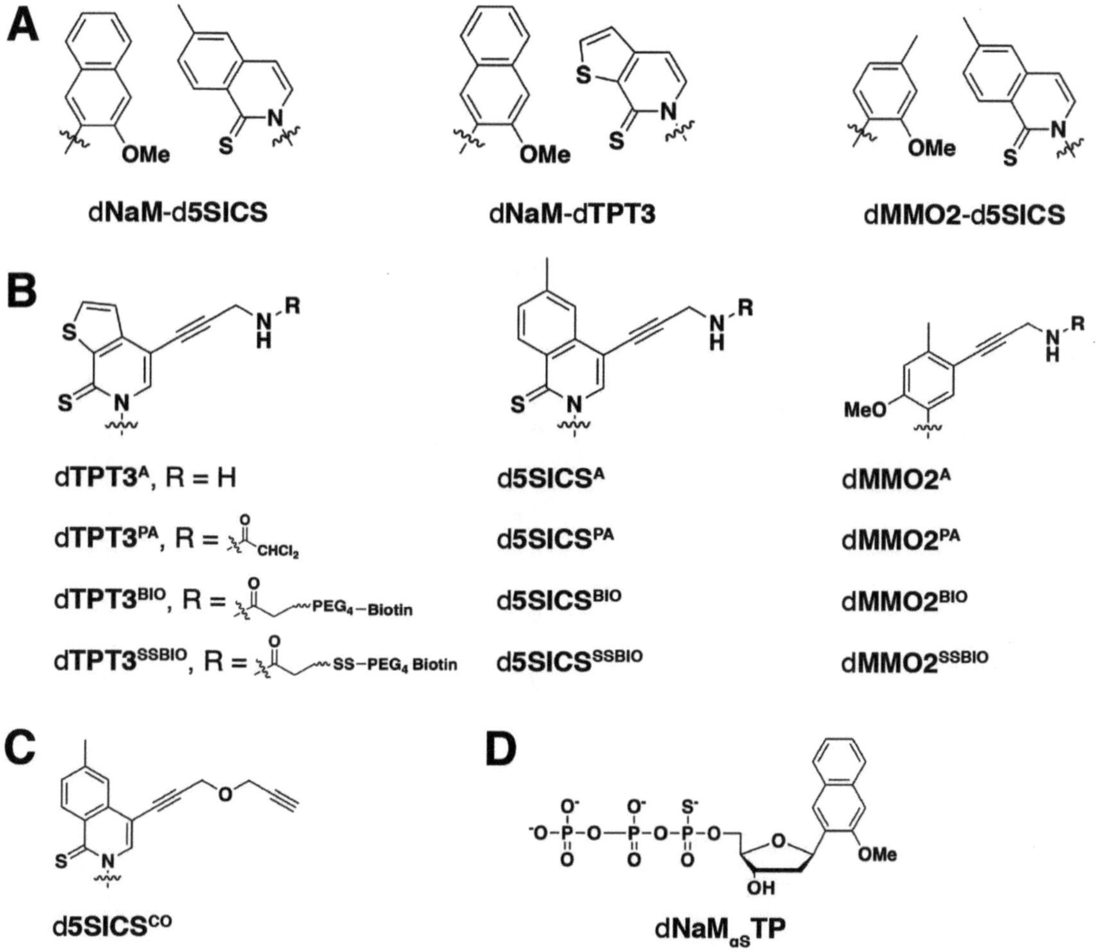


**Fig. 1** The structure of the **dNaM**-**d5SICS**, **dNaM**-**dTPT3**, and **dMMO2**-**d5SICS** UBPs (**a**) and their analogs with attached linkers for nucleic acid labeling using amine (**b**), alkyne (**c**), or thiotriphosphate (**d**) moieties

amino acids into proteins [41], creating the first SSO that stores and retrieves increased information. Moreover, analogs of d**NaM**, d**5SICS**, and d**TPT3** that are modified with linkers for the attachment of a functionality of interest have also been developed and used to produce DNA and RNA that are site-specifically modified with different functionalities of interest (Fig. 2) [42–45]. These linker-modified UBPs dramatically expand the potential applications of PCR and the resulting oligonucleotides. Here we describe basic PCR protocols for the amplification of DNA containing UBPs, including where one of the unnatural nucleotides bears a linker for functional group attachment (either before or after amplification). The protocols are generally focused on the d**NaM**-d**5SICS** UBP, because it has been most extensively explored in vitro, and its corresponding component phosphoramidites and

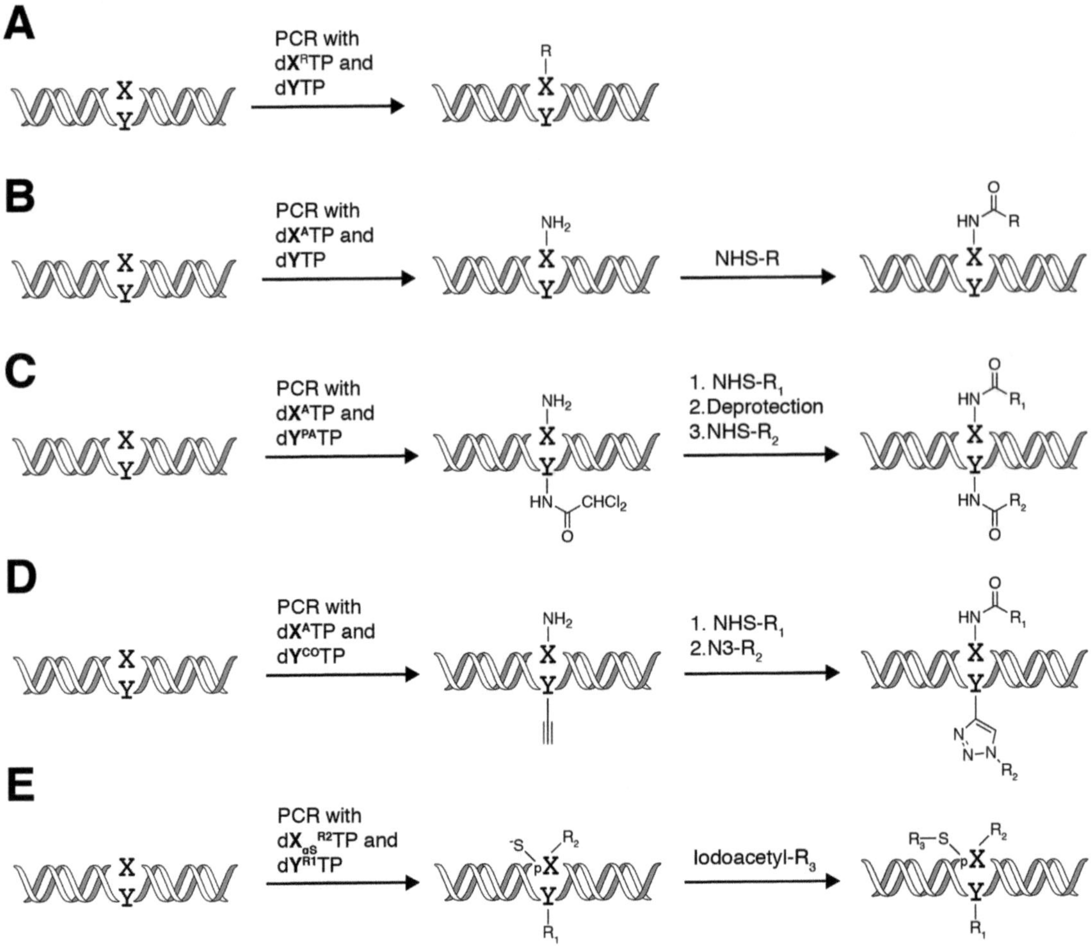


**Fig. 2** Strategies for site-specific dsDNA labeling using a UBP (with the unnatural nucleotides represented as **X** and **Y**): (**a**) direct incorporation of a functional group (R) into DNA; (**b**) post-PCR conjugation strategy; (**c**) double labeling using amino and protected unnatural nucleotides; (**e**) double labeling using amino and alkyne unnatural nucleotides; (**d**) combination of both pre- and post-amplification labeling

triphosphates (d**NaM**TP and d**5SICS**TP) are commercially available. However, the same protocols may be applied to replication of DNA containing d**NaM**-d**TPT3**, and any exceptions for optimal results are noted. Additionally, we provide two assays for the analysis of UBP retention during amplification.

## 2 Materials

### *2.1 Templates and Primers*

To amplify DNA containing a UBP, one of the nucleotides must first be incorporated into a template or primer using standard solid-phase chemical synthesis with the corresponding phosphoramidites (*see* **Note 1**). The syntheses of unnatural phosphoramidites are similar to natural phosphoramidite syntheses [38], and d**5SICS** and d**NaM** phosphoramidites are commercially available from Berry & Associates, Inc. (*see* **Note 2**). All phosphoramidites should be stored at −20 °C, protected from moisture, and dissolved in anhydrous acetonitrile immediately prior to use. For details of unnatural DNA synthesis, please refer to ref. 40, and *see* also **Note 3**. Templates may be synthesized directly if not prohibitively long; otherwise they may be produced from shorter oligonucleotides using standard methods such as overlap extension PCR [46] or megaprimer PCR [47].

If the primer region contains one or more UBPs, then the annealing temperature should be lowered slightly, relative to a dA-dT pair (~3 to 5 °C lower for a typical 18-mer primer). As always, when calculating the melting temperatures, primer and template concentration as well as $Mg^{2+}$ and triphosphate concentration should be taken into account. We have had success using both NEB's $T_m$ Calculator (http://tmcalculator.neb.com) and OligoAnalyzer 3.1 (https://www.idtdna.com/calc/analyzer) for primer design. The former tool possesses preset solution conditions for all NEB buffers, while the latter allows for manual description of solution conditions and predicted secondary structure information. If necessary, touchdown PCR [48] can be used to achieve the highest specificity.

### 2.2 Deoxyribonucleotides

The purity of the d**NaM**TP, d**5SICS**TP, and d**TPT3**TP is important for obtaining high efficiency and fidelity during amplification. Triphosphates may be prepared under standard phosphorylation conditions [49], starting from nucleosides synthesized as described [43, 44, 50] or obtained from Berry & Associates, Inc. (*see* **Note 2**). Alternatively, triphosphates can be obtained directly from MyChem LLC (*see* **Note 2**). Reproducibly, we have had the best results with triphosphates that have been purified by anion-exchange chromatography (DEAE-Sephadex). Phosphorylation must be carried out carefully, and the purity of the final products should be verified using C18 RP HPLC and MALDI-TOF or $^{31}P$ NMR methods before proceeding. As with natural dNTPs, unnatural triphosphates can be stored in dd$H_2O$ but should be aliquoted and stored at −20 °C.

### 2.3 $Mg^{2+}$

As with PCR of fully natural DNA, the concentration of $Mg^{2+}$ should be carefully optimized. For PCR amplification with *Taq* or One*Taq* enzymes, we recommend using 3 mM $Mg^{2+}$ with a concentration of 200/100 μM of natural/unnatural triphosphates, respectively (*see* **Note 4**). With increased exonuclease activity (i.e., when using Deep Vent, Vent, Phusion, etc.), we recommend using 6 mM $Mg^{2+}$ with a concentration of 700/100 μM of natural/

unnatural triphosphates, respectively. $Mg^{2+}$ concentration can be optimized in 1 mM increments if required.

### 2.4 Thermostable DNA Polymerase

DNA polymerases have likely evolved to possess optimal ratios of polymerase and 3′ → 5′ exonuclease activity based on the relative rates of natural correct pair and mispair synthesis and extension. As these relative rates are different for the natural and unnatural base pairs, especially for linker-modified UBPs, the optimum ratio of polymerase and exonuclease activity is expected to be somewhat different. Indeed, we have found that the efficiency and fidelity with which DNA containing a UBP may be optimized by altering this ratio in PCR reactions that contain varying relative concentrations of two polymerases, one exonuclease proficient and one exonuclease deficient [40, 51]. Conveniently, we found that for **dNaM-d5SICS** and **dNaM-dTPT3**, the commercially available One*Taq*, a mixture of exonuclease proficient Deep Vent and exonuclease deficient *Taq*, is at least nearly optimal. While further optimization is likely possible with different ratios of polymerases, the reliability of a commercially available product justified its detailed exploration. We found that One*Taq* amplifies DNA containing a UBP with high efficiency and fidelity (here defined as the retention of UBP per doubling) and, most importantly, amplifies the DNA with limited sequence biases [52]. Such unbiased amplification is critically important for any application requiring the amplification of randomly selected sequences, for example, during in vitro evolution experiments, where only a few copies of functional sequence must be efficiently amplified in the presence of a great excess of nonfunctional members.

One*Taq* is less optimal for the amplification of DNA containing linker-modified UBPs. Fidelity may be easily maximized in these cases by adding a small amount of Deep Vent(exo+) and increasing the concentration of dNTPs (*see* Subheading 3.6).

### 2.5 SYBR Green (For qPCR Only)

Although not required for PCR amplification, we routinely add SYBR Green I nucleic acid gel stain at 1× concentration and run qPCR on the CFX Connect system. This allows for direct monitoring of the amplification and facilitates optimization. Importantly, when working with DNA libraries, e.g., for in vitro selection, qPCR prevents overamplification and complications associated with correct duplex reannealing (i.e., if a library of dsDNA is denatured, reannealing will not be sequence-specific).

### 2.6 DNA Size Analysis and Purification

We routinely use agarose or native polyacrylamide gels for the analysis of PCR amplicons. The UBP does not significantly affect the gel migration of dsDNA; thus the same guidelines for selection of gel percentage should be followed as for natural dsDNA. For example, for short DNA fragments 100–200 bp in length, we recommend 2–4% agarose or 6% polyacrylamide gels. For visualization, we recommend SYBR Gold nucleic acid stain from Invitrogen. It allows detection of as little as 0.1 ng of product. Ethidium bromide can successfully detect as little as 10 ng of dsDNA. Radioactive labeling of DNA with α-[$^{32}$P]dATP is our method of choice when detection of subpicogram quantities of dsDNA or ssDNA is required.

If a given PCR generates a single amplicon, any standard PCR purification kit is sufficient for DNA purification. While multiple commercial kits can be used, we routinely employ the DNA Clean & Concentrator-5, because it allows for concentration of DNA in as little as 5 μL of eluant. Given the cost of working with UBPs, we generally aim to work at the smallest scale possible. If a given PCR generates multiple amplicons, agarose or polyacrylamide gel purification can be employed to isolate the desired amplicon. Again multiple commercial gel purification kits can be used, but we routinely employ the Zymoclean Gel DNA Recovery Kit.

For high concentrations of the purified DNA product (above 10 ng/μL), quantification via UV-vis spectrophotometry is sufficient. For lower concentrations, we commonly employ the Qubit dsDNA HS Assay, which allows for accurate quantification of dsDNA down to 10 pg/μL.

## 3 Methods

### 3.1 Setting Up PCR

As an example that should be sufficient as a starting point for the optimization of any given PCR, we describe the One*Taq* PCR amplification of the 149-bp DNA template shown below (primer regions underlined):

*Template (sequence of **dTPT3** strand is shown)*

5′-dCACACAGGAAACAGCTATGACCCGGGTTATTA
CATGCGCTGGCACTTGCCCGTA
CGGCGGTTGCACT**TPT3**GTGATGGG-
GACCGGCTTCTTGG AGCCCATGGTATATCTCCTTCT-
TAAAGTTAACCCTATAGTGAGTC GTATTAATTTC.

*Primers for PCR and biotin shift PCR*

Primer1: 5′-dCACACAGGAAACAGCTATGAC (21mer).

Primer2: 5′-dGAAATTAATACGACTCACTATAGG (24mer).

*Primers for sequencing*

Primer1-polyT: 5′-d$T_{59}$CACACAGGAAACAGCTATGAC (80mer).

Primer2-polyT: 5′-d$T_{61}$GAAATTAATACGACTCACTATAGG (85mer).

The following components are mixed together:

| | Stock concentration | Final concentration | Volume, μL |
|---|---|---|---|
| Deionized water[a] | – | – | 25.3 |
| One*Taq* Standard Reaction Buffer[b] | 5× | 1× | 10.0 |
| $MgSO_4$[c] | 100 mM | 3 mM | 0.6 |
| dNTPs (N = A, C, G, T) | 25 mM each | 0.2 mM | 0.4 |
| SYBR Green I[d] | 10× | 0.5× | 2.5 |
| **dTPT3**TP[e] | 2 mM | 0.1 mM | 2.5 |
| **dNaM**TP | 2 mM | 0.1 mM | 2.5 |
| Primers | 10 μM each | 1 μM | 5.0 |
| DNA template | 1 ng/μL | 0.02 ng/μL | 1.0 |
| One*Taq*[f] | 5 units/μL | 0.02 units/μL | 0.2 |

[a]Final volume of PCR is 50 μL. We routinely use water from a Millipore Milli-Q Advantage A10 system

[b]One*Taq* reaction buffer: 20 mM Tris–HCl, 22 mM KCl, 1.8 mM $MgCl_2$, 0.05% Tween-20, 22 mM $NH_4Cl$, 0.06% IGEPAL® CA-630, pH 8.9 at 25 °C

[c]Since 1× One*Taq* reaction buffer contains 1.8 mM $Mg^{2+}$, supplementation with 1.2 mM $MgSO_4$ is required to achieve 3 mM final concentration

[d]Required only for qPCR monitoring. Already diluted 1000 times from 10,000× stock solution in deionized water. Allow full thawing of the tube, vortex vigorously, and spin down before use

[e]*See* **Note 5**

[f]We typically run more than one PCR and prepare a master-mix solution, so the addition of 0.2 μL of enzyme is not a problem. We suggest running at least one extra sample, a negative control without the DNA template to detect contamination. However, if only one PCR is run or if an even smaller volume is required, we recommend diluting the enzyme in 1× One*Taq* reaction buffer immediately before the addition

The order of reagent addition is not important with the exception of the enzyme, which should be added last. We also typically add the template immediately before adding polymerase to prevent contamination of stock solutions. The use of aliquoted reagents usually improves the efficiency and reproducibility of PCR amplification and also decreases the risk of contamination. We have found that mixing all reagents on ice does not affect the efficiency or fidelity of PCR amplification; however, this may be beneficial in more complicated or challenging cases, for example, when a very small amount of a large DNA template is utilized, and thus incubation with a proofreading enzyme at room temperature may result in degradation.

Due to the presence of the Tween-20 detergent in the One*Taq* reaction buffer, vortexing or excessive mixing of the PCR reaction should be avoided. We recommend gentle mixing by pipette after the addition of each reagent and/or gentle inversion of the reaction tube followed by brief centrifugation.

### 3.2 PCR Cycling

*Initial denaturation*. Initial denaturation is only required when amplifying genomic, plasmid, or long dsDNA (>1000 bp). For shorter fragments of DNA, as in our example (149 bp), the initial denaturation does not affect PCR efficiency or fidelity and thus is not required.

*Denaturation step*. Both of the enzymes in the One*Taq* mixture (*Taq* and Deep Vent) have high thermal stability, so denaturation is usually performed at 96 °C for 10 s.

*Annealing step*. For the given 149 bp template and primer pair, PCR with DNA containing the UBP works with a wide range of annealing temperatures (48–64 °C). Using NEB's $T_m$ Calculator, we design primers with $T_m$ values around 60 °C in OneTaq Standard Reaction Buffer (*see* Subheading 2.1). Primers with melting temperatures below 40 °C or above 70 °C should be avoided. Using a 10 s annealing step at the $T_m$ Calculator's recommended annealing temperature typically results in specific production of the desired amplicon. For longer templates and more complicated or challenging amplifications, the annealing temperature should be experimentally optimized (a gradient option is very helpful in these cases). PCR specificity can be also improved by using touchdown PCR [48].

*Extension step*. We generally use the extension temperature recommended by the enzyme manufacturer (68 °C for One*Taq*). Though d**NaM**-d**TPT3** is the most efficiently replicated UBP reported to date, it is still thought to be extended less efficiently than its natural counterparts [53]. Extension times should be optimized for a given sequence, and we have typically found 1–4 min to be optimal.

*Number of cycles*. The efficiency of One*Taq* PCR [54] amplification is usually above 95% relative to the same sequence with a dA-dT pair at the position occupied by the UBP [40]. Thus, as long as PCR amplification remains in the exponential phase (i.e., a fluorescence plateau is not observed by qPCR), DNA is amplified approximately 1000-fold every 10 PCR cycles. If enough PCR cycles are run, then a typical yield is 0.2–1.0 μg of DNA from 50 μL PCR sample. With our example, starting with 1 ng of DNA, 12 cycles are sufficient to generate 0.5–0.6 μg of DNA after purification without the production of side products.

### 3.3 DNA Purification

Before purification, PCR products are typically visualized on 6% native polyacrylamide with SYBR Gold; however, a more traditional agarose gel with ethidium bromide also works as well. In this manner the presence and approximate purity of the product may be confirmed. No changes to the standard PAGE technique are required [55]. For SYBR Gold staining, refer to the manufacturer's website (https://www.thermofisher.com/order/catalog/product/S11494).

In cases such as the example provided, amplification specificity is usually high, and only the desired band is typically detected on the gel. Thus, DNA may be purified using the DNA Clean & Concentrator-5 according to the protocol available from the manufacturer's website (http://www.zymoresearch.com/) with 5 volumes (250 μL) of DNA binding buffer for the binding step and 20 μL of water for the elution step. Ensure that the pH of water used is above 5.0.

DNA concentration may be quantified by the Qubit dsDNA HS Assay and for the given example is usually 10–15 ng/μL. UV measurement on a NanoDrop instrument tends to overestimate the concentration of DNA by a factor of 20–50% (or even more for low concentrations of DNA).

### 3.4 Site-Specific Labeling of DNA Using Linker-Modified UBPs

For site-specific labeling of DNA, we synthesized and evaluated a variety of d**NaM**, d**5SICS**, and d**TPT3** analogs with linkers, which may be used to attach different functional groups, either before amplification or after [43, 44, 56] (Fig. 1). Additionally, many of these corresponding ribonucleotides are compatible with T7 RNA polymerase-mediated transcription, and they have been used to site-specifically label and study RNAs of interest [42]. The linker-modified unnatural deoxy and ribonucleoside triphosphates are not yet commercially available; however they may be easily synthesized [39, 43, 44, 56].

Two general approaches for DNA or RNA labeling may be employed, direct incorporation of an unnatural linker-derivatized triphosphate already coupled to the desired label (pre-amplification labeling, Fig. 2a) or modification of an unnatural linker-derivatized nucleotide already incorporated into an amplified product (post-amplification labeling, Fig. 2b). If post-amplification labeling with two different functional groups is desired, the derivatization of one nucleotide with a removable protecting group may be employed (Figs. 1b and 2c). Alternatively, linkers with orthogonal reactive centers such as alkynes or azides for Click chemistry may be employed (Figs. 1c and 2d). Incorporation of α-phosphorothioate within an unnatural triphosphate (i.e., d**5SICS**$_{\alpha S}$TP, Figs. 1d and 2e) also provides a convenient route to an additional site-specific labeling option [44]. These strategies are not mutually exclusive and, in fact, may be employed in tandem for generality or convenience, allowing for the site-specific modification of DNA with up to three different groups (Fig. 2e). For many applications it may be useful to remove part of an attached group, for example, when the group is an affinity tag such as biotin, allowing for release of the captured DNA. In these cases, linkers including S-S cleavable linkers may be used (Fig. 1b).

For pre-amplification labeling, the unnatural triphosphates may be modified using standard methodologies. However, the efficient and high-fidelity amplification of DNA using the resulting triphosphates requires case-by-case consideration. While kinetic data suggests that the replication of DNA containing linker-modified unnatural nucleotides is typically less efficient than that containing their unmodified unnatural counterparts, the same model template described in Subheading 3.1 possessing linker-modified nucleotides can typically be amplified by PCR with fidelities in excess of 99% per doubling. If necessary, extension time, exonuclease-positive polymerase ratios, and magnesium and triphosphate conditions may be optimized to improve replication fidelity. No other changes to purification, sequencing, or biotin shift analysis of labeled DNAs are necessary.

We have found that the incorporation of linker-modified nucleotides bearing free amines typically proceeds with lower fidelity, especially in the case of d**5SICS**$^{A}$TP. Thus, for post-amplification conjugation, we recommend carrying out PCR with the protected variants d**5SICS**$^{PA}$TP and/or d**MMO2**$^{PA}$TP. After amplification, the dichloroacetyl protecting group(s) may be easily removed by treatment with 0.1 M sodium hydroxide or 30% aqueous ammonia, for 1 and 3 h, respectively (*see* **Note 6**). The deprotected amine may then be coupled to NHS reagents, which are widely available with different functional groups. It is important to remember that functional groups linked to nucleotides embedded in the middle of a long stretch of DNA, duplex or single stranded, may be less accessible than those positioned at or near the end. We recommend varying linker lengths if a given functional group is inaccessible for chemistry or binding in its position within a duplex.

### 3.5 Determination of UBP Retention by Sequencing

#### 3.5.1 Sanger Sequencing

We routinely use Sanger sequencing (3730xl DNA analyzer) to quantify the level at which the UBP is retained in the DNA during amplification. The 5′-poly-dT-tailed sequencing primer strategy is used to improve the quality of the sequencing reads [57]. This allows for the use of the standard sequencing reagents instead of the specialized and more costly dye terminator reaction kits designed for short PCR products. Sequencing is accomplished via the following steps:

1. After quantification of the DNA concentration, dilute the sample to a concentration of 0.25 ng/μL.
2. In a well of a 96-well plate, mix 3 μL of the template (0.75 ng) in total with 1 μL of 3 μM sequencing primer (3 pmol).
3. Add 3.75 μL water and 2.25 μL cycle sequencing mix of the BigDye Terminator v3.1 Cycle Sequencing Kit (*see*

manufacturer's website http://www.ibt.lt/sc/files/BDTv3.1_Protocol_04337035.pdf for details), and mix well by pipetting.

4. Run 25 cycles of PCR (initial denaturation 98 °C, 1 min; 98 °C, 10 s; 60 °C, 15 s; 68 °C, 1.5 min) on a 9800 Fast Thermal Cycler.
5. Remove the residual dye terminators from the reaction with Agencourt CleanSEQ according to the manufacturer's protocol. Elute the product off the beads with deionized water, and sequence directly on a 3730 DNA Analyzer.
6. Analyze data (*see* below). In our laboratory, sequencing traces are collected using Applied Biosystems Data Collection software v3.0 and analyzed with the Applied Biosystems Sequencing Analysis v5.2 software.

### *3.5.2 Analysis of Sanger Sequencing Traces*

The retention of the UBP in an amplification product may be determined because in the absence of the unnatural triphosphates, the sequencing reaction abruptly terminates at an unnatural nucleotide in a template (Fig. 3, top). In contrast, fully natural amplicons produced through mutation of the UBP generate full-length sequencing traces that can be easily quantified using the chromatogram (Fig. 3, bottom). Thus, the extent of this read-through can be quantified by comparison of the peak amplitudes before and after

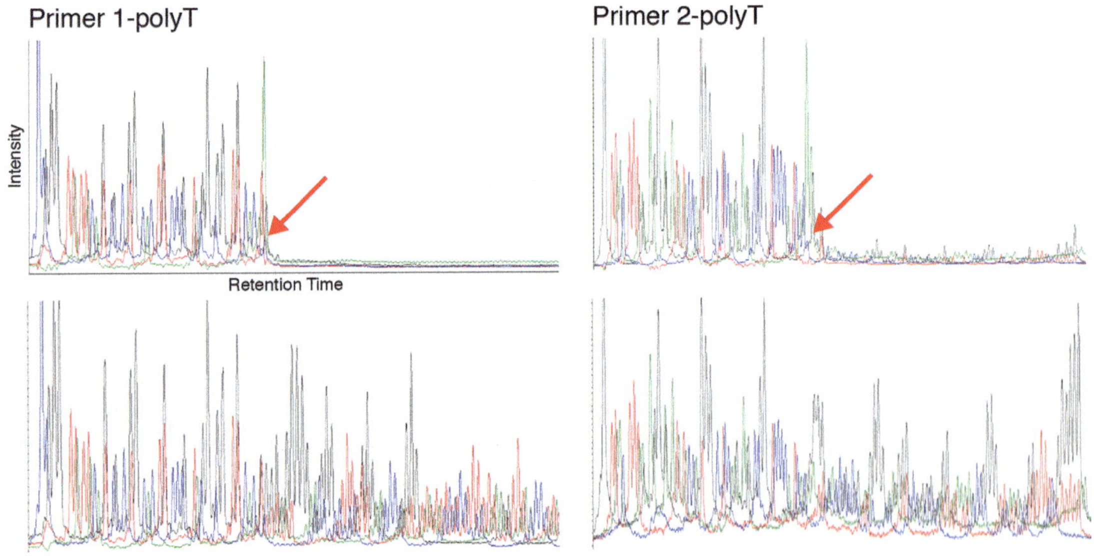


**Fig. 3** Sequencing profiles of high- (top) and low-fidelity (bottom) replication with DNA containing UBPs using primer 1-polyT or primer 2-polyT. The upper profiles correspond to sequencing of the PCR product resulting from amplification of DNA containing d**NaM**-d**5SICS** with both d**NaM**TP and d**5SICS**TP. The lower profiles correspond to the same reactions amplified with only d**NaM**TP so that the UBP is lost during amplification. The position of the unnatural nucleotide is indicated with a red arrow

an unnatural nucleotide. Raw sequencing traces are analyzed by first adjusting the start and stop points for the Applied Biosystems Sequencing Analysis v5.2 software and then determining the average signal intensity individually for each channel (A, C, G, and T) for peaks within the defined points (35–45 nucleotides in length). This is done separately for the parts of the sequencing trace before (section $L$) and after (section $R$) the unnatural nucleotide. Calibration experiments are carried out with control mixtures containing defined percentages of different templates with and without the UBP (the mixtures contained 50–100% of the template with the UBP; *see* [51] for a complete description). The $R/L$ ratio over the percentage of the natural template is plotted and fit by linear regression, and the corresponding equations for normalization of the $R/L$ ratio are used to account for read-through in the unamplified control templates as well as for the amplitude decay, which would otherwise result in the overestimation of fidelity. Finally, the retention of the UBP ($F$) is calculated as $1 - (R/L)_{norm}$ and the retention of the UBP per doubling (fidelity, $f$) is calculated as $F^{\frac{1}{\log_2 A}}$, where $A$ is an amplification and $\log_2 A$ is the number of doublings. Note that this assay (as well as the biotin shift assay described in the following section) does not account for the possibility that the UBP has switched strands during amplification; however, kinetic data suggests that this is an unlikely event [38, 43, 53].

### 3.6 Determination of UBP Retention by Biotin Shift PCR

#### 3.6.1 Biotin Shift PCR

While Sanger sequencing measures both UBP position and retention, use of biotin-modified UBPs enables the estimation of retention with fewer equipment requirements and higher sample throughput. We routinely estimate the UBP content of a given DNA through amplification with d**5SICS**TP and d**MMO2**$^{\mathbf{BIO}}$TP (Fig. 1) and subsequent PAGE analysis of amplicons incubated with streptavidin. As described above, linker modification alters polymerase recognition; therefore, PCR of DNA with d**5SICS**TP and d**MMO2**$^{\mathbf{BIO}}$TP (biotin shift PCR) is performed under the following reaction conditions:

| | Stock concentration | Final concentration | Volume, µL |
|---|---|---|---|
| Deionized water[a] | – | – | 5.4135 |
| One*Taq* Standard Reaction Buffer[b] | 5× | 1× | 3.0 |
| $MgSO_4$[c] | 100 mM | 4 mM | 0.33[d] |
| dNTPs (N = A, C, G, T) | 25 mM each | 0.4 mM | 0.6 |
| SYBR Green I[e] | 10× | 1× | 1.5 |
| d**5SICS**TP | 2 mM | 65 µM | 0.975 |
| d**MMO2**$^{\mathbf{BIO}}$TP | 2 mM | 65 µM | 0.975 |

(continued)

| | Stock concentration | Final concentration | Volume, μL |
|---|---|---|---|
| Primers | 25 μM each | 1 μM | 0.6 |
| DNA template | 0.1–10 nM | 0.01–1 nM | 1.5 |
| One*Taq* | 5 units/μL | 0.018 units/μL | 0.054 |
| Deep Vent[f] | 2 units/μL | 0.007 units/μL | 0.0525 |

[a]Final volume of PCR is 15 μL. We routinely use water from a Millipore Milli-Q Advantage A10 system

[b]One*Taq* reaction buffer: 20 mM Tris–HCl, 22 mM KCl, 1.8 mM $MgCl_2$, 0.05% Tween-20, 22 mM $NH_4Cl$, 0.06% IGEPAL® CA-630, pH 8.9 at 25 °C

[c]Since 1× One*Taq* reaction buffer contains 1.8 mM $Mg^{2+}$, supplementation with 2.2 mM $MgSO_4$ is required to achieve 4 mM final concentration

[d]This reaction is normally prepared as a master mix for several reactions. Thus, larger volumes are required for all reagents. If necessary, we recommend diluting reagents keep all pipetting volumes at 0.5 μL or greater

[e]SYBR Green I is required for monitoring reaction progress when performing biotin shift PCR (*see* below). Already diluted 1000 times from 10,000× stock solution in deionized water. Allow full thawing of the tube, vortex vigorously, and spin down before use

[f]While One*Taq* already contains Deep Vent DNA polymerase, additional exonuclease activity is required for proper amplification of DNA with d**MMO2**$^{\mathbf{BIO}}$TP. The additional Deep Vent fulfills this requirement

Cycling conditions are as described above: 96 °C. 10 s for denaturing, annealing temperature estimated using NEB's $T_m$ calculator and optimized experimentally if necessary, and a 68 °C, 4 min extension. All biotin shift PCRs are performed with a qPCR system (CFX Connect). Fluorescence is monitored at the end of each thermal cycle, and reactions are stopped after reaching the end of the exponential phase (typically 14–18 cycles, Fig. 4). Reaction monitoring is required as we have observed decreased biotin labeling in reactions cycled well past the exponential phase of amplification (presumably due to depletion of the UBPs late in the reaction and corresponding selection for fully natural amplicons).

#### *3.6.2 PAGE Analysis of DNA-Streptavidin Binding*

To assess biotin labeling of amplicons and correspondingly estimate template DNA UBP content, 1 μL of a biotin shift PCR is mixed with 5 μg of streptavidin (2.5 μL at 2 μg/μL), incubated at room temperature for 10 min, mixed with 1 μL 6× purple loading dye, and loaded onto a 6% native polyacrylamide gel. Given that this assay aims to resolve streptavidin-DNA species from amplicons without a UBP, we design biotin PCR primers such that amplicons are 50–200 bp long. This ensures that streptavidin-DNA species will migrate significantly slower than amplicons without a UBP in the gel (Fig. 4c). After electrophoresis, the polyacrylamide gel is stained by shaking the gel in 50 mL 1× TBE + 1× SYBR Gold for 20 min. Stained gels are imaged with a UV transilluminator on a Molecular Imager Gel Doc XR System with a 520DF30 62 mm

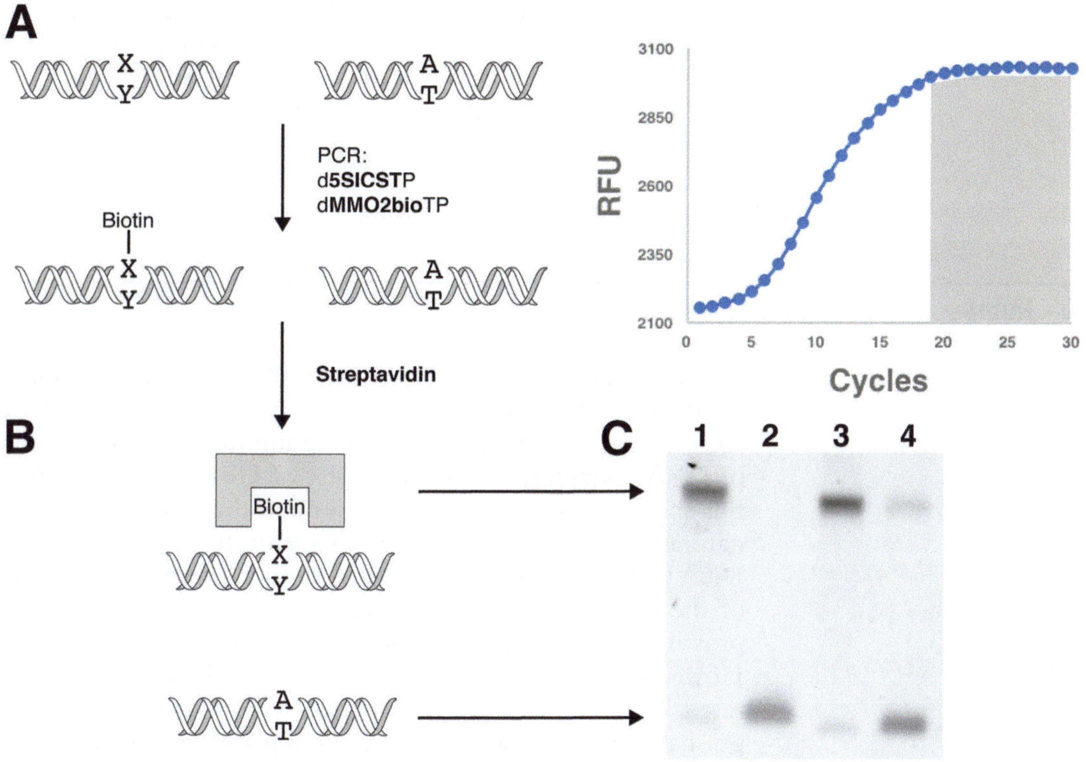


**Fig. 4** Workflow for biotin shift analysis of UBP containing DNA. (**a**) Populations of DNA produced through replication of UBP containing DNA by either PCR or in vivo replication can be analyzed by biotin shift PCR. Biotin shift PCRs should not be cycled past the exponential amplification phase, because the efficiency of biotin labeling decreases in the shaded region. (**b**) Incubation of biotin shift PCR products with streptavidin allows for PAGE resolution of biotinylated amplicons from fully natural amplicons. (**c**) A representative polyacrylamide gel is provided from the biotin shift analysis of a 63 base segment of UBP containing DNA from the following samples: 1. UBP-containing chemically synthesized oligonucleotide with streptavidin, 2. UBP containing chemically synthesized oligonucleotide without streptavidin, 3. A UBP containing plasmid that was well replicated in vivo with streptavidin, and 4. A UBP containing plasmid that was poorly replicated in vivo without streptavidin

filter. Image exposure time is limited to prevent saturation of the CCD for pixels in the bands of interests.

Gel images are analyzed using Quantity One software v4.6.9. Bands are identified manually and quantified with the "volume rectangle tool" and local background subtraction. Background subtracted volumes ($V$) for each band are then used to calculate a percent shift value ($S$) for every sample as $S = 100^* V_{SA\text{-}DNA}/(V_{SA\text{-}DNA} + V_{DNA})$. While the sequence context bias of d**NaM**-d**5SICS** is limited, significant sequence context bias has been observed for d**MMO2$^{BIO}$**-d**5SICS**. Biotin shift PCR amplification of DNA containing a UBP in a difficult to replicate sequence will not produce 100% biotin-labeled amplicons (*see* ref. 52 for a complete discussion). Therefore, the percent shift value for a given sample is only a qualitative measurement of UBP content. To get a quantitative measurement of UBP retention in a given sample, the percent shift value must be normalized to the percent shift value of a chemically synthesized oligonucleotide mimic of the desired template (Fig. 4c, lane 1). Thus, UBP retention ($R$) is reported as $R = 100 * S_{\mathrm{Sample}}/S_{\mathrm{Oligo}}$. Fidelity of UBP replication can be estimated as described above using sample retention values.

## 4 Notes

1. We have successfully demonstrated PCR with two **dNaM**-d**5SICS** UBPs separated by zero, one, or six natural base pairs. While we have not explicitly examined the amplification of DNA containing a higher density of UBPs, we expect that this should be straightforward as long as they are separated by natural base pairs. The major challenge in the amplification of DNA with multiple UBPs adjusted to each other is the analysis of the amplicon because Sanger sequencing does not always provide single nucleotide resolution at the position of the UBP. However, the template with two d**NaM**-d**5SICS** pairs separated by six nucleotides is amplified with a virtually indistinguishable efficiency and fidelity as a template with a single UBP, even with extension times of only 1 min [40].
2. Phosphoramidites and nucleosides of d**NaM** and d**5SICS** are available from Berry & Associates, Inc. (Dexter, MI). Phosphoramidite product numbers are as follow: d**NaM** CEP = BA 0343 and d**5SICS** CEP = BA 0344. Nucleoside product numbers are as follow: d**NaM** CEP = FC 8110 and d**5SICS** CEP = FC 8120. Triphosphates of d**NaM**, d**5SICS**, and d**MMO2** are available from MyChem, LLC (San Diego, CA). Triphosphate products numbers are as follow: d**NaM**TP = M-1013, d**5SICS**TP = M-1014, and d**MMO2**TP = M-1012.
3. We have found that chemically synthesized DNA templates exhibit some sequence biases that are not observed with templates that themselves were produced by amplification [40], which likely results from incomplete nucleotide deprotection after chemical synthesis. Thus, we recommend using Ultra-MILD phosphoramidites with phenoxyacetyl anhydride capping, using the manufacturer's suggested protocol. Moreover, to ensure full deprotection after chemical synthesis, conc. aqueous ammonia with overnight treatment at 37 °C is used. Pre-amplification of a chemically synthesized template is also an option to avoid the complications associated with chemically synthesized templates. For this, we typically run between six and eight PCR cycles under the conditions described above, purify the DNA, and then use it as a template for the next round of PCR.

4. According to a large body of kinetic data [38, 50], the rate-limiting step of the replication of DNA containing a UBP is not the incorporation of the unnatural nucleotide triphosphate itself, but rather incorporation of the next natural nucleotide after the unnatural one (which we refer to as UBP extension). Due to the relatively high $K_M$ value of the extension step, sufficient concentrations of dNTPs, not the unnatural triphosphates are crucial for efficient amplification. This is especially true for the case when only exonuclease proficient enzymes are utilized because of the competition between polymerization and $3' \rightarrow 5'$ exonuclease activities.
5. For PCR production of small amplicons (≤200 bp) d**TPT3**TP is added to the reaction at 0.1 mM final concentration. This allows for high-fidelity amplification of UBP containing DNA. However, at high concentrations, d**TPT3**TP can be misincorporated opposite natural nucleotides [53]. While this misinsertion is unlikely to yield mutated amplicons, due to poor extension of the mispairs, they can reduce the efficiency of amplification with long templates (≥200 bp). Thus, for the amplification of long amplicons, we recommend using d**TPT3**TP at a final concentration of 5 μM. If inefficient amplification or poor retention is observed with this condition, a minimum d**TPT3**TP concentration that provides sufficient amplification and retention should be determined experimentally.
6. Both 0.1 M sodium hydroxide and concentrated aqueous ammonia solutions will denature DNA. Thus, if dsDNA is required, it can easily be obtained via annealing in the appropriate buffer (e.g., 10 mM sodium phosphate, pH 7.5, 100 mM NaCl, and 0.1 mM EDTA) through heating to 85–90 °C and slow cooling to room temperature over 1 h. This approach, however, cannot be used for reannealing a library of DNA sequences.

## Acknowledgments

We thank Dr. Jodie Chin, Kirandeep Dhami, Henry Quach, Dr. Phillip Ordoukhanian, and Dr. Thomas Lavergne for helpful discussions and assistance with editing this manuscript.

### References

1. Battersby TR, Ang DN, Burgstaller P, Jurczyk SC, Bowser MT, Buchanan DD, Kennedy RT, Benner SA (1999) Quantitative analysis of receptors for adenosine nucleotides obtained via in vitro selection from a library incorporating a cationic nucleotide analog. J Am Chem Soc 121:9781–9789
2. Brakmann S, Lobermann S (2001) High-density labeling of DNA: preparation and characterization of the target material for single-

molecule sequencing. Angew Chem Int Ed 40:1427–1429

3. Gourlain T, Sidorov A, Mignet N, Thorpe SJ, Lee SE, Grasby JA, Williams DM (2001) Enhancing the catalytic repertoire of nucleic acids. II. Simultaneous incorporation of amino and imidazolyl functionalities by two modified triphosphates during PCR. Nucleic Acids Res 29:1898–1905
4. Hocek M, Fojta M (2011) Nucleobase modification as redox DNA labelling for electrochemical detection. Chem Soc Rev 40:5802–5814
5. Jager S, Rasched G, Kornreich-Leshem H, Engeser M, Thum O, Famulok M (2005) A versatile toolbox for variable DNA functionalization at high density. J Am Chem Soc 127:15071–15082
6. Kuwahara M, Nagashima J, Hasegawa M, Tamura T, Kitagata R, Hanawa K, Hososhima S, Kasamatsu T, Ozaki H, Sawai H (2006) Systematic characterization of 2-′-deoxynucleoside- 5′-triphosphate analogs as substrates for DNA polymerases by polymerase chain reaction and kinetic studies on enzymatic production of modified DNA. Nucleic Acids Res 34:5383–5394
7. Mehedi Masud M, Ozaki-Nakamura A, Kuwahara M, Ozaki H, Sawai H (2003) Modified DNA bearing 5(methoxycarbonylmethyl)-2′-deoxyuridine: preparation by PCR with thermophilic DNA polymerase and post-synthetic derivatization. Chembiochem 4:584–588
8. Perrin DM, Garestier T, Helene C (1999) Expanding the catalytic repertoire of nucleic acid catalysts: simultaneous incorporation of two modified deoxyribonucleoside triphosphates bearing ammonium and imidazolyl functionalities. Nucleosides Nucleotides 18:377–391
9. Held HA, Benner SA (2002) Challenging artificial genetic systems: thymidine analogs with 5-position sulfur functionality. Nucleic Acids Res 30:3857–3869
10. Hollenstein M, Hipolito CJ, Lam CH, Perrin DM (2009) A self-cleaving DNA enzyme modified with amines, guanidines and imidazoles operates independently of divalent metal cations (M-2). Nucleic Acids Res 37:1638–1649
11. Lee SE, Sidorov A, Gourlain T, Mignet N, Thorpe SJ, Brazier JA, Dickman MJ, Hornby DP, Grasby JA, Williams DM (2001) Enhancing the catalytic repertoire of nucleic acids: a systematic study of linker length and rigidity. Nucleic Acids Res 29:1565–1573
12. Roychowdhury A, Illangkoon H, Hendrickson CL, Benner SA (2004) 2′-deoxycytidines carrying amino and thiol functionality: synthesis and incorporation by Vent (exo-) polymerase. Org Lett 6:489–492
13. Sakthivel K, Barbas CF III (1998) Expanding the potential of DNA for binding and catalysis: highly functionalized dUTP derivatives that are substrates for thermostable DNA polymerases. Angew Chem Int Ed 37:2872–2875
14. Tasara T, Angerer B, Damond M, Winter H, Dorhofer S, Hubscher U, Amacker M (2003) Incorporation of reporter molecule-labeled nucleotides by DNA polymerases. II. High-density labeling of natural DNA. Nucleic Acids Res 31:2636–2646
15. Ellington AD, Szostak JW (1990) In vitro selection of RNA molecules that bind specific ligands. Nature 346:818–822
16. Tuerk C, Gold L (1990) Systematic evolution of ligands by exponential enrichment: RNA ligands to bacteriophage T4 DNA polymerase. Science 249:505–510
17. Atanasova P, Weitz RT, Gerstel P, Srot V, Kopold P, van Aken PA, Burghard M, Bill J (2009) DNA-templated synthesis of ZnO thin layers and nanowires. Nanotechnology 20:365302
18. Hoffmann RC, Atanasova P, Dilfer S, Bill J, Schneider JJ (2011) Templating of polycrystalline ZnO with DNA and its performance in field-effect transistors. Phys Status Solidi A 208:1983–1988
19. Lazareck AD, Cloutier SG, Kuo TF, Taft BJ, Kelley SO, Xu JM (2006) DNA-directed synthesis of zinc oxide nanowires on carbon nanotube tips. Nanotechnology 17:2661–2664
20. Fritzsche W, Bier FF (2008) DNA-based nanodevices: international symposium on DNA-based nanodevices. American Institute of Physics, Melville, NY
21. Chen T, Hongdilokkul N, Liu Z, Thirunavukarasu D, Romesberg FE (2016) The expanding world of DNA and RNA. Curr Opin Chem Biol 34:80–87
22. Chen T, Romesberg FE (2014) Directed polymerase evolution. FEBS Lett 588:219–229
23. Watson JD, Crick FH (1953) Molecular structure of nucleic acids; a structure for deoxyribose nucleic acid. Nature 171:737–738
24. Franklin RE, Gosling RG (1953) Evidence for 2-chain helix in crystalline structure of sodium deoxyribonucleate. Nature 172:156–157
25. Piccirilli JA, Krauch T, Moroney SE, Benner SA (1990) Enzymatic incorporation of a new base pair into DNA and RNA extends the genetic alphabet. Nature 343:33–37

26. Switzer CY, Moroney SE, Benner SA (1993) Enzymatic recognition of the base pair between isocytidine and isoguanosine. Biochemistry 32:10489–10496
27. Yang Z, Chen F, Chamberlin SG, Benner SA (2010) Expanded genetic alphabets in the polymerase chain reaction. Angew Chem Int Ed 49:177–180
28. Moran S, Ren RXF, Rumney S, Kool ET (1997) Difluorotoluene, a nonpolar isostere for thymine, codes specifically and efficiently for adenine in DNA replication. J Am Chem Soc 119:2056–2057
29. Betz K, Malyshev DA, Lavergne T, Welte W, Diederichs K, Dwyer TJ, Ordoukhanian P, Romesberg FE, Marx A (2012) KlenTaq polymerase replicates unnatural base pairs by inducing a Watson-Crick geometry. Nat Chem Biol 8:612–614
30. McMinn DL, Ogawa AK, Wu Y, Liu J, Schultz PG, Romesberg FE (1999) Efforts toward expansion of the genetic alphabet: DNA polymerase recognition of a highly stable, self-pairing hydrophobic base. J Am Chem Soc 121:11585–11586
31. Kimoto M, Kawai R, Mitsui T, Yokoyama S, Hirao I (2008) An unnatural base pair system for efficient PCR amplification and functionalization of DNA molecules. Nucleic Acids Res 37:e14
32. Yamashige R, Kimoto M, Takezawa Y, Sato A, Mitsui T, Yokoyama S, Hirao I (2011) Highly specific unnatural base pair systems as a third base pair for PCR amplification. Nucleic Acids Res 40:2793–2806
33. Zhang L, Yang Z, Sefah K, Bradley KM, Hoshika S, Kim MJ, Kim HJ, Zhu G, Jimenez E, Cansiz S, Teng IT, Champanhac C, McLendon C, Liu C, Zhang W, Gerloff DL, Huang Z, Tan W, Benner SA (2015) Evolution of functional six-nucleotide DNA. J Am Chem Soc 137:6734–6737
34. Biondi E, Lane JD, Das D, Dasgupta S, Piccirilli JA, Hoshika S, Bradley KM, Krantz BA, Benner SA (2016) Laboratory evolution of artificially expanded DNA gives redesignable aptamers that target the toxic form of anthrax protective antigen. Nucleic Acids Res 44:9565–9577
35. Zhang L, Yang Z, Le Trinh T, Teng IT, Wang S, Bradley KM, Hoshika S, Wu Q, Cansiz S, Rowold DJ, McLendon C, Kim MS, Wu Y, Cui C, Liu Y, Hou W, Stewart K, Wan S, Liu C, Benner SA, Tan W (2016) Aptamers against cells overexpressing glypican 3 from expanded genetic systems combined with cell engineering and laboratory evolution. Angew Chem Int Ed 55:12372–12375
36. Kimoto M, Yamashige R, Matsunaga K, Yokoyama S, Hirao I (2013) Generation of high-affinity DNA aptamers using an expanded genetic alphabet. Nat Biotechnol 31:453–457
37. Matsunaga KI, Kimoto M, Hirao I (2017) High-affinity DNA aptamer generation targeting von Willebrand Factor A1-domain by genetic alphabet expansion for systematic evolution of ligands by exponential enrichment using two types of libraries composed of five different bases. J Am Chem Soc 139:324–334
38. Seo YJ, Hwang GT, Ordoukhanian P, Romesberg FE (2009) Optimization of an unnatural base pair toward natural-like replication. J Am Chem Soc 131:3246–3252
39. Seo YJ, Matsuda S, Romesberg FE (2009) Transcription of an expanded genetic alphabet. J Am Chem Soc 131:5046–5047
40. Malyshev DA, Dhami K, Quach HT, Lavergne T, Ordoukhanian P, Torkamani A, Romesberg FE (2012) Efficient and sequence-independent replication of DNA containing a third base pair establishes a functional six-letter genetic alphabet. Proc Natl Acad Sci U S A 109:12005–12010
41. Zhang Y, Ptacin JL, Fischer EC, Aerni HR, Caffaro CE, San Jose K, Feldman AW, Turner CR, Romesberg FE (2017) A semi-synthetic organism that stores and retrieves increased genetic information. Nature 551 (7682):644–647
42. Lavergne T, Lamichhane R, Malyshev DA, Li Z, Li L, Sperling E, Williamson JR, Millar DP, Romesberg FE (2016) FRET characterization of complex conformational changes in a large 16S ribosomal RNA fragment site-specifically labeled using unnatural base pairs. ACS Chem Biol 11:1347–1353
43. Li L, Degardin M, Lavergne T, Malyshev DA, Dhami K, Ordoukhanian P, Romesberg FE (2014) Natural-like replication of an unnatural base pair for the expansion of the genetic alphabet and biotechnology applications. J Am Chem Soc 136:826–829
44. Seo YJ, Malyshev DA, Lavergne T, Ordoukhanian P, Romesberg FE (2011) Site-specific labeling of DNA and RNA using an efficiently replicated and transcribed class of unnatural base pairs. J Am Chem Soc 133:19878–19888
45. Uter NT, Perona JJ (2006) Active-site assembly in glutaminyl-tRNA synthetase by tRNA-mediated induced fit. Biochemistry 45:6858–6865

46. Heckman KL, Pease LR (2007) Gene splicing and mutagenesis by PCR-driven overlap extension. Nat Protoc 2:924–932
47. Burke E, Barik S (2003) Megaprimer PCR: application in mutagenesis and gene fusion. Methods Mol Biol 226:525–532
48. Korbie DJ, Mattick JS (2008) Touchdown PCR for increased specificity and sensitivity in PCR amplification. Nat Protoc 3:1452–1456
49. Ludwig J, Eckstein F (1989) Rapid and efficient synthesis of nucleoside 5′-0-(1-thiotriphosphates), 5′-triphosphates and 2′,3′-cyclophosphorothioates using 2-chloro-4H-1,3,2-benzodioxaphosphorin-4-one. J Org Chem 54:631
50. Lavergne T, Malyshev DA, Romesberg FE (2012) Major groove substituents and polymerase recognition of a class of predominantly hydrophobic unnatural base pairs. Chem Eur J 18:1231–1239
51. Malyshev DA, Seo YJ, Ordoukhanian P, Romesberg FE (2009) PCR with an expanded genetic alphabet. J Am Chem Soc 131:14620–14621
52. Malyshev DA, Dhami K, Lavergne T, Chen T, Dai N, Foster JM, Correa IR Jr, Romesberg FE (2014) A semi-synthetic organism with an expanded genetic alphabet. Nature 509:385–388
53. Morris SE, Feldman AW, Romesberg FE (2017) Synthetic biology parts for the storage of increased genetic information in cells. ACS Synth Biol 6(10):1834–1840. https://doi.org/10.1021/acssynbio.1027b00115
54. New England Biolabs. (2018) Polymerase Overview. PCR Reagents, Version 10.0, New England Biolabs, Ipswich, MA, p 2
55. Andrus A, Kuimelis RG (2001) Polyacrylamide gel electrophoresis (PAGE) of synthetic nucleic acids. Curr Prot Nucleic Acids Chem Chapter 10:Unit 10.4
56. Li Z, Lavergne T, Malyshev DA, Zimmermann J, Adhikary R, Dhami K, Ordoukhanian P, Sun Z, Xiang J, Romesberg FE (2013) Site-specifically arraying small molecules or proteins on DNA using an expanded genetic alphabet. Chem Eur J 19:14205–14209
57. Binladen J, Gilbert MT, Campos PF, Willerslev E (2007) 5′-tailed sequencing primers improve sequencing quality of PCR products. BioTechniques 42:174–176

# Chapter 14

# Flexible Nucleic Acids (FNAs) as Informational Molecules: Enzymatic Polymerization of fNTPs on DNA Templates and Nonenzymatic Oligomerization of RNA on FNA Templates

**Maryline Chemama and Christopher Switzer**

## Abstract

The methodology enabling enzymatic and nonenzymatic information transfer with FNAs is described. This methodology includes the chemical synthesis of fNTPs and fN phosphoramidites, in addition to protocols for the enzymatic and nonenzymatic transfer of information.

**Key words** FNA, RNA, DNA, Enzymatic, Nonenzymatic, Polymerization, Oligomerization, Template-directed, Prebiotic, Pre-RNA, Synthetic biology

## 1 Introduction

Polymerases are natural protein enzymes responsible for replication and transcription of DNA. These enzymes are template-dependent, catalyzing the synthesis of a product strand in the presence of a primer, nucleoside triphosphates, and divalent ions. The fidelity of enzyme-assisted template-directed polymerization is extremely high, with errors occurring once every $10^3$–$10^6$ coupling events dependent on the absence or presence of an error-correcting 3′-5-′-exonuclease domain [1]. Over the course of molecular evolution, nonenzymatic polymerization may have preceded enzymatic polymerization. Models of nonenzymatic polymerization parallel the enzymatic polymerization process except that nucleoside triphosphates are replaced by more reactive nucleotides, such as phosphoroimidazolides, to enable activation by divalent ions alone in the absence of a protein catalyst [2]. Nonenzymatic polymerization exhibits fidelity on the order of one error in $10^2$ coupling events for the incorporation of G [3] and as a rule requires the use of C-rich templates [4]. However, there have been recent inroads that suggest sequence-independent nonenzymatic replication may be attainable [5, 6].

Nathaniel Shank (ed.), *Non-Natural Nucleic Acids: Methods and Protocols*, Methods in Molecular Biology, vol. 1973,
https://doi.org/10.1007/978-1-4939-9216-4_14, © Springer Science+Business Media, LLC, part of Springer Nature 2019

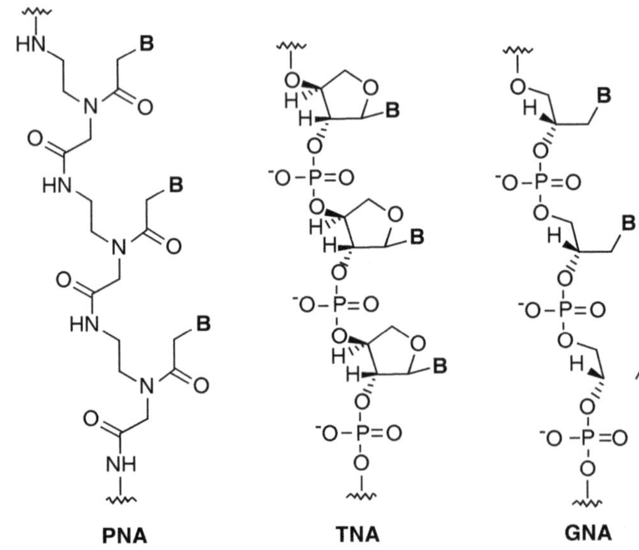


**Fig. 1** Examples of acyclic DNA analogs

The structures of DNA and RNA are exquisitely suited to their roles as informational molecules. Moreover, RNA incorporates molecular features that can be recreated in prebiotic simulation reactions [7]. Although an RNA-first origin of informational polymers on Earth has merits, other scenarios involving less chemically complex alternatives to DNA and RNA remain to be fully explored [7–10]. Three examples of simplified alternatives that have been the subject of past studies are depicted in Fig. 1: PNA [11, 12], TNA [13], and GNA [14]. PNA, a radical departure from RNA, is devoid of any stereogenic atoms and bears an amide backbone. TNA incorporates the seemingly unlikely carbohydrate backbone candidate L-threose and contains one less stereocenter than RNA. Finally, GNA is a 4′-nor variant of TNA bearing a single stereogenic carbon atom. All three of these polymers mirror to varying degrees the functional properties of DNA and RNA.

Perhaps the most structurally conservative simplification of DNA and RNA results from formal 2′-carbon atom deletion [15]. The resulting alternative nucleic acid structure has been dubbed flexible nucleic acid (FNA; Fig. 2) [16]. Similar to the alternative structures already mentioned, FNA exhibits some of the properties of the parent polymers. For example, FNA can participate in duplex formation as a copolymer with DNA [17, 18] or as a homopolymer depending on the sequence [18–20], despite an atactic composition owing to the presence of both (*R*)- and (*S*)-configurations at the lone stereogenic carbon atom. Beyond base-pairing within a helix, atactic FNA templates have been shown to promote the oligomerization of RNA nonenzymatically [17]. Further, DNA templates have successfully

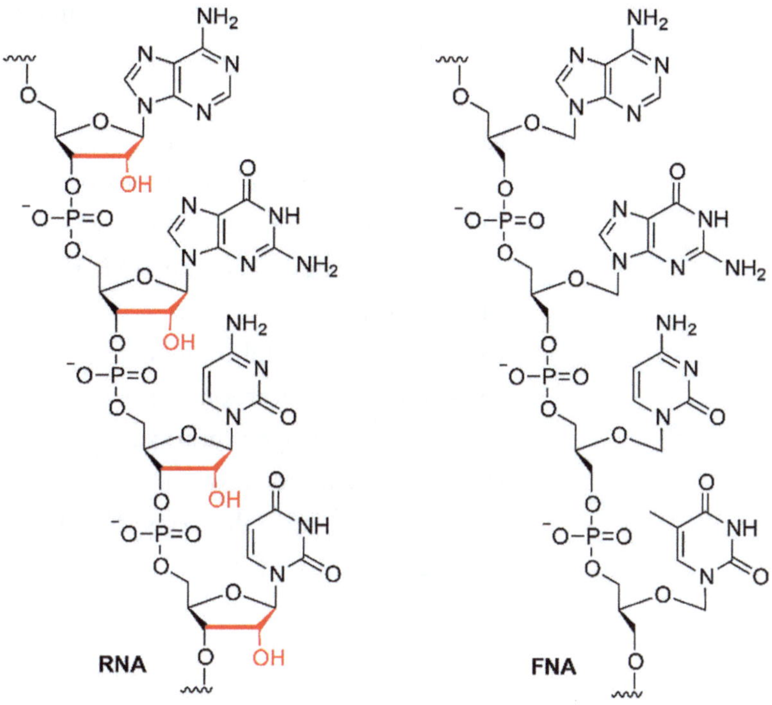


**Fig. 2** Structures of RNA and FNA

directed the enzymatic oligomerization of up to seven monomers of either the (*R*)- or (*S*)-antipodes of homochiral fNTPs [16]. This latter observation could be relevant to the development of an informational polymer that avoids enantiomer cross-inhibition during polymerization [21].

Here we detail the chemical synthesis of fNTPs and an FNA phosphoramidite. Subsequently, the use of these materials in the context of enzymatic and nonenzymatic oligomerization is described.

## 2 Materials

Solvents were dried using standard methods and distilled before use. Unless otherwise specified, materials were purchased from commercial suppliers and used without further purification.

### 2.1 Synthesis of fNTPs

1. (*R*)- and (*S*)-glycidyl tosylate.
2. Benzyl alcohol.
3. Adenine.
4. Cytosine.
5. Guanine.
6. Thymine.

7. Trimethylsilyl trifluoromethanesulfonate.
8. *N*,*O*-bistrimethylsilylacetamide.
9. Pd black.
10. DOWEX 50×8–400 [H+].
11. Tetrabutylammonium hydroxide.
12. *p*-Toluenesulfonyl chloride.
13. Boron trifluoride etherate ($BF_3$•$Et_2O$).
14. *s*-Trioxane.
15. Pentasodium triphosphate.
16. Ammonium bicarbonate powder ($NH_4HCO_3$).
17. Sodium hydrogen carbonate ($NaHCO_3$), certified ACS.
18. Sodium chloride (NaCl).
19. Sodium sulfate anhydrous ($Na_2SO_4$).
20. 1,2-Dichloroethane.
21. Cyclohexene.
22. Ultrapure $H_2O$.
23. 3 or 4 Å molecular sieve powder.
24. Celite® 575, powder.
25. DEAE Sephadex A-25 ion exchange column.
26. Silica gel 60 Geduran®, 40–63 μm.
27. Thin-layer chromatography (TLC) was carried out on aluminum sheets coated with silica gel $60F_{254}$ (with fluorescent indicator).
28. Nitrogen (or argon) gas, dry.
29. Round-bottom flasks.
30. 100-mL 2-necked round-bottom flask.
31. Magnetic stir plate and stir bar.
32. Rotary evaporator.
33. 254-nm UV lamp.
34. Balloon.
35. Desiccator.
36. High vacuum oil pump.
37. SpeedVac concentrator.
38. TEAA buffer, 50 mM, pH 7.0 is prepared by stirring at 0 °C 1.43 mL glacial acetic acid with 250 mL $H_2O$ then adding dropwise at 0 °C 3.48 mL of triethylamine. Adjust the mixture to pH to 7.0 using acetic acid or triethylamine, and then add water to a final volume of 500 mL.

### 2.2 Cytosine FNA (fC) Synthesis

1. Cytosine FNA.
2. Trimethylsilyl chloride.
3. Benzoyl chloride.
4. Ammonium hydroxide 28% ($NH_4OH$).
5. Hydrochloric acid 37% (HCl).
6. Dimethoxytrityl chloride 98%.
7. Sodium sulfate anhydrous ($Na_2SO_4$).
8. Diisopropylammonium tetrazolide.
9. 2-Cyanoethyl *N,N*-diisopropylchlorophosphoramidite.

### 2.3 Synthesis of Guanosine 5′-Phosphoro-2-Methylimidazolide (2-MeImpG)

1. Guanosine 5′-monophosphate.
2. 2-Methylimidazole.
3. Dimethyl sulfoxide (DMSO), distilled from $CaH_2$.
4. *N,N*-Dimethylformamide (DMF), water <50 ppm.
5. Triphenylphosphine ($Ph_3P$).
6. 2,2′-Dipyridyl disulfide.
7. Triethylamine (TEA), distilled from KOH and stored over KOH.
8. 11:2:7 (v/v/v) *n*-propanol/ammonium hydroxide/water.
9. Acetone.
10. Diethyl ether, anhydrous.
11. Saturated anhydrous sodium perchlorate ($NaClO_4$) in acetone.
12. Phosphorus pentoxide ($P_2O_5$).
13. Sodium hydroxide (NaOH).

### 2.4 Synthesis of DNA Natural and Unnatural Primers and Templates

1. DNA synthesizer using commercially available reagents, including nucleoside-derivatized controlled pore glass (CPG) columns and phosphoramidites; the base identity of the nucleoside-derivatized CPG column used to prepare a given oligonucleotide corresponds to that of the 3′-end.
2. Unnatural FNA-C (fC) phosphoramidite (*see* Subheading 2.2 for Materials).
3. Anhydrous acetonitrile.
4. Anhydrous helium or nitrogen gas.
5. Ammonium hydroxide 28% ($NH_4OH$).
6. 3 or 4 Å molecular sieves powder.
7. Screw-cap glass vial (preferably fitted with rubber O-ring).
8. Heating block 37 °C, 55 °C, and 90 °C.
9. 1.5-mL microcentrifuge tube.

10. Micro-Spin™ filter tube 0.45 μm PVDF.
11. Oligonucleotide desalting by gel filtration is carried out on a Sephadex G-25 column.
12. Polyacrylamide gel electrophoresis (PAGE) is performed in an Adjustable Vertical Electrophoresis System, with vertical glass plates connected to a voltage source.
13. PAGE reagents: acrylamide, bis-acrylamide, urea, *N*,*N*,*N*′,*N*-′-tetramethylethylenediamine, and ammonium persulfate.
14. TBE buffer 10×: tris(hydroxymethyl)aminomethane hydrochloride, boric acid, ethylenediaminetetraacetic acid.
15. Loading buffer: bromophenol blue, xylene cyanol.
16. Thin-layer chromatography (TLC) was carried out on aluminum sheets coated with silica gel $60F_{254}$ (with fluorescent indicator).
17. 254-nm UV lamp.
18. Razor blade.
19. 15-mL culture tubes.
20. Rotomix shaker.
21. SpeedVac evaporator.

### 2.5 Primer Phosphorylation

1. T4 Polynucleotide Kinase and dilution buffer.
2. 10× kinase buffer (700 mM Tris–HCl, pH 7.6, 100 mM $MgCl_2$, 50 mM DTT).
3. [$\gamma$-$^{32}$P]ATP, 6000 Ci/mmol, 10 mCi/mL.
4. Adenosine 5′-triphosphate (EMD).
5. TEAA buffer: glacial acetic acid; triethylamine.
6. 1.5-mL microcentrifuge tubes.
7. 15-mL culture tubes.
8. Plastic wrap.
9. Autoradiographic film.
10. Developer.
11. Fixer.
12. Razor blade.
13. Vortex mixer.
14. Microcentrifuge.
15. Sephadex G-25 NAP columns, preswollen in $H_2O$.
16. Rotomix 50,800 shaker.
17. Geiger counter.
18. SpeedVac evaporator.

19. Phosphorimager screen.
20. Imaging scanner.
21. Image-quantifying software.

### *2.6 Enzymatic Oligomerization Assay*

1. Tth pyrophosphatase.
2. Therminator Polymerase.
3. 10× Therminator pol buffer: 100 mM KCl, 100 mM $(NH_4)_2SO_4$, 200 mM Tris–HCl (pH 8.8), 20 mM $MgSO_4$, 1% Triton X-100; 10× Therminator pol buffer is included with commercial sources of the polymerase as a separate container.

### *2.7 Nonenzymatic Oligomerization Assay*

1. TEAA buffer: glacial acetic acid; triethylamine.
2. 2,6-Lutidine.
3. Ribonuclease T1.
4. $tRNA^{Phe}$.
5. 2 M NaCl.
6. 1 M $MgCl_2$.
7. Tris-EDTA buffer (10 mM Tris·Cl, pH 7.4; 1 mM EDTA, pH 8.0); Tris-EDTA buffer was prepared by mixing 1 mL of 1 M Tris–HCl (pH 7.4) with 200 μL of 0.5 M EDTA (pH 8.0) and adding deionized water to a final volume of 100 mL. Store at room temperature.
8. 2,6-Lutidine•HCl 1 M (pH 8.0) was prepared by dissolving 10.7 g of 2,6-lutidine in 50 mL of $ddH_2O$ and adding 3.09 mL of HCl dropwise at 0 °C. Adjust the pH to 8.0 using 2,6-lutidine or HCl, and add deionized water to a final volume of 100 mL.
9. 10× TBE is prepared by mixing 108 g of Tris Base, 55 g of boric acid and 40 mL of EDTA (pH 8.0). Adjust the volume to 1 L with deionized water.
10. Loading buffer is prepared by making a solution containing 7 M urea, 0.1% (w/v) bromophenol blue and 0.1% (w/v) xylene cyanol solution in 1× TBE buffer. Store at −20 °C.

## 3 Methods

### *3.1 Synthesis of fNTPs*

Alcohol **2** (Fig. 3) was obtained from commercially available (*S*)- and (*R*)-glycidyl tosylates. After chloromethylation of compound **2**, the Vorbrüggen reaction between the silylated nucleobases and chloromethyl ether **3** afforded benzyl-protected nucleosides **4**. Removal of the benzyl-protecting group by treatment with Pd/cyclohexene followed by substitution of the tosylate with butylammonium triphosphate yielded fNTPS **6** [16].

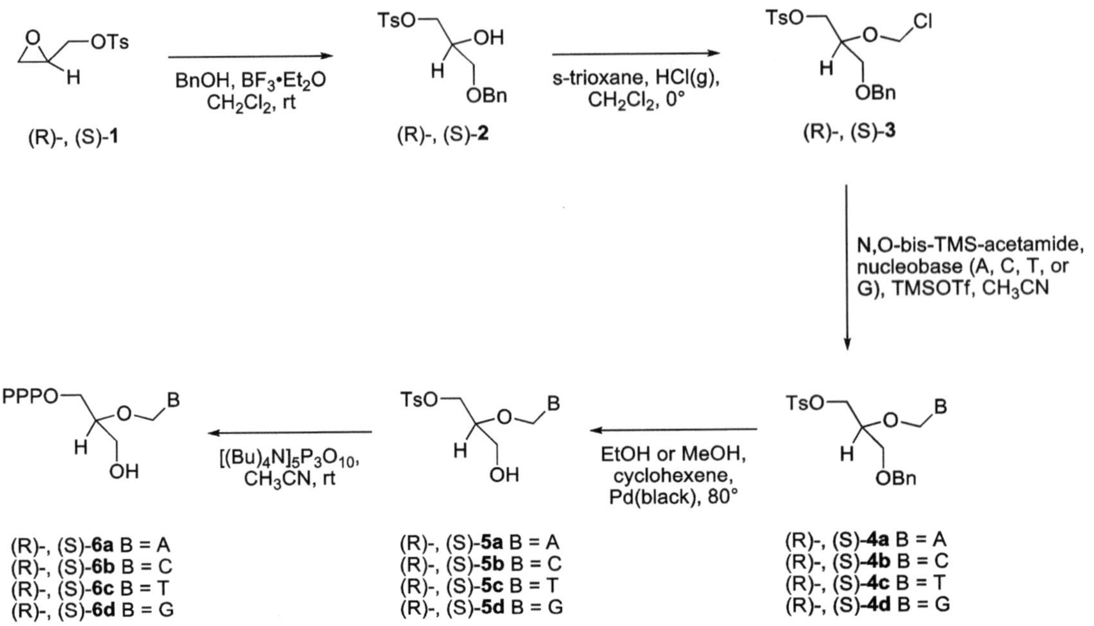


**Fig. 3** Synthesis of fNTPs (R)-**6a–d** and (S)-**6a–d** from (R)- and (S)-glycidyl tosylate **1**, respectively. *B* nucleobase (*A* adenine, *G* guanine, *C* cytosine, *T* thymine), *BnOH* benzyl alcohol, *$BF_3$•$Et_2O$* boron trifluoride etherate, *TMSOTf* trimethylsilyl trifluoromethanesulfonate, *N,O-bis-TMS-acetamide* N,O-bistrimethylsilylacetamide

#### *3.1.1 Procedure for 1-O-Tosyl-3-O-Benzylglycerol **2***

1. Suspend 5.3 g of glycidyl tosylate (23.2 mmol) and freshly activated 3 Å molecular sieve powder (1.5 g) in 50 mL of anhydrous $CH_2Cl_2$ in a 250-mL round-bottom flask with magnetic stirring under a nitrogen atmosphere (*see* **Note 1**).
2. Add 4.8 mL of freshly distilled benzyl alcohol (46.4 mmol), and cool the mixture at 0 °C in an acetone bath with a cryostat cold finger.
3. Add 150 μL of freshly distilled $BF_3$•$Et_2O$ (1.16 mmol), and stir overnight at room temperature under a nitrogen atmosphere (*see* **Note 2**).
4. Wash the mixture with saturated aqueous $NaHCO_3$ (25 mL), followed by aqueous brine (25 mL).
5. Dry the organic phase over anhydrous $Na_2SO_4$, and filter off the drying agent.
6. Evaporate the mixture to dryness under reduced pressure using a rotary evaporator.
7. Purify the residue by column chromatography (EtOAc/Hexanes = 3:7 → 2:3 → 1:1) (*see* **Note 3**).
8. Collect fractions containing the product, and evaporate to dryness to obtain the product **2** as a solid wax (6.03 g; 80%) (*see* **Note 4**).

9. Characterize the compound by $^1$H NMR and mass spectrometry: (*S*)-1-*O*-tosyl-3-*O*-benzylglycerol, (*R*)-1-*O*-tosyl-3-*O*-benzylglycerol $^1$H NMR (300 MHz, $CDCl_3$) δ 7.84–7.77 (m, 2H), 7.39–7.25 (m, 7H), 4.51 (s, 2H), 4.16–3.97 (m, 3H), 3.58–3.47 (m, 2H), 2.45 (s, 3H); ZAB-SE FAB+: calculated for $C_{17}H_{21}O_5S$ 337.1110 ($MH^+$), found 337.1118.

#### 3.1.2 Procedure for 1-O-Tosyl-2-Chloromethyl-3-O-Benzylglycerol 3

1. Place 5.6 g (16.6 mmol) of 1-*O*-tosyl-3-*O*-benzylglycerol **2** and 9.0 g of s-trioxane (100.0 mmol) in a flame-dried 100-mL 2-necked round-bottom flask with magnetic stirring under a nitrogen atmosphere.
2. Add 28 mL of 1,2-dichloroethane, and equilibrate the mixture to −4 °C in a salt water ice bath.
3. Bubble HCl (g) through the mixture for 5 h at −4 °C under an argon atmosphere (*see* **Note 5**).
4. Quench the reaction by adding 30 mL saturated aqueous $NaHCO_3$.
5. Wash the organic layer with brine (25 mL).
6. Dry the organic phase over anhydrous $Na_2SO_4$, and filter off the drying agent.
7. Evaporate the solution to dryness under reduced pressure using a rotary evaporator to obtain compound **3** as an oil which is used directly without further purification (*see* **Note 6**).

#### 3.1.3 General Vorbrüggen Reaction Procedure to Obtain Nucleosides 4 a–d

1. Combine nucleobase (2.2 mmol) and *N,O*-bistrimethylsilylacetamide [1.1 mL; 4.4 mmol (A, C, and T) or 1.6 mL; 6.6 mmol (G)] in 10 mL of anhydrous $CH_3CN$ in a 25-mL round-bottom flask with magnetic stirring under a nitrogen atmosphere.
2. Reflux the mixture until homogeneous (*see* **Note 7**).
3. Cool the mixture to room temperature.
4. Add a solution of 770 mg of 1-*O*-tosyl-2-chloromethyl-3-*O*-benzylglycerol **3** (2.0 mmol) in 8 mL of $CH_3CN$ followed by 18 μL of trimethylsilyltrifluoromethane sulfonate (0.10 mmol), and stir at room temperature for 20 min under a nitrogen atmosphere.
5. Quench the reaction by the addition of methanol, followed by 3 g of $NaHCO_3$(s), and stir for 20 additional minutes.
6. Filter the mixture over Celite, and evaporate the filtrate to dryness using a rotary evaporator.
7. Final products **4 a–d** were difficult to separate from impurities and were used without further purification in the next step (yield over 2 steps ≈30% for T and C, ≈5% for A and G (*see* **Note 8**).

#### 3.1.4 General Debenzylation Procedure to Obtain Nucleosides **5 a–d**

1. Dissolve [(1-*O*-Tosyl-3-benzyloxy-2-propoxy)methyl]nucleobase **4 a–d** (0.46 mmol) in 4.2 mL cyclohexene and 8.4 mL EtOH (A, C, and T) or MeOH (G) in a 25-mL round-bottom flask with magnetic stirring under a nitrogen atmosphere.
2. Add 172 mg of Pd (black) (1.6 mmol), and heat the mixture at a gentle reflux (*see* **Note 9**).
3. Filter the solution over Celite, and concentrate the filtrate to dryness using a rotary evaporator (*see* **Note 10**).
4. Purify the crude material by column chromatography (*see* **Note 11**).
5. The following yields were obtained: **5a** 69% (A), **5b** 72% (C), **5c** 92% (T), and **5d** 27% (G).
6. (*S*)-9-[(1-*O*-tosyl-3-hydroxy-2-propoxy)methyl]adenine, (*R*)-9-[(1-*O*-tosyl-3-hydroxy-2- propoxy)methyl]adenine **5a**: $^1$H NMR (400 MHz, D6-DMSO) δ 8.20 (s, 1H), 8.15 (s, 1H), 7.65 (d, *J* = 8.09 Hz, 2H), 7.41 (d, *J* = 8.25 Hz, 2H), 7.29 (s, 2H), 5.55 (s, 2H), 4.87 (t, *J* = 5.49 Hz, 1H), 4.09 (dd, *J* = 10.21, 2.60 Hz, 1H), 3.96–3.83 (m, 2H), 2.41 (s, 3H); ZAB-SE FAB+: calculated for $C_{16}H_{20}N_5O_5S$ 394.1185 ($MH^+$), found 394.1195.
7. (*S*)-1-[(1-*O*-tosyl-3-hydroxy-2-propoxy)methyl]cytosine, (*R*)-1-[(1-*O*-tosyl-3-hydroxy-2-propoxy)methyl]cytosine **5b**: $^1$H NMR (300 MHz, D6-DMSO) δ 7.75 (d, *J* = 8.33 Hz, 2H), 7.52 (d, *J* = 7.25 Hz, 1H), 7.47 (d, *J* = 8.29 Hz, 2H), 7.21 (d, *J* = 19.32 Hz, 2H), 5.68 (d, *J* = 7.25 Hz, 1H), 5.06 (s, 2H), 4.83 (t, *J* = 5.55 Hz, 1H), 4.15–4.06 (m, 1H), 3.98–3.90 (m, 1H), 3.79 (dq, *J* = 5.80, 5.77, 3.17 Hz, 1H), 2.42 (s, 3H); ZAB-SE FAB+: calculated for $C_{15}H_{20}N_3O_6S$ 370.1073 ($MH^+$), found 370.1077.
8. (*S*)-1-[(1-*O*-tosyl-3-hydroxy-2-propoxy)methyl]thymine, (*R*)-1-[(1-*O*-tosyl-3-hydroxy-2-propoxy)methyl]thymine **5c**: $^1$H NMR (400 MHz, $CDCl_3$) δ 9.08 (s, 1H), 7.76 (d, *J* = 8.25 Hz, 2H), 7.35 (d, *J* = 8.57 Hz, 2H), 7.15–7.13 (m, 1H), 5.21 (q, *J* = 10.55, 10.47 Hz, 2H), 4.14–4.02 (m, 3H), 3.70–3.57 (m, 2H), 2.65 (s, 2H), 2.45 (s, 3H), 1.93 (s, 3H); ZAB-SE FAB+: calculated for $C_{16}H_{21}N_2O_7S$ 385.1070 ($MH^+$), found 385.1067.
9. (*S*)-9-[(1-*O*-tosyl-3-hydroxy-2-propoxy)methyl]guanine, (*R*)-9-[(1-*O*-tosyl-3-hydroxy-2- propoxy)methyl]guanine **5d**: 1H NMR (400 MHz, D6-DMSO) δ 10.82 (s, 1H), 8.04 (s, 1H), 7.68 (d, *J* = 8.11 Hz, 2H), 7.43 (d, *J* = 8.32 Hz, 2H), 6.20 (s, 2H), 5.56 (s, 2H), 4.83 (d, *J* = 5.36 Hz, 1H), 4.08 (d, *J* = 8.13 Hz, 1H), 3.89 (dd, *J* = 16.53, 6.82 Hz, 2H), 2.42 (s, 3H); TOF-MS: calculated for $C_{16}H_{20}N_5O_6S$ 410.1134 ($MH^+$), found 410.1145.

*3.1.5 Butylammonium Triphosphate*

1. Combine 2.0 g of pentasodium triphosphate (5.4 mmol) with the minimum amount of $H_2O$ required to dissolve it in a 5-mL round-bottom flask.
2. Pass the mixture through a column of 16.0 mL DOWEX 50×8-400 [$H^+$] (suspended in $H_2O$) eluting with $H_2O$ to afford the acidic fractions.
3. Combine the fractions and titrate to pH 7.00 with 5–10% tetrabutylammonium hydroxide (*see* **Note 12**).
4. Lyophilize the neutralized aqueous solution to afford a white, hygroscopic solid (252 mg, yield 69%) (*see* **Note 13**).
5. The analytical data of the resulting compound is: $^1$H NMR (300 MHz, D6-DMSO) δ 3.28–3.15 (m, 8H), 1.63–1.49 (m, 8H), 1.39–1.24 (m, 8H), 0.92 (t, $J$ = 7.25 Hz, 12H); $^{31}$P NMR (162 MHz, $D_2O$) δ −7.00 (d, $J_{P,P}$ = 29.83 Hz, 2P), −22.36 (t, $J_{P,P}$ = 30.04 Hz, 1P).

*3.1.6 General Procedure for Preparation of Triphosphates **6 a–d***

1. Suspend 1-*O*-tosylated nucleoside **5 a–d** (0.11 mmol) in 1.0 mL $CH_3CN$ in a 5-mL round-bottom flask with magnetic stirring (*see* **Note 14**).
2. Add 480 mg of butylammonium triphosphate (0.33 mmol), and stir at room temperature for 1–2 days (*see* **Note 15**).
3. Remove the solvent using a rotary evaporator.
4. Purify the crude residue on a DEAE Sephadex A-25 ion exchange column with an aqueous $NH_4HCO_3$ gradient (0.05 M → 0.5 M, *see* **Note 16**).
5. Lyophilize the product three times from $H_2O$ to obtain a white powder with the following yields: **6a** 50% (A), **6b** 67% (C), **6c** 53% (T), and **6d** 9% (G).
6. (*S*)-9-[(3-Hydroxy-2-propoxy)methyl]adenine-1-triphosphate, (*R*)-9-[(3-hydroxy-2-propoxy)methyl]adenine-1-triphosphate **6a**: $^1$H NMR (400 MHz, $D_2O$) δ 8.36 (s, 1H), 8.27 (s, 1H), 5.83 (q, $J$ = 11.29, 11.27 Hz, 2H), 4.17–3.91 (m, 3H), 3.60 (ddd, $J$ = 18.29, 12.46, 4.89 Hz, 2H); $^{31}$P NMR (162 MHz, $D_2O$) δ −7.13 to −8.93 (m, 1P), −9.90 to −10.66 (m, 1P), −21.14 to −22.67 (m, 1P); TOF-MS: calculated for $C_9H_{17}N_5O_{12}P_3$ 480.0087 ($MH^+$), found 480.0080.
7. [(3-Hydroxy-2-propoxy)methyl]cytosine-1-triphosphate, (*R*)-1-[(3-hydroxy-2- propoxy)methyl]cytosine-1-triphosphate, (±)-1-[(3-hydroxy-2-propoxy)methyl]cytosine-1-triphosphate **6b**: $^1$H NMR (400 MHz, $D_2O$) δ 7.63–7.57 (m, 1H), 5.96–5.87 (m, 1H), 5.21 (d, $J$ = 2.51 Hz, 2H), 4.01–3.72 (m, 3H), 3.54 (ddd, $J$ = 18.07, 12.32, 4.97 Hz, 2H); $^{31}$P NMR (162 MHz, $D_2O$) δ −5.83 to −7.54 (m, 2P), −20.52 to −21.70 (m, 1P); MALDI-MS: calculated for $C_8H_{17}N_3O_{13}P_3$ 455.9974 ($MH^+$), found 455.9978.

8. (*S*)-1-[(3-hydroxy-2-propoxy)methyl]thymine-1-triphosphate, (*R*)-1-[(3-hydroxy-2-propoxy)methyl]thymine-1-triphosphate **6c**: $^1$H NMR (300 MHz, $D_2O$) δ 7.60 (s, 1H), 5.31 (s, 2H), 4.15–3.85 (m, 3H), 3.66 (ddd, *J* = 18.33, 12.41, 4.99 Hz, 2H), 1.87 (s, 3H); $^{31}$P NMR (121 MHz, $D_2O$) δ −5.26 to −6.79 (m, 2P), −20.31 to −21.40 (m, 1P); MALDI-MS: calculated for $C_9H_{18}N_2O_{14}P_3$ 470.9971 ($MH^+$), found 470.9966.
9. (*S*)-9-[(3-hydroxy-2-propoxy)methyl]guanine-1-triphosphate, (*R*)-9-[(3-hydroxy-2-propoxy)methyl]guanine-1-triphosphate **6d**: $^1$H NMR (400 MHz, D2O) δ 8.30 (s, 1H), 5.84 (dd, *J* = 24.38, 11.38 Hz, 2H), 4.16–3.88 (m, 3H), 3.67–3.52 (m, 2H); $^{31}$P NMR (162 MHz, $D_2O$) δ −8.34 to −12.16 (m, 2P), −20.98 to −24.49 (m, 1P); TOF-MS: calculated for $C_9H_{17}N_5O_{13}P_3$ 496.0036 ($MH^+$), found 496.0029.

### 3.2 Cytosine FNA (fC) Synthesis

Acyclic nucleoside 7 was synthesized as reported [17], and then converted to phosphoramidite **10** in analogy with the synthesis of standard phosphoramidites (Fig. 4).

#### 3.2.1 4-N-Benzoyl-[(1′,3′-Dihydroxy-2′-Propoxy)Methyl]Cytosine (8)

1. Suspend 430 mg of glycerol nucleoside 7 (2.0 mmol) in 20 mL of pyridine in a 50-mL round-bottom flask with magnetic stirring at 0 °C using an ice bath under a nitrogen atmosphere.
2. Add 1.3 mL of trimethylsilyl chloride (10 mmol) at 0 °C.
3. After 30 min stirring, add 1.2 mL of benzoyl chloride (10 mmol).
4. Bring the reaction mixture to room temperature, and stir for 3 h.
5. Cool to 0 °C with an ice bath, and quench with 2 mL of cold $H_2O$ and 2 mL of concentrated $NH_4OH$ following a 15-min wait between additions.
6. Remove the solvent using a rotary evaporator.
7. Dissolve the crude residue in 50 mL of $CH_2Cl_2$, wash with 10% HCl (3 × 10 mL), and dry over anhydrous $Na_2SO_4$.
8. Filter off the drying agent, and concentrate the filtrate under reduced pressure.

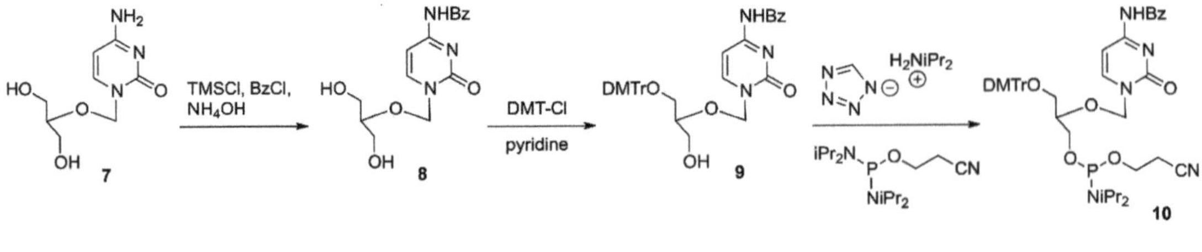


**Fig. 4** Synthesis of the FNA cytosine phosphoramidite

9. Purify the crude residue by column chromatography on silica gel eluting with 5–10% $MeOH/CH_2Cl_2$ to afford **8** (441 mg, 69% yield).

10. The analytical data of compound **8** are: $^1H$ NMR (DMSO-D6) δ 3.35–3.4 (m, 4H), 3.59 (m, 1H), 4.65 (t, 2H), 5.34 (s, 2H), 7.31 (d, 1H, *J* = 6.75 Hz), 7.52 (t, 2H, *J* = 7.74, 7.33 Hz), 7.63 (t, 1H, *J* = 7.33, 7.29 Hz), 8.01 (d, 2H, *J* = 7.43 Hz), 8.20 (d, 1H, *J* = 6.94 Hz), 11.21 (s, 1H). MS ($EI^+$): *m/z* 319. Exact mass ($EI^+$): calculated for $C_{15}H_{17}N_3O_5$ ($MH^-$), 319.116821; found, 319.1159.

*3.2.2 4-N-Benzoyl-[(18-O-(4,4′-Dimethoxytrityl)-2′-Propoxy)Methyl]-Cytosine (**9**)*

1. Dissolve 435 mg of nucleoside **8** (1.4 mmol) in 5 mL pyridine in a 10-mL round-bottom flask with magnetic stirring at 0 °C using an ice bath.

2. Add 475 mg of dimethoxytrityl chloride (1.4 mmol), and stir for 3 h at room temperature.

3. Add 50 mL of $CH_2Cl_2$, and wash the mixture with a saturated aqueous $NaHCO_3$ solution (3 × 10 mL).

4. Dry the organic layer over anhydrous $Na_2SO_4$, and filter off the drying agent.

5. Evaporate the solution to dryness under reduced pressure using a rotary evaporator.

6. Purify the crude residue by column chromatography on silica gel eluting with 2–5% $MeOH/CH_2Cl_2$ to give **9** (238 mg, 30% yield).

7. The analytical data of compound **9** are $^1H$ NMR ($CDCl_3$): δ 2.63 (s, 1H), 3.22 (d, 2H), 3.65 (m, 2H), 3.78 (s, 6H), 3.90 (m, 1H), 5.42 (q, 2H), 6.81 (d, 4H), 7.20–7.31 (m, 8H), 7.4 (d, 2H, *J* = 7.48 Hz), 7.51 (t, 2H, *J* = 7.68, 7.31 Hz), 7.61 (d, 1H, *J* = 7.26 Hz), 7.68 (t, 1H, *J* = 7.57, 7.85 Hz), 7.80 (d, 1H, *J* = 7.15 Hz) 7.91 (d, 2H, *J* = 7.30 Hz), 8.61 (s, 1H). MS (FAB−): *m/z* 620. HRMS ($FAB^-$): calculated for $C_{36}H_{35}N_3O_7$ ($MH^-$), 620.239676; found, 620.2378.

*3.2.3 4-N-Benzoyl-1-[1′-O-(4,4′-Dimethoxytrityl)-2′-Propoxy)Methyl]-Cytosine 2-Cyanoethyl N,N-Diisopropyl-phosphoramidite (**10**)*

1. Suspend 235 mg of nucleoside **9** (0.38 mmol) and 34 mg of diisopropylammonium tetrazolide (0.20 mmol) in 5 mL of acetonitrile in a 10-mL round-bottom flask with magnetic stirring under nitrogen atmosphere.

2. Add 0.13 mL of 2-cyanoethyl *N,N*-diisopropylchlorophosphoramidite (0.4 mmol), and stir for 5 h at room temperature.

3. Add 20 mL of $CH_2Cl_2$ (20 mL), and wash the mixture with a saturated aqueous $NaHCO_3$ solution (1 × 10 mL).

4. Dry the organic layer over anhydrous $Na_2SO_4$, and filter off the drying agent.

5. Evaporate the solution to dryness under reduced pressure using a rotary evaporator.
6. Purify the crude residue by column chromatography on silica gel eluting with 50% EtOAc/1% pyridine/49% hexanes to give **10** (128 mg, 40% yield).
7. The analytical data of compound **10** are $^{1}$H NMR ($CDCl_3$): δ 1.08–1.18 (m, 12H), 2.6 (m, 2H), 3.22 (m, 2H), 3.52 (m, 2H), 3.61–3.74 (m, 4H), 3.77 (s, 6H), 4.01 (m, 1H), 5.45 (q, 2H), 6.82 (m, 4H), 7.16–7.31 (m, 9H), 7.41 (d, 2H), 7.5 (t, 2H), 7.6 (t, 1H), 7.92 (t, 1H). $^{31}$P NMR ($CDCl_3$): δ 148.52, 148.34. MS (FAB$^+$): $m/z$ 844. HRMS (FAB$^+$): calculated for $C_{45}H_{52}N_5O_8PNa$ (MNa$^+$), 844.345122; found, 844.3482.

### *3.3 DNA Synthesis: Primer and Template Synthesis*

Primer and templates (Fig. 5) were synthesized on a DNA synthesizer using the standard 1.0-μmol-scale DNA synthesis protocol (trityl-off mode) with commercially available reagents.

#### *3.3.1 Automated DNA Synthesis*

#### *3.3.2 Deprotection and Cleavage from the Solid Support*

1. Remove the solid support from the column by cutting one side of the column.
2. In a screw-cap glass vial equipped with a rubber O-ring, suspend the CPG bearing the oligomers in 1.0 mL concentrated ammonium hydroxide.
3. Place the sample in a heat block or oven for 16 h at 55 °C.
4. Let the vial cool to room temperature and carefully open it.
5. Remove of the solid support by filtration with a centrifuged filter tube.
6. Lyophilize the sample to dryness.

#### *3.3.3 PAGE Purification*

1. Prepare a vertical denaturing 20% polyacrylamide gel (*see* **Note 17**).
2. Place the gel in the gel electrophoresis apparatus, using 1× TBE buffer in both top and bottom buffer reservoirs, and rinse wells thoroughly with buffer.
3. Prerun the gel for 1 h at 27 mA until it reaches the temperature of 45 °C.
4. Using a syringe, rinse wells with 1× TBE buffer just prior to loading samples to remove urea.

primer: $^{5'}$dCGC CAG GGT TTT CCC AGT CAC GAC G$^{3'}$
template: $^{3'}$dGCG GTC CCA AAA GGG TCA GTG CTG C **XX XXX XX**$^{5'}$

**Fig. 5** Primer and template sequences; bold text refers to template coding regions (X=A, T, C or G)

5. Prepare oligonucleotide samples by mixing the crude oligonucleotide in water with the loading buffer. Heat the samples for 5 min at 90 °C and load samples in separate adjacent wells.
6. Run the gel at 27 mA and observe migration of marker dyes (on a 20% gel bromophenol blue migrates as an 8-nt linear DNA and xylene cyanol as a 28-nt linear DNA) to determine running time of electrophoresis.
7. Disassemble the gel apparatus. Wrap the gel with transparent plastic wrap and visualize the bands under UV light by sliding a silica TLC sheet under the gel.
8. Cut out the desired bands with a razor blade, and transfer them to a 15-mL culture tube. Crush the gel thoroughly with a glass rod, add 6 mL of a 50 mM TEAA solution (pH 7.0), and shake overnight using a rotary shaker.
9. Filter by centrifugation and keep the filtrate. Resuspend the gel in 5 mL of dd$H_2O$ and centrifuge and filter again. Combine the filtrates and lyophilize the sample to dryness.
10. Dissolve the oligonucleotide in 1 mL of dd$H_2O$.
11. Desalt the purified product with a Sephadex G-25 NAP column by applying the solution to the column preswollen with $H_2O$. Elute the column with 10 mL $H_2O$, collecting 1-mL fractions in a 1.5-mL microcentrifuge tube.
12. Detect fractions containing oligonucleotide by UV absorbance and lyophilize the combined fractions to dryness using a SpeedVac evaporator.
13. Quantify the amount of purified oligonucleotide by UV absorbance at 260 nm at 70 °C.

#### 3.3.4 Radiolabel the Oligonucleotide Primer

1. In a 1.5-mL microcentrifuge tube, mix the following:
   100 pmol template in 5 μL of dd$H_2O$.
   1 μL T4 polynucleotide kinase (3 U/μL).
   1 μL 10× kinase buffer.
   2 μL [γ-$^{32}$P]ATP (1 μCi/μL).
2. Vortex, centrifuge 1 min at 2910 × *g* at room temperature, and incubate the mixture for 30 min at 37 °C.
3. Add 1 μL of a 660 μM ATP solution to the mixture. Vortex, centrifuge 1 min at 2910 × *g* at room temperature, and then incubate the mixture for an additional 30 min at 37 °C.

#### 3.3.5 Purify Labeled Primer Oligonucleotide Template by PAGE

1. Add 10 μL loading buffer to the above mixture. Vortex, centrifuge 1 min at 2910 × *g* at room temperature, and place the mixture in a heating block preheated to 90 °C for 5 min.

2. Load the solution on a 20 × 52-cm, 7Murea/20% denaturing polyacrylamide gel and electrophorese at 27 mA and observe the migration of marker dyes.
3. Dismantle the gel apparatus and plates, wrap the gel with a transparent plastic wrap, and expose the gel to autoradiographic film.
4. Develop the film using developer and fixer in a dark room.
5. Overlay the film on the gel. Cut out the desired oligonucleotide template bands from the gel using the film as a guide.
6. Place the gel slices containing labeled template in a 15-mL culture tube and crush thoroughly with a glass rod.
7. Incubate the mixture in 6 mL of 50 mM TEAA buffer (pH 7.0) by shaking overnight using a rotary shaker.
8. Filter by centrifugation, transfer the filtrate into a new 15-mL culture tube, and lyophilize the solution.
9. Dissolve the residue in 1 mL dd$H_2O$.
10. Apply the solution to a Sephadex G-25 NAP column preswollen in dd$H_2O$. Elute the column with 10 mL dd$H_2O$, collecting 1-mL fractions in 1.5-mL microcentrifuge tubes.
11. Detect radioactive fractions using a Geiger counter and lyophilize the combined fractions to dryness.
12. Add dd$H_2O$ to afford a ~10 μM template solution (assuming 30% recovery from the gel).

### 3.4 Synthesis of FNA Cytosine-Bearing Hairpin Oligonucleotides

DNA hairpin templates incorporating FNA cytosine at the positions indicated (Fig. 6) were synthesized on a solid phase support beginning from a riboG-derivatized CPG column [17].

#### 3.4.1 Automated DNA Synthesis

Hairpin templates (Fig. 6) were synthesized on a DNA synthesizer using the standard 1.0-μmol-scale DNA synthesis protocol (trityl-off mode) on a CPG solid support derivatized with a commercially available riboguanosine nucleoside bearing a 3′-*O*-acetyl-protecting group along with other commercially available reagents and phosphoramidites (*see* **Note 18**).

#### 3.4.2 5′-Radiolabeling Procedure

1. In a 15-mL culture tube, mix the following:
   100 pmol oligonucleotide.
   
   1 mL T4 polynucleotide kinase.
   
   1 mL 10× kinase buffer.
   
   1 mL [γ-$^{32}$P]ATP (2 μCi).
   
   6 mL $H_2O$.

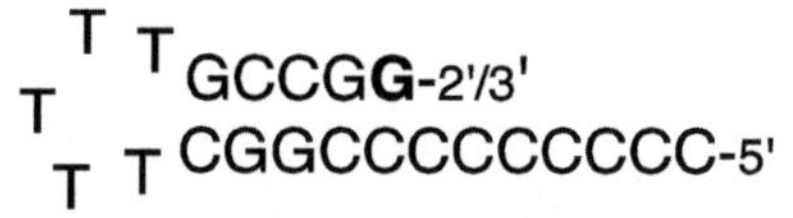


**11**

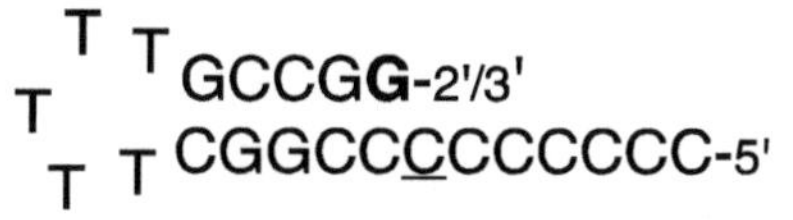


**12**

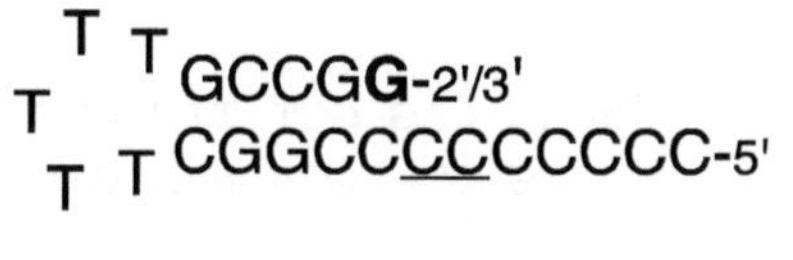


**13**

**14**

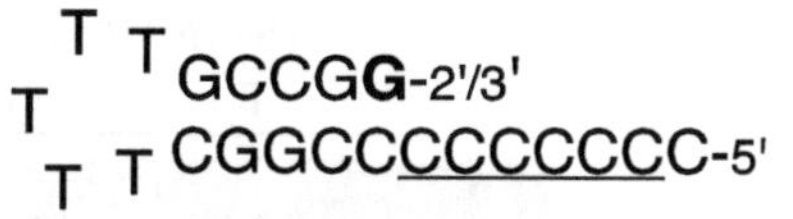


**15**

**Fig. 6** Hairpin templates used in oligomerization reactions. Underlined regions refer to FNA cytosine residues. Terminal 3′-riboG residues are italicized. DNA regions are in standard type

2. Vortex and then centrifuge for 1 min at 2910 × *g* at room temperature; incubate the sample for 30 min at 37 °C.
3. Add 1 μL of 660 μM ATP to the mixture. Vortex, centrifuge 1 min at 2910 × *g* at room temperature, and then incubate the mixture for an additional 30 min at 37 °C.
4. Lyophilize the solution.
5. Dissolve the residue in 1 mL dd$H_2O$.
6. Apply the solution to a Sephadex G-25 NAP column preswollen in dd$H_2O$. Elute the column with 10 mL dd$H_2O$, collecting 1-mL fractions in 1.5-mL microcentrifuge tubes.
7. Detect radioactive fractions using a Geiger counter and lyophilize the combined fractions to dryness.

### 3.5 Polymerase-Directed Primer Extension

Despite a diminished ability of FNAs to cross-pair with DNA or with themselves [17–20] compared to natural nucleic acids, fNTPs can be used as substrates in enzymatic polymerization by DNA polymerases. Moreover, both (*S*)- and (*R*)-FNA monomers allow serial chain extension of up to 7 nucleotides (full-length product in the examples; Figs. 7 and 8) [16].

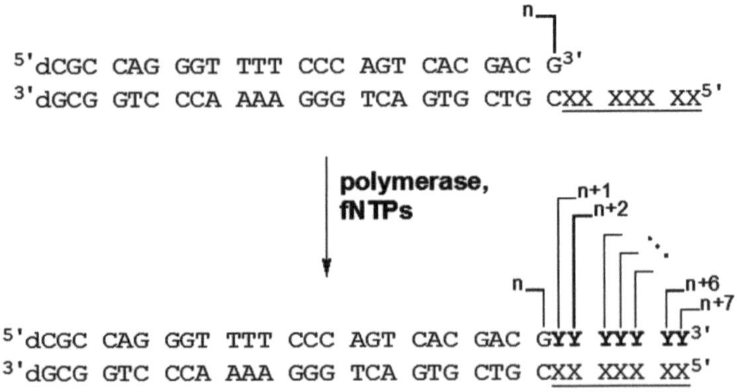


**Fig. 7** Template-directed primer extension with fNTPs. Underlined regions refer to template regions. Incorporated fNMP residues are **bold**. DNA regions are in standard type, X/Y = A/T, T/A, C/G, or G/C

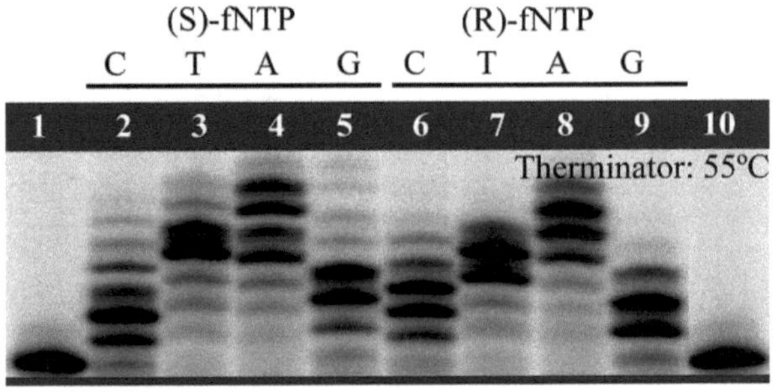


**Fig. 8** Primer extension using templates **1–4** with $^{32}$P-5′-labeled primer, templates, and (S)-, and (R)-fNTPs as indicated, 2 U Therminator DNA polymerase, 1 U Tth pyrophosphatase, 55 °C, 24 h. Lanes 1 and 10 contain radiolabeled primer only. Parts of this figure reproduced from ref. 16 with permission. Copyright 2008 Am. Chem. Soc.

*3.5.1 Perform DNA Polymerase Reaction*

1. Combine $^{32}$P-5′-labeled primer and template [1.2 μM primer, 1.5 μM template], 10× Therminator pol buffer, 200 μM fNTP (racemic, (*S*)- or (*R*)- as indicated), and 1 U Tth pyrophosphatase in a 1.5-mL centrifuge tube and heat at 90 °C for 5 min.
2. After briefly cooling to room temperature, add 2 U of Therminator DNA polymerase.
3. Cover the samples with a layer of light mineral oil and incubate at 75 °C for 24 h.

*3.5.2 Analyze Oligomerization Products by PAGE*

1. Quench the reaction with an equal volume of 8 M urea in 1× TBE, and heat at 90 °C for 10 min.
2. Immediately load the sample on an analytical 20% (29:1) denaturing PAGE, and observe migration of marker dyes.

3. Expose the gel to a phosphorimager screen overnight, and analyze the results using a phosphorimage scanner and image-quantifying software.

```
primer:    5'dCGC CAG GGT TTT CCC AGT CAC GAC G3'
template:  3'dGCG GTC CCA AAA GGG TCA GTG CTG CXX XXX XX5'

template 1: 3'dGCG GTC CCA AAA GGG TCA GTG CTG CGG GGG GG5'
template 2: 3'dGCG GTC CCA AAA GGG TCA GTG CTG CAA AAA AA5'
template 3: 3'dGCG GTC CCA AAA GGG TCA GTG CTG CTT TTT TT5'
template 4: 3'dGCG GTC CCA AAA GGG TCA GTG CTG CCC CCC CC5'
```

### 3.6 Nonenzymatic Oligomerization of DNA by FNA Templates

Nonenzymatic oligomerization of 2-methylimidazole-activated ribonucleotides proceeds on DNA hairpins appended with templates incorporating atactic FNA (Figs. 6 and 9). In particular, atactic FNA-cytidylate units participate in the template-directed oligomerization of activated 2-MeImpG (guanosine 5′-phosphoro-2-methylimidazole). However, these templates show a diminished activity compared to *isotactic* 2′-deoxyribocytidine templates [17].

#### 3.6.1 Synthesis of Guanosine 5′-Phosphoro-2-Methylimidazolide (2-MeImpG)

This protocol is based upon the synthesis of mononucleoside 5′-phosphoro-2-methylimidazolides (Fig. 10) previously reported [22].

1. Dissolve 0.50 g of guanosine 5′-monophosphate (1.38 mmol) in 10 mL of DMSO in a 50-mL round-bottom flask with magnetic stirring at room temperature under nitrogen.

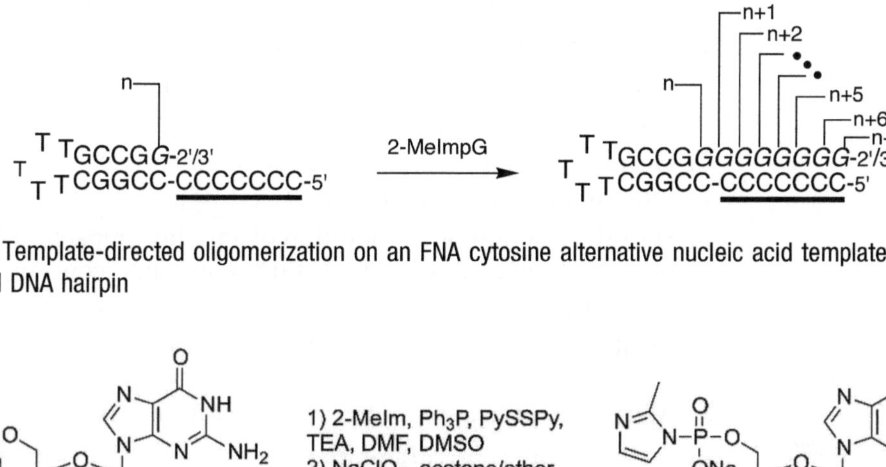


**Fig. 9** Template-directed oligomerization on an FNA cytosine alternative nucleic acid template grafted to a natural DNA hairpin

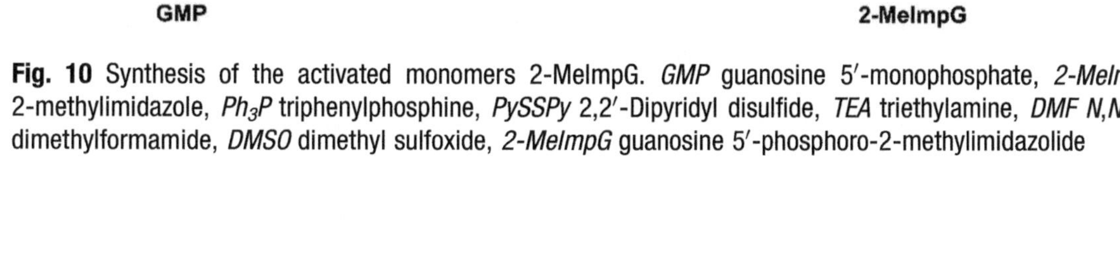


**Fig. 10** Synthesis of the activated monomers 2-MeImpG. *GMP* guanosine 5′-monophosphate, *2-MeIm* 2-methylimidazole, *$Ph_3P$* triphenylphosphine, *PySSPy* 2,2′-Dipyridyl disulfide, *TEA* triethylamine, *DMF* *N,N*-dimethylformamide, *DMSO* dimethyl sulfoxide, *2-MeImpG* guanosine 5′-phosphoro-2-methylimidazolide

2. Add sequentially 10 mL of DMF, 0.5 mL of TEA (3.59 mmol), 0.70 g of $Ph_3P$ (2.67 mmol), and 0.80 g of aldrithiol-2 (3.63 mmol) (*see* **Note 19**). Stir the mixture for 5 h at room temperature under a nitrogen atmosphere (*see* **Note 20**).
3. Add the reaction mixture dropwise to a vigorously stirring solution of 400 mL of acetone, 250 mL of anhydrous diethyl ether, 30 mL of TEA, and 2 mL of saturated $NaClO_4$ in acetone and continue stirring for 20 min.
4. Collect the precipitate by centrifuging 5 min at 13,600 × *g* at room temperature, and wash the precipitate with 1:1 (v/v) acetone/anhydrous diethyl ether (3 × 30 mL) and then with anhydrous diethyl ether (2 × 30 mL).
5. Desiccate the product over $P_2O_5$ and NaOH under high vacuum overnight to obtain the activated 2-MeImpG (0.57 g, yield 92%) (*see* **Note 21**).
6. The analytical data of **2-MeImpG** are $^1$H NMR (300 MHz, $D_2O$): δ 7.76 (s, 1H), 6.99 (s, 1H), 6.62 (s, 1H), 5.77–5.59 (m, 1H), 4.82–4.53 (m, 1H), 4.36–4.24 (m, 1H), 4.17–4.09 (m, 1H), 3.96 (d, 2H, $J = 4.16$ Hz), 2.22 (s, 3H). $^{31}$P-NMR (120 MHz, $CDCl_3$): δ −6.8. HRMS (ESI): calculated for $C_{14}H_{18}N_7O_7NaP$, 450.0903; found 450.0896.

### *3.6.2 Oligomerization Reaction Conditions*

Template reactions were carried out in Eppendorf tubes in a total volume of 20 μL using the general procedure described [17].

1. In a 1.5-mL Eppendorf tube, combine 5′-$^{32}$P-labeled-oligonucleotide (10 pmol), 10 μL NaCl (2 M),and 4 μL $MgCl_2$ (1 M).
2. Lyophilize the resulting mixture to dryness.
3. Dissolve the residue in 8 μL 2,6-lutidine·HCl buffer (0.5 M, pH 8.0) and 9 μL dd$H_2O$.
4. Heat the samples for 5 min at 55 °C, then cool for 5 min at room temperature and for 5 min at 0 °C.
5. Start the reaction by the addition of a freshly prepared solution of 2-MeImpG (3 μL; 21.3 OD/rxn; final concentration, ≈100 mM).
6. Incubate the reaction up to 10 days at 5 °C.

### *3.6.3 Analyze Oligomerization Products by PAGE*

1. Add 10 μL of loading buffer to the reaction mixture.
2. Apply 15 μL of each samples on a 7 M urea/20% denaturing polyacrylamide gel, and observe migration of marker dyes.
3. Expose the gel to a phosphorimager screen, and analyze the results (Fig. 11) using phosphorimage scanner and image-quantifying software.

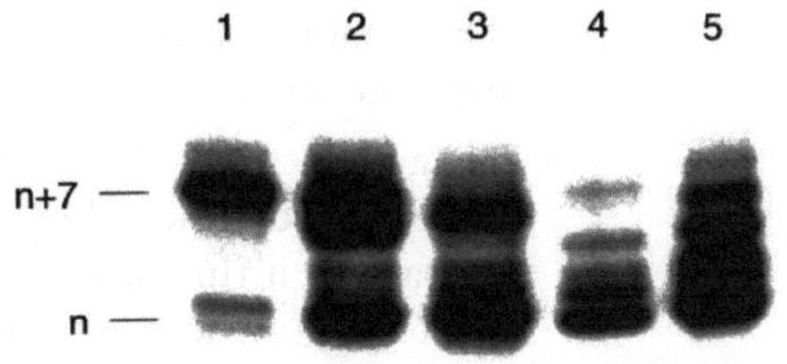


**Fig. 11** Autoradiogram after 20% PAGE analysis of 2-MeImpG oligomerization on templates 11–15 (Fig. 6), lanes 1–5, respectively, after 10 days

*3.6.4 Ribonuclease T1 Digestion*

1. In a 1.5-mL microcentrifuge tube, mix the gel-purified oligomerization product (≈1 pmol) in 9 μL Tris-EDTA buffer along with 1 μL tRNA$^{Phe}$ (10 mg/mL).
2. Vortex and then centrifuge for 1 min at 2910 × *g* at room temperature; incubate for 10 min at 55 °C in a heating block.
3. Add 1 μL RNase T1 (100 U/μL). Vortex and then centrifuge 1 min at 2910 × *g* at room temperature; incubate for 1 h at 55 °C in a heating block.
4. Quench by addition of 5 μL loading buffer.
5. Analyze the result by electrophoresing on a 7 M urea/20% denaturing polyacrylamide gel running until the markers reached the desired positions.
6. Bands were quantified using a phosphorimager and image-quantifying software.

## 4 Notes

1. Vessel needs to be flame-dried and charged with nitrogen before use.
2. Be careful to add $BF_3$•$Et_2O$ directly to the solution.
3. Mix silica with the nonpolar component of the eluent solvent, and stir with a glass rod until a homogeneous and fluid gel is attained. Pour the gel rapidly into a chromatography glass column. Let the solvent flow through the column until the silica bed volume is stabilized and the silica is well packed using low-pressure airflow. Stop solvent the flow just before the top of the column dries. Dilute the product to be purified in a minimum volume of the eluent, and apply the solution carefully on the top of the column. Once the material penetrates into the silica gel, add 5 mL of eluent and let it penetrate in the silica gel once again. Repeat this operation several times before adding more eluent. Collect the effluent in glass tubes and analyze by TLC. Combine the fractions containing the pure product, and concentrate to dryness under reduced pressure.

4. The final product needs to be kept under high vacuum overnight at 50 °C to remove the excess of benzyl alcohol.
5. To generate dry HCl (g), add concentrated sulfuric acid to potassium chloride. Dry the HCl gas produced in this process by passing it through sulfuric acid before allowing it to bubble directly into the reaction flask. A Pasteur pipette may be used for bubbling the gas; do not use a metal needle.
6. TLC (hexanes/EtOAc 1:1) showed no change in the starting material but $^1$H NMR indicated complete conversion of starting material to chloromethyl ether. This compound is unstable, and $^1$H NMR showed significant decomposition after a week's time at room temperature; it should be stored under an inert atmosphere at −20 °C.
7. Reflux time is 30 min for A, C, or T and 2 h for G.
8. TLC characterization yielded the following Rf values: T, 0.48 (MeOH/$CH_2Cl_2$ 1:19); C, 0.61 (MeOH/$CH_2Cl_2$ 1:9); A, 0.59 (MeOH/$CH_2Cl_2$ 8:92); G, 0.59 (MeOH/$CH_2Cl_2$ 1:9).
9. Reflux the mixture gently (90–100 °C) for 30 min (C and T) or overnight (A and G). Reaction progress was monitored by TLC (MeOH/$CH_2Cl_2$ 1:9) and also by visual inspection of the Pd (black) catalyst that turned from a fine powder to a clumpy solid upon reaction completion.
10. Filter warm over Celite.
11. T, Rf: 0.29 (MeOH/$CH_2Cl_2$ 6:94); C, Rf: 0.18 (MeOH/$CH_2Cl_2$ 1:9); A, Rf: 0.34 (MeOH/$CH_2Cl_2$ 1:9); G, Rf: 0.31 (MeOH/$CH_2Cl_2$ 15:85).
12. Using a pH meter.
13. The compound was stored at −78 °C under argon.
14. The starting material was coevaporated twice with freshly distilled pyridine under high vacuum with a dry ice/acetone trap and then dried under high vacuum overnight.
15. The reaction progress was monitored by TLC (11:7:1 propanol/$NH_4OH$/$H_2O$).
16. Prepare the medium by weighing out the required amount of dry powder, and suspend it in the binding buffer. Pack the Sephadex media in the column. Add the buffer and collect the effluent in glass tubes and analyze using a UV spectrometer. Combine the fractions containing the pure product, and concentrate to dryness with a SpeedVac evaporator.
17. Assemble glass plates (20 × 52 cm) using 1.5-mm uniform thickness PTFE spacers. Mix 42.4 g of urea, 50 mL of acrylamide/bis-acrylamide 29:1 solution, 10 mL of 10× TBE, 8 mL of deionized water, 600 μL of 10% APS solution (1 g of ammonium persulfate is dissolved in 10 mL $H_2O$; this

solution has to be prepared just prior to use), and 60 μL TEMED immediately before pouring the gel between plates. Degas the mixture using an aspirator vacuum for several minutes. Insert the flat side of a 1.5-mm gel comb into the solution at the top of the gel plates, being careful to avoid bubbles. Allow the system to polymerize for 1–2 h. Remove the gel comb. *Caution: acrylamide and bis-acrylamide are hazardous chemicals. Work in a well-ventilated fume hood taking extreme care in minimizing contact.*

18. The chemically synthesized modified cytosine phosphoramidite (FNA-C) was incorporated into oligonucleotides with efficiencies comparable to the commercially available natural phosphoramidites. Cleavage from the solid support and deprotection was accomplished in concentrated $NH_4OH$ for 15 h at 55 °C. Oligonucleotides were purified by preparative gel electrophoresis and quantified by UV absorbance at 260 nm and 70 °C. The extinction coefficients for the natural nucleotides used for the calculation were as follows: dCMP, 7050; TMP, 8840; dGMP, 12,010; and FNA-CMP, taken as dCMP.
19. Ensure a homogeneous mixture before adding each reagent.
20. Check by TLC eluting with 11:2:7 (v/v/v) *n*-propanol/ammonium hydroxide/water to confirm the completion of the reaction. TLC (11:2:7 v/v/v *n*-propanol/ammonium hydroxide/water): Rf = 0.67.
21. Store the pure activated monomers obtained indefinitely at −80 °C under nitrogen.

## Acknowledgments

This work was supported by grants from the NASA Astrobiology Institute and the NASA Exobiology Program.

### References

1. Echols H, Goodman MF (1991) Fidelity mechanisms in DNA replication. Annu Rev Biochem 60:477–511
2. Orgel LE (2004) Prebiotic chemistry and the origin of the RNA world. Crit Rev Biochem Mol Biol 39:99–123
3. Orgel LE (1992) Molecular replication. Nature 358:203–209
4. Hill AR, Orgel LE (1993) The limits of template-directed synthesis with nucleoside-5-′-phosphoro(2-methyl)imidazolides. Orig Life Evol Biosph 23:285–290
5. Prywes N, Blain JC, Del Frate F, Szostak JW (2016) Nonenzymatic copy of RNA templates containing all four letters is catalyzed by activated oligonucleotides. Elife 5:17756
6. Li L, Prywes N, Tam CP, O'Flaherty DK, Lelyveld VS, Izgu EC, Pal A, Szostak JW (2017) Enhanced nonenzymatic RNA copying with 2-aminoimidazole activated nucleotides. J Am Chem Soc 139:1810–1813
7. Ruiz-Mirazo K, Briones C, de la Escosura A (2014) Prebiotic systems chemistry: new perspectives for the origins of life. Chem Rev 114:285–366

8. Engelhart AE, Hud NV (2010) Primitive genetic polymers. Cold Spring Harb Perpect Biol 2:a002196
9. Cleaves HJ, Meringer M, Goodwin J (2015) 227 views of RNA: is RNA unique in its chemical isomer space? Astrobiology 15:538–558
10. Joyce GF (2002) The antiquity of RNA-based evolution. Nature 418:214–221
11. Nielsen PE, Egholm M, Berg RH, Buchardt O (1991) Sequence-selective recognition of DNA by strand displacement with a thymine-substituted polyamide. Science 254:1497–1500
12. Nelson KE, Levy M, Miller SL (2000) Peptide nucleic acids rather than RNA may have been the first genetic molecule. Proc Natl Acad Sci U S A 97:3868–3871
13. Schoning KJ, Scholtz P, Guntha S, Wu X, Krishnamurthy R, Eschenmoser A (2000) Chemical Etiology of Nucleic Acid Structure: The α-Threofuranosyl-(3′→2′) Oligonucleotide System. Science 290:1347–1351
14. Zhang LL, Peritz A, Meggers E (2005) A Simple Glycol Nucleic Acid. J Am Chem Soc 127:4174–4175
15. Joyce GF, Schwartz AW, Miller SL, Orgel LE (1987) The case for an ancestral genetic system involving simple analogues of the nucleotides. Proc Natl Acad Sci U S A 84:4398–4402
16. Heuberger BD, Switzer C (2008) A pre-RNA candidate revisited: both enantiomers of flexible nucleoside triphosphates are DNA polymerase substrates. J Am Chem Soc 130:412–413
17. Chaput JC, Switzer C (2000) Nonenzymatic oligomerization on templates containing phosphodiester-linked acyclic glycerol nucleic acid analogues. J Mol Evol 51:464–470
18. Merle Y, Bonneil E, Merle L, Sagi J, Szemzo A (1995) Acyclic oligonucleotide analogs. Int J Biol Macromol 17:239–246
19. Merle L, Spach G, Merle Y, Sagi J, Szemzo A (1993) Some biochemical-properties of an acyclic oligonucleotide analog – a plausible ancestor of DNA. Orig Life Evol Biosph 23:91–103
20. Schneider KC, Benner SA (1990) Oligonucleotides containing flexible nucleoside analogs. J Am Chem Soc 112:453–455
21. Joyce GF, Visser GM, van Boeckel CAA, van Boom JH, Orgel LE, van Westrenen J (1984) Chiral selection in poly(C)-directed synthesis of oligo(G). Nature 310:602–604
22. Joyce GF, Inoue T, Orgel LE (1984) Non-enzymatic template-directed synthesis on RNA random copolymers. J Mol Biol 176:279–306

# Chapter 15

# Artificial Nucleosides as Diagnostic Probes to Measure Translesion DNA Synthesis

**Jung-Suk Choi and Anthony Berdis**


## Abstract

The misreplication of damaged DNA, a biological process termed translesion DNA synthesis (TLS), produces a large number of adverse effects on human health. This chapter describes the application of an artificial nucleoside/nucleotide system that functions as a biochemical probe to quantify TLS activity under in vitro and in vivo conditions. For in vitro studies, the artificial nucleotide, 3-ethynyl-5-nitroindolyl-2′-deoxyriboside triphosphate (3-Eth-5-NITP), is used as it is efficiently inserted opposite an abasic site, a highly pro-mutagenic DNA lesion produced by several types of DNA-damaging agents. The placement of the ethynyl moiety allows the incorporated nucleoside triphosphate to be selectively tagged with azide-containing fluorophores via "click" chemistry. This reaction provides a facile way to quantify the extent of nucleotide incorporation opposite this and other noninstructional DNA lesions. The corresponding nucleoside, 3-Eth-5-NIdR, can be used to monitor TLS activity in hematological and adherent cancer cells treated with compounds that produce noninstructional DNA lesions. As described above, visualizing the replication of these lesions is achieved using copper-catalyzed "click" chemistry to tag the ethynyl moiety present on the nucleotide with fluorogenic probes. This technique represents a new diagnostic approach to quantify TLS activity inside cells. In addition, the application of this "clickable" nucleoside provides a chemical probe to identify cells that become drug resistant by the facile replication of noninstructional DNA lesions produced by DNA-damaging agents.


**Key words** DNA damage, DNA polymerization, Translesion DNA synthesis, Nucleoside analogs, Chemotherapy

## 1 Introduction

DNA is often referred to as the "molecule of life" since it contains the genetic blueprint encoding for all organisms on this planet. In this case, the information provided by the spatial arrangement of the four natural nucleobases (adenine, cytosine, guanine, and thymine) in DNA provides a fundamental code that ultimately allows for RNA and protein synthesis. As a result, maintaining the integrity of the genetic blueprint is essential for a cell to function properly. Unfortunately, the functional groups present on these nucleobases are highly reactive and susceptible to modifications

Nathaniel Shank (ed.), *Non-Natural Nucleic Acids: Methods and Protocols*, Methods in Molecular Biology, vol. 1973,
https://doi.org/10.1007/978-1-4939-9216-4_15, © Springer Science+Business Media, LLC, part of Springer Nature 2019

caused by a large number of endogenous and exogenous DNA-damaging agents [1–3]. These reactions lead to the formation of DNA lesions that can adversely change the proper coding information contained in DNA. To help prevent catastrophic consequences caused by mutagenesis, cells possess a number of DNA repair pathways that can correct these formed DNA lesions to revert the damaged DNA back to its original state [4–6]. However, some DNA lesions escape repair and can be inappropriately replicated by DNA polymerases in a process termed translesion DNA synthesis [7–9]. The process of translesion DNA synthesis involves the activity of numerous different DNA polymerases and is highly regulated under normal cellular conditions [10, 11]. Despite this tight regulation, dysfunctional translesion DNA synthesis occurs quite frequently and plays an important role in several pathological conditions including neurological disorders and cancer [12–14]. This process can also influence how patients respond to DNA-damaging agents used during chemotherapy [15]. The ability of DNA polymerases to effectively replicate the DNA lesions formed by these anticancer agents is often reflected in acquired drug resistance as well as the generation of more aggressive cancers caused by an increased frequency in mutagenic events [16–18].

To better understand translesion DNA synthesis at the molecular and cellular level, we developed a series of artificial nucleosides and nucleotides that can visualize the replication of damaged DNA [19–22]. Visualization is achieved using "click" chemistry to covalently attach azide-containing fluorophores to the ethynyl group present on the nucleoside analog after its incorporation opposite damaged DNA. Microscopy techniques can be used with adherent and hematological cancer cells to evaluate if the extent of nucleotide incorporation into genomic DNA is enhanced after treatment with DNA-damaging agents. These approaches can be used with high throughput techniques such as flow cytometry and plate reader assays to quantify translesion DNA synthesis across a variety of cancer cell lines with multiple DNA-damaging agents. Collectively, the diagnostic capabilities of these synthetic nucleoside analogs represent a new paradigm in basic science endeavors to study translesion DNA synthesis at the cellular level in addition to translational efforts using personalized medicine approaches to identify cancer cells that acquire drug resistance through unregulated translesion DNA synthesis activity.

## 2 Materials

Prepare all solutions using ultrapure water prepared by purifying deionized water to attain a sensitivity of 18 MΩ-cm at 25 °C. Prepare all reagents at 25 °C unless otherwise noted. Store all reagents at either 4 °C or 25 °C, respectively, as indicated. All materials should be of the highest quality available and obtained from commercial sources. Personal protective equipment (PPE) should be worn during all of the procedures described below. Follow all waste disposal regulations when disposing of chemical, biological, and radioactive materials.

### 2.1 Denaturing Polyacrylamide Gel

1. 10× Tris-Borate-EDTA (TBE) Buffer: 1 M Tris base, 1 M boric acid, and 20 mM EDTA (pH 8.0).
2. 1× TBE Buffer.
3. 20% denaturing gel solution (*see* **Note 1**).
4. Ammonium persulfate (APS) (*see* **Note 2**).
5. *N*,*N*,*N*′,*N*′-Tetramethylethylenediamine (TEMED) (*see* **Note 3**).
6. Silane solution: 10 mL dichlorodimethylsilane and 90 mL of cyclohexane (*see* **Note 4**).

### 2.2 Polymerization Reactions

1. 5× Reaction buffer: 250 mM Tris-acetate, 5 mg/mL bovine serum albumin, 50 mM DTT, and 25 mM $MgCl_2$ at pH 7.5 (*see* **Note 5**).
2. 100 mM magnesium chloride solution.
3. 200 mM EDTA solution.
4. DNA substrate.
5. Nucleotide substrate (dNTP) and/or 3-Eth-5-NITP (100 μM).
6. 1× Assay buffer: 50 mM TrisOAc, 1 mg/mL bovine serum albumin, 10 mM DTT, and 5 mM $MgCl_2$ at pH 7.5.
7. DNA polymerase.
8. Small microfuge tube (0.65 mL).
9. Gel-loading dye solution: 90% ultrapure formamide, 1× TBE, 0.1% xylene cyanol, and 0.1% bromophenol blue.
10. Gel electrophoresis equipment.
11. Gel imager and accompanying software to analyze image and quantify data.

### 2.3 In Vitro "Click" Chemistry Reactions

1. 100 μM $CuSO_4$ solution.
2. 100 μM Alexa Fluor 488 azide dye solution (*see* **Note 6**).
3. 10 mM ascorbate solution.

### 2.4 Cell Culture Equipment and Reagents

1. Cancer cell lines (*see* **Note** 7).
2. Biological safety cabinet.
3. Incubator that can maintain a humidified atmosphere of 5% $CO_2$ at 37 °C.
4. RPMI 1640 Media.

5. Fetal bovine serum (FBS).
6. L-Glutamine.
7. Penicillin/streptomycin.
8. 1× phosphate-buffered saline (1× PBS).
9. T-75 flasks.
10. 12-well tissue culture plates.
11. Micropipette (10, 20, 200, and 1000 μL) (*see* **Note 8**).
12. Serological pipette (5, 10, and 25 mL).
13. Aspirator.
14. Light microscope.
15. Centrifuge (5000 × *g*, 5 min).
16. Trypan blue solution (*see* **Note 9**).
17. Hemocytometer (*see* **Note 10**).
18. Cell counter.

#### 2.5 Click Chemistry to Visualize Cellular Translesion DNA Synthesis

1. Dimethyl sulfoxide (DMSO).
2. 5-Ethynyl-2′-deoxyuridine (EdU) (*see* **Note 11**).
3. Alexa Fluor™ 488 azide (*see* **Note 12**).
4. Copper solution (10 mM).
5. 100 mM 3-Eth-5-NIdR (*see* **Note 13**).
6. "Click-reaction" additive buffer.
7. "Click-reaction" saponin-based permeabilization and wash reagent.
8. 1× phosphate-buffered saline (pH 7.5).
9. Fixative solution (4% formaldehyde in 1× PBS).
10. Permeabilization reagent (0.5% Triton X-100 in 1× PBS, 1% bovine serum albumin in 1× PBS).
11. Coverslips and microscope slides.
12. Mounting media.
13. Fluorescence microscope (*see* **Note 14**).

## 3 Methods

Perform all procedures and reactions at 25 °C unless otherwise specified.

#### 3.1 Polyacrylamide Gel Electrophoresis

1. Clean both glass plates and spacers thoroughly with soap and deionized water.
2. After drying the plates, apply a thin coat of silane solution using paper towels to spread the solution across one side of the smaller glass plate.

3. Rinse both plates with ethanol and set them aside to dry. The glass plates must be free of grease spots to prevent air bubbles from forming in the gel.
4. Assemble the glass plates with spacers in gel caster.
5. Prepare the 20% polyacrylamide gel solution by adding 200 μL of a 10% APS solution (prepared as described in Subheading 2.1) to 60 mL of denaturing gel solution (prepared as described in Subheading 2.1).
6. Add 15 μL of TEMED to the 20% polyacrylamide gel solution. Mix the solution briefly with a glass stir rod, and carefully pour the polyacrylamide solution into the assembled glass plate apparatus.
7. Immediately insert the comb into the gel. Do not allow air bubbles to become trapped under the teeth of the comb. The tops of the teeth should be slightly higher than the top of the smaller glass plate.
8. Use large binder clips to clamp the comb in place.
9. Use the remaining acrylamide gel solution to fill the gel mold completely. Make sure that no acrylamide solution is leaking from the gel mold.
10. Wait 30–60 min at room temperature to allow the acrylamide to polymerize. The gel can be run after polymerization is complete (*see* **Note 15**).
11. When ready to proceed with electrophoresis, remove the polymerized gel from the gel caster, and carefully pull the comb from the polymerized gel. Carefully remove excess polyacrylamide material from the top of the gel.
12. Insert the gel into sequencing gel box. Add 1× TBE running buffer to the top and bottom reservoirs.
13. Use a Pasteur pipette or a syringe to flush out the wells once more with 1× TBE solution.
14. Mix the DNA samples with the appropriate amount of gel-loading buffer.
15. Load the mixture into the wells using a micropipette equipped with a drawn-out plastic tip. Do not attempt to expel all of the sample from the loading device since this produces air bubbles that can displace the DNA sample from the well.
16. Connect the electrodes to a power pack, turn on the power, and begin the electrophoresis run. Denaturing polyacrylamide gels on small oligonucleotides (<30-mers) are usually run using 1× TBE using the following conditions: volts = 2200 V, watts = 70 W, amperage = 30–34 mA (*see* **Note 16**).
17. Run the gel until the marker dyes (bromophenol blue and xylene cyanol) have migrated the desired distance.

18. Turn off the electric power, disconnect the leads, and discard the electrophoresis buffer from the reservoirs.
19. Detach the glass plates. Lay the glass plates on the bench, and use a spacer or plastic wedge to lift a corner of the upper glass plate.
20. Check that the gel remains attached to the lower (larger) plate. Pull the upper (smaller) plate smoothly away and remove the spacers.
21. For gels containing radioactive material, expose the gel to phosphorimager screen, and quantify using densitometry software supplied by the manufacturer.

### 3.2 Polymerization Reactions

The following section describes quantitative assays that should be performed first to demonstrate that the artificial nucleotide is a bona fide substrate for the specific DNA polymerase(s) under investigation. Assays are typically performed using pseudo-first-order reaction conditions in which the concentration of DNA and nucleotide substrate are in molar excess versus the concentration of DNA polymerase. However, reactions can also be performed using single-turnover reaction conditions in which the concentration of DNA polymerase and nucleotide substrate are maintained in molar excess versus the concentration of DNA substrate.

1. DNA substrate (500 nM) is preincubated with a fixed concentration of nucleotide substrate (~10 μM) in assay buffer. DNA polymerase (10 nM) is added last to initiate the polymerization reaction (*see* **Note 17**).
2. At times ranging from 5 to 600 s, quench a 5 μL aliquot of the reaction by placing it into a small microfuge tube (0.65 mL) containing 5 μL of 200 mM EDTA.
3. Place 10 μL of gel-loading dye into microfuge tube containing quenched polymerization reaction sample.
4. Run samples using 20% denaturing polyacrylamide gel electrophoresis as described above in Subheading 3.1.
5. Obtain gel images using a phosphorimager instrument, and quantify product formation using the software supplied by the manufacturer of the instrument.
6. Calculate product formation using the following relationship: [amount of elongated DNA ($DNA_{n+1}$)]/[amount of initial DNA ($DNA_n$) + amount of elongated DNA ($DNA_{n+1}$)] * concentration of DNA substrate initially used in reaction.
7. Plot product as a function of DNA polymerase and/or DNA lesion to compare the efficiency of nucleotide incorporation.

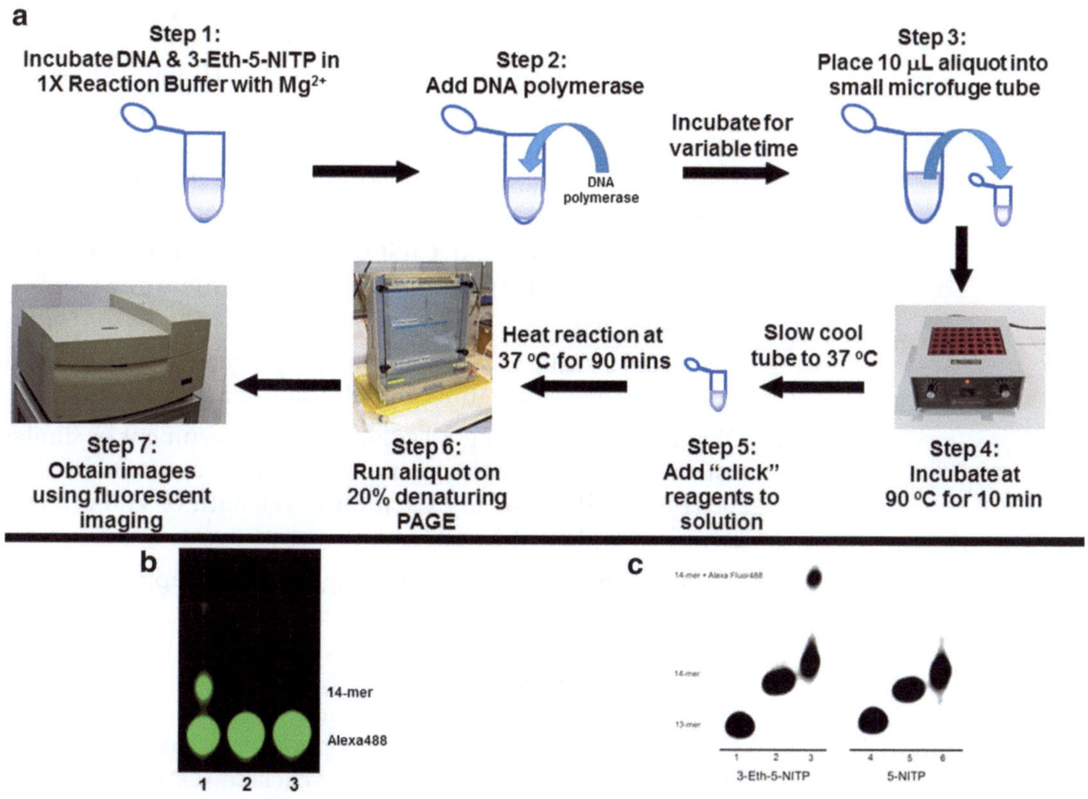


**Fig. 1** (**a**) Flow chart illustrating key steps used to perform in vitro "click" chemistry reactions to monitor translesion DNA synthesis catalyzed by DNA polymerases. Please *see* text for complete experimental details. (**b**) Representative denaturing gel electrophoresis validating that 3-Eth-5-NITP which is incorporated opposite an abasic site can react with an azide-containing fluorophore to visualize translesion DNA synthesis. Please *see* text for complete experimental details. (**c**) Autoradiogram demonstrating the selectivity for the covalent attachment of azide-containing fluorophores to 3-Eth-5-NITP when incorporated opposite an abasic site. The left panel shows "clicking" of 3-Eth-5-NITP opposite an abasic site (top band, lane 3). In contrast, the right panel does not show a change in mobility associated with "clicking" of the azide-containing dye to 5-NITP when inserted opposite an abasic site (top band, lane 6). Please *see* text for complete experimental details

### 3.3 In Vitro "Click" Chemistry Reactions

Figure 1a provides a representative flow chart highlighting key steps in the experimental process described below.

1. Perform polymerization assays as described above by adding DNA polymerase (10 nM) to a preincubated solution containing 10 μM 3-Eth-5-NITP and DNA (500 nM) in 1× reaction buffer and 10 mM $Mg^{2+}$. After initiation of the reaction, incubate at room temperature for the appropriate time interval. In most cases, a reaction time of 10 min is sufficient for adequate incorporation of 3-Eth-5-NITP opposite a noninstructional DNA lesion.

2. Quench the reaction through heat inactivation by transferring a small aliquot of the reaction to a microfuge and placing the microfuge in a heat block set at 90 °C for 10 min. After this time period, remove the sample from the heat block, and allow it to cool to 37 °C (*see* **Note 18**).
3. Subject a 10 μL aliquot of the heat-inactivated reaction to the "clicking" reaction by adding a cocktail containing Alexa Fluor 488 azide (10 μM) and $CuSO_4$ (10 μM) in a Tris-buffered saline solution with 10 mM of ascorbic acid. Allow the reaction to proceed at 37 °C for 90 min. Protect the reaction from light.
4. Run samples using 20% denaturing polyacrylamide gel electrophoresis as described above in Subheading 3.1.
5. Obtain gel images using a Typhoon 9400 instrument or similar instrument capable of detecting fluorogenic compounds. For Alexa Fluor 488, use an excitation wavelength of 488 nm and an emission filter of 526 nm.
6. Quantify product formation using software supplied by the manufacturer of the instrument. Figure 1b provides representative-denaturing gel electrophoresis validating the ability of 3-Eth-5-NITP to react with an azide-containing fluorophore. Lane 1 shows the "clicking" reaction of 3-Eth-5-NITP incorporated opposite abasic-containing DNA. The upper band represents "clicked" DNA containing 3-Eth-5-NIMP, while the lower band represents unreacted Alexa Fluor 488 azide. Lane 2 shows the presence of a single fluorogenic species in "clicked" reactions of DNA and 3-Eth-5-NITP without DNA polymerase. Lane 3 shows Alexa Fluor 488 azide alone. Figure 1c shows an autoradiogram of "clicked" DNA. Lane 1 represents radiolabeled primer (13-mer) in the absence of polymerase and nucleotide substrate. Lane 2 represents the incorporation of 3-Eth-5-NITP opposite an abasic site. Lane 3 represents the "clicking" reaction DNA containing 3-Eth-5-NITP opposite an abasic site. Lane 4 represents radiolabeled primer (13-mer) in the absence of polymerase and nucleotide substrate. Lane 5 represents the incorporation of 5-NITP opposite an abasic site. Lane 6 shows the "clicking" reaction for abasic site containing DNA with 5-NITP incorporated opposite the DNA lesion. The upper band represents "clicked" primer containing 3-Eth-5-NIMP, while the lower band represents unreacted 3-Eth-5-NIMP in the primer.

### 3.4 Cell-Based Assays to Measure Translesion DNA Synthesis

#### 3.4.1 Cell Culture Experiments

1. Culture MOLT4 cells in a humidified atmosphere of 5% $CO_2$ at 37 °C. Maintain cells in ATCC-formulated RPMI-1640 media with the following components added to the base medium. Fetal bovine serum (FBS) to a final concentration of 10%, L-glutamine to a final concentration of 5%, and penicillin/streptomycin to a final concentration of 0.5% (*see* **Note 19**).
2. Measure cell density by counting cells with a hemocytometer (*see* **Note 20**).
3. Dilute cells by calculating the required dilution factor (*see* **Note 21**).
4. Dilute cells in the correct amount of culture media. Make sure that cells are well mixed by using a pipette to suspend any cells that may have settled in the flask during the cell counting process.
5. Pipette 1 mL of cells into 12-well plate.
6. Treat cells with 0.1% DMSO (vehicle), 10 μg/mL 3-Eth-5-NIdR alone, DNA-damaging alone, and 10 μg/mL 3-Eth-5-NIdR combined with DNA-damaging agent. Incubate cells at 37 °C for 48 h in an incubator with 5% $CO_2$ (*see* **Note 22**).
7. After 48 h, harvest cells by centrifuge (5000 × *g* for 5 min) and wash with 1× PBS.
8. Fix the cells by resuspension with 0.5 mL cold 70% methanol. Store overnight at −20 °C.

### 3.5 "Click" Chemistry to Visualize Cellular Translesion DNA Synthesis

Allow vials to warm to room temperature before opening unless otherwise specified

1. Prepare 50 mL of 1× Click saponin-based permeabilization, and wash reagent by adding 5 mL of 10× Click saponin-based permeabilization and wash reagent to 45 mL of 1% BSA in PBS.
2. Make a 10× stock solution of the Click-EdU buffer additive by adding 2 mL of deionized water to the vial, and mix until the Click-EdU buffer additive is fully dissolved.
3. Wash the cells once with 3 mL of 1% BSA in PBS, pellet the cells by centrifugation, and remove the supernatant. Resuspend cells with 0.3 mL of saponin-based permeabilization and wash buffer.
4. Prepare Click reaction cocktail by adding 1× PBS, copper additive, Alexa Fluor™ 488 azide, and reaction buffer additive (*see* **Note 23**).
5. Add 0.5 mL of Click reaction cocktail to each sample tube and mix well by vortexing the tube for 15 s.
6. Incubate the reaction mixture for 90 min at 25 °C. Wrap sample tube in aluminum foil to protect from light.

7. Centrifuge cells and wash cells twice with 1× Click buffer.
8. Dislodge cell pellets using 0.5 mL solution of 10 μg/mL PI and RNAase A in saponin-based permeabilization buffer. Incubate cells for 15 min with 1 μg/mL DAPI prior to analysis.
9. Resuspend cells in the correct amount of mounting solution. Transfer 10 μL of cells onto a microscope slide and gently cover with coverslip.
10. Obtain images using an EVOS Advanced microscope (20× or 40× magnification). Images provided in Fig. 2b shows representative microscopy images of MOLT4 cells treated with 100 μM temozolomide (TMZ) and 10 μg/mL 3-Eth-5-NIdR. The merged image of DAPI and Alexa Fluor 488 shows nuclear co-localization, thus validating the enzymatic incorporation of 3-Eth-5-NITP opposite damaged DNA. Microscopy analyses monitoring the incorporation of 3-Eth-5-NIdR opposite abasic sites generated by TMZ

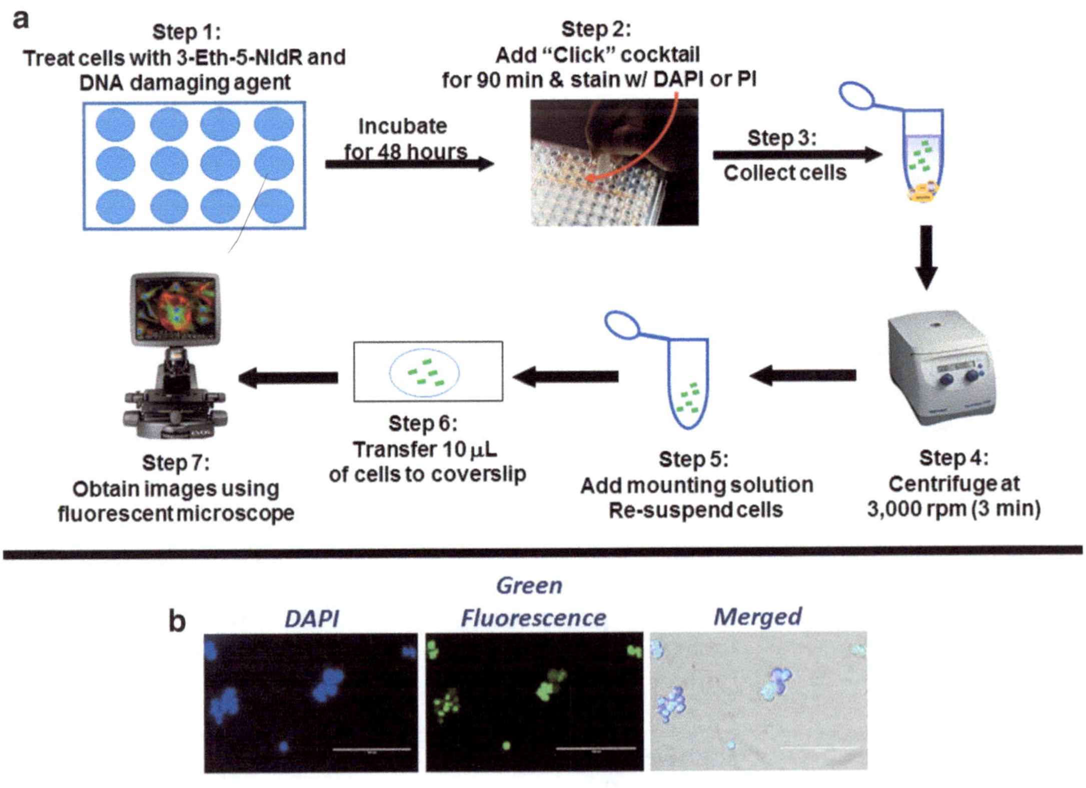


**Fig. 2** Cell-based assays to measure translesion DNA synthesis. (**a**) Flow chart illustrating key steps used to perform "click" chemistry to visualize cellular translesion DNA synthesis. Please *see* text for complete experimental details. (**b**) Representative microscopy images of MOLT4 leukemia cells treated with 100 μM temozolomide (TMZ) and 10 μg/mL 3-Eth-5-NIdR. The merged image of DAPI and Alexa Fluor 488 shows nuclear co-localization, thus validating the enzymatic incorporation of 3-Eth-5-NITP opposite damaged DNA

treatment. Cells were co-treated with 100 μM TMZ, 10 μg/mL 3-Eth-5-NIdR, and a combination of 100 μM TMZ and 10 μg/mL 3-Eth-5-NIdR for 3 days. Cells treated with a combination of 100 μM TMZ and 10 μg/mL 3-Eth-5-NIdR display significantly higher levels of green fluorescence that co-localizes in the nucleus compared to cells treated with 100 μM TMZ or 10 μg/mL 3-Eth-5-NIdR alone.

## 4 Notes

1. Avoid direct handling of acrylamide and bis-acrylamide powder as these reagents are potentially harmful. Always use PPE (lab coats, gloves, eye goggles, and respiratory mask) when handling these materials. The final solution may be stored at 4 °C in the dark and is stable for several weeks after filtering. Ready-to-use solutions of acrylamide/bis-acrylamide are also commercially available.
2. For best results, APS solutions should be made fresh everyday.
3. Store at 4 °C.
4. Dichlorodimethylsilane is a potentially dangerous compound that should be handled with extreme caution. It is a flammable liquid and vapor may cause flash fire. This agent is corrosive and very water-reactive. Exposure can cause eye and skin burns. May cause irritation of the respiratory tract with burning pain in the nose and throat, coughing, wheezing, shortness of breath, and pulmonary edema. Inhalation may be fatal as a result of spasm, inflammation, edema of the larynx and bronchi, chemical pneumonitis, and pulmonary edema. Perform the transfer of solutions in a chemical hood and store at 4 °C.
5. Note that alterative reaction buffers can be used. The composition of these reaction buffers depends upon the type of DNA polymerase being investigated.
6. Store 1 mL aliquots at −20 °C.
7. The protocols described here use the leukemia cell line, MOLT4, as a model. However, other hematological and adherent cell lines can be used with minor modifications to these protocols.
8. Use aerosol barrier tips to prevent aerosol contamination with bacteria, viruses, DNA, and organic material. Aerosol barrier tips do not obstruct airflow.
9. Trypan blue is an azo dye that is used to selectively stain dead cells blue. Live cells do not take up the dye and thus remain clear.

10. The hemocytometer is a device usually used to count hematological cells. The gridded area of the hemocytometer consists of nine 1 × 1 mm (1 $mm^2$) squares. The number of cells in the each gridded area can be determined by direct counting using a microscope, and visually distinguishable cells (live versus dead) can be differentially counted.
11. After use, store any remaining stock solution at ≤−20 °C. This solution is stable for up to 1 year.
12. Alexa Flour 488 can be stored at 2–8 °C. Protect from light.
13. 3-Eth-5-NIdR can be purchased from Red5 Pharmaceuticals, LLC. The nucleoside can be stored at 2–8 °C. Protect from light.
14. The protocols described here use EVOS Advanced microscope with 20×, 40× and/or 60× magnification. However, other fluorescent microscope systems can be used to give identical results.
15. Alternatively, the gel can be run the following day. In this case, cover the comb and the top of the gel with dampened paper towels soaked in deionized water. Cover the dampened paper towels in plastic wrap, place binder clips to secure the plastic wrap, and store the gel at room temperature overnight.
16. Note that differential heating in the center of the gel may cause distortion of the DNA bands. Run duration depends upon several factors including the percentage of polyacrylamide used, ionic strength of the buffer, and gel thickness. A typical run lasts 2–4 h. Check on the temperature of the gel to make sure that the gel does not overheat as this can cause the glass plates to crack. A gel thermometer can help to monitor the temperature during the run.
17. Polymerization reactions using human DNA polymerases such as pol delta, pol epsilon, pol eta, pol iota, pol kappa, pol lambda, and pol mu are performed using an assay buffer consisting of 50 mM TrisOAc, 1 mg/mL bovine serum albumin, 10 mM DTT, and 5 mM $MgCl_2$ at pH 7.5.
18. Do not use EDTA as the metal-chelating agent will inhibit the $Cu^{2+}$-dependent "click" reaction.
19. Warm buffers and media to 37 °C.
20. Calculate the cell density using the following formula: Cell density = (Total number of cells counted/Number of gridded areas measured) * 20,000 cells/mL.
21. Dilution factor = $N$ (cell plating number/mL)/counted cells/mL.
22. In the protocols described here, temozolomide was used as the DNA-damaging agent to produce abasic sites. However, other

DNA-damaging agents such as doxorubicin, cisplatin, chlorambucil, carmustine, and ionizing radiation can be used.

23. Prepare cocktail within 15 min prior to the next step.

## References


1. Greenberg MM (2012) The formamidopyrimidines: purine lesions formed in competition with 8-oxopurines from oxidative stress. Acc Chem Res 45:588–597
2. Cadet J, Douki T, Ravanat JL (2008) Oxidatively generated damage to the guanine moiety of DNA: mechanistic aspects and formation in cells. Acc Chem Res 41:1075–1083
3. Neeley WL, Essigmann JM (2006) Mechanisms of formation, genotoxicity, and mutation of guanine oxidation products. Chem Res Toxicol 19:491–505
4. Krokan HE, Bjørås M (2013) Base excision repair. Cold Spring Harb Perspect Biol 5: a012583
5. Spivak G (2015) Nucleotide excision repair in humans. DNA Repair (Amst) 36:13–18
6. Jasin M, Rothstein R (2013) Repair of strand breaks by homologous recombination. Cold Spring Harb Perspect Biol 5:a012740
7. Goodman MF, Tippin B (2000) Sloppier copier DNA polymerases involved in genome repair. Curr Opin Genet Dev 10:162–168
8. Goodman MF, Woodgate R (2013) Translesion DNA polymerases. Cold Spring Harb Perspect Biol 5:a010363
9. Sale JE (2013) Translesion DNA synthesis and mutagenesis in eukaryotes. Cold Spring Harb Perspect Biol 5:a012708
10. Sutton MD, Walker GC (2001) Managing DNA polymerases: coordinating DNA replication, DNA repair, and DNA recombination. Proc Natl Acad Sci U S A 98:8342–8349
11. Zhao L, Washington MT (2017) Translesion synthesis: insights into the selection and switching of DNA polymerases. Genes (Basel) 8(1). pii: E24
12. Zhao J, Yu S, Zheng Y, Yang H, Zhang J (2017) Oxidative modification and its implications for the neurodegeneration of Parkinson's disease. Mol Neurobiol 54:1404–1418
13. Makridakis NM, Reichardt JK (2012) Translesion DNA polymerases and cancer. Front Genet 3:174
14. Kunz BA, Straffon AF, Vonarx EJ (2000) DNA damage-induced mutation: tolerance via translesion synthesis. Mutat Res 451:169–185
15. Calderón-Montaño JM, Burgos-Morón E, Orta ML, López-Lázaro M (2014) Effect of DNA repair deficiencies on the cytotoxicity of drugs used in cancer therapy – a review. Curr Med Chem 21:3419–3454
16. Nicolay NH, Helleday T, Sharma RA (2012) Biological relevance of DNA polymerase β and translesion synthesis polymerases to cancer and its treatment. Curr Mol Pharmacol 5:54–67
17. Roy U, Schärer OD (2016) Involvement of translesion synthesis DNA polymerases in DNA interstrand crosslink repair. DNA Repair (Amst) 44:33–41
18. Haynes B, Saadat N, Myung B, Shekhar MP (2015) Crosstalk between translesion synthesis, Fanconi anemia network, and homologous recombination repair pathways in interstrand DNA crosslink repair and development of chemoresistance. Mutat Res Rev Mutat Res 763:258–266
19. Salehan MR, Morse HR (2013) DNA damage repair and tolerance: a role in chemotherapeutic drug resistance. Br J Biomed Sci 70:31–40
20. Devadoss B, Lee I, Berdis AJ (2013) Spectroscopic analysis of polymerization and exonuclease proofreading by a high-fidelity DNA polymerase during translesion DNA synthesis. Biochim Biophys Acta 1834:34–45
21. Motea EA, Lee I, Berdis AJ (2012) A non-natural nucleoside with combined therapeutic and diagnostic activities against leukemia. ACS Chem Biol 7:988–998
22. Motea EA, Lee I, Berdis AJ (2012) Development of a 'clickable' non-natural nucleotide to visualize the replication of non-instructional DNA lesions. Nucleic Acids Res 40:2357–2367

# Chapter 16

# FRET Assay for Ligands Targeting the Bacterial A-Site RNA

**Renatus W. Sinkeldam and Yitzhak Tor**

## Abstract

A robust, fluorescence-based analysis and discovery platform is described for bacterial A-site binders. The assay relies on an incorporated isomorphic fluorescent uridine analog, which substitutes the A-site's U1406 and serves as a FRET donor to an A-site bound coumarin-labeled aminoglycoside that serves as the FRET acceptor. Binding efficiency of unlabeled A-site ligands can be determined by competition experiments, where the acceptor-labeled aminoglycoside is displaced. The replacement efficiency is gauged by the concentration-dependent loss of the sensitized FRET acceptor's signal with concomitant restoration of the donor's emission. Plotting the relative emission intensity of both the donor and acceptor as a function of ligand concentration followed by fitting of the data points to a dose-response curve yields $IC_{50}$ values, one possible measure of the antibiotic potency of new A-site binders.

**Key words** Fluorescent nucleosides, FRET assay, Bacterial A-site, A-site binders, Aminoglycosides, Fluorescence spectroscopy

## 1 Introduction

Many naturally occurring antibiotics target the ribosomal RNA and meddle with translation [1–5]. The aminoglycoside antibiotics, for example, have evolved to corrupt bacterial protein production by targeting the ribosomal decoding site and lowering its fidelity [6–10]. These potent antibiotics bind a loop within the 16S rRNA and impede the conformational dynamics of A1492 and A1493, two key adenosines in the A-site (Fig. 1) [5, 11–13]. It is suggested that by stabilizing a conformation similar to the one induced by the cognate acyl–tRNA–mRNA complex, the ribosome's ability to discriminate between cognate and non-cognate tRNAs drops and translational fidelity is compromised [14–19].

The autonomous nature of the ribosomal A-site allows one to implement a hierarchical screening approach for the discovery of new antibiotics, which starts with binding assays using small RNA constructs as the primary selection filter for active compounds. Such assays, contrary to traditional approaches based on functional screens, facilitate the identification of target-specific ligands as the

Nathaniel Shank (ed.), *Non-Natural Nucleic Acids: Methods and Protocols*, Methods in Molecular Biology, vol. 1973, https://doi.org/10.1007/978-1-4939-9216-4_16, © Springer Science+Business Media, LLC, part of Springer Nature 2019

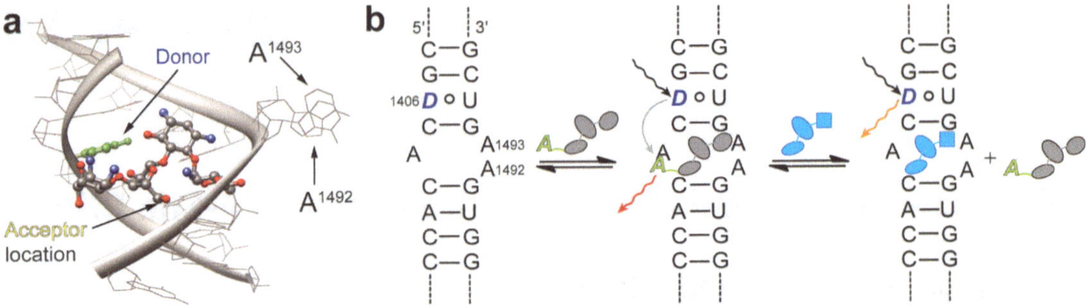


**Fig. 1** (**a**) A model of neomycin bound to the bacterial A-site (PDB 2ET4) revealing how the flipped in/out dynamics of A1492 and A1493 is arrested upon binding of neomycin inducing loss of protein synthesis fidelity and (**b**) a cartoon of the FRET system with an emissive U analog (**D**) as the FRET donor and an A-site ligand labeled with coumarin (**A**) as the FRET acceptor. Excitation of the donor chromophore, following formation of the FRET complex (middle cartoon), results in attenuated donor emission and enhanced sensitized acceptor emission. A titration experiment with an unlabeled competitor ligand can displace the labeled ligand resulting in loss of the FRET signal and concomitant restoration of the donor emission

first step. A number of tools have been developed to assess ligand binding to the ribosomal decoding site [20, 21]. Singly labeled fluorescent A-site constructs, which contain emissive and responsive nucleoside analogues, have shown great promise [22, 23]. Unfortunately, their fluorescence response can frequently be antibiotic-dependent [22, 24]. To overcome this limitation and devise a robust analysis and discovery platform for A-site binders, we developed a FRET assembly, where detection of ligand binding is dependent on the interaction between two chromophores acting as a Förster resonance energy transfer (FRET) pair (Fig. 1). Experimentally, one can then monitor binding and displacement with "FRET accuracy" (Fig. 1).

To realize such an analysis and discovery assay, we substitute one of the native nucleosides in the A-site (U1406), which is proximal to the binding site, with an emissive isomorphic nucleobase analog (**D**) acting as a FRET donor (Fig. 1). We conjugated an appropriate FRET acceptor to aminoglycosides, the cognate binders of the A-site, at positions that are not essential for RNA binding (Fig. 2). Aminoglycosides have been fluorescently labeled before, but typically by modifying the amines that are also essential for specific RNA-binding (*see* [25–27]). Several criteria were taken into consideration in assembling this emissive RNA–ligand complex, including the spectral overlap between donor emission and acceptor absorption and the donor–acceptor distance. We found that the emission spectrum of the fluorescent modified nucleoside (**D**) used as substitute for U1406 overlaps with the absorption of coumarin (**A**) used to label the aminoglycosides [28]. Based on reported crystal structures of aminoglycoside–A-site complexes, the distance between the donor and acceptor was estimated to be 15 Å, well within the experimentally determined critical Förster

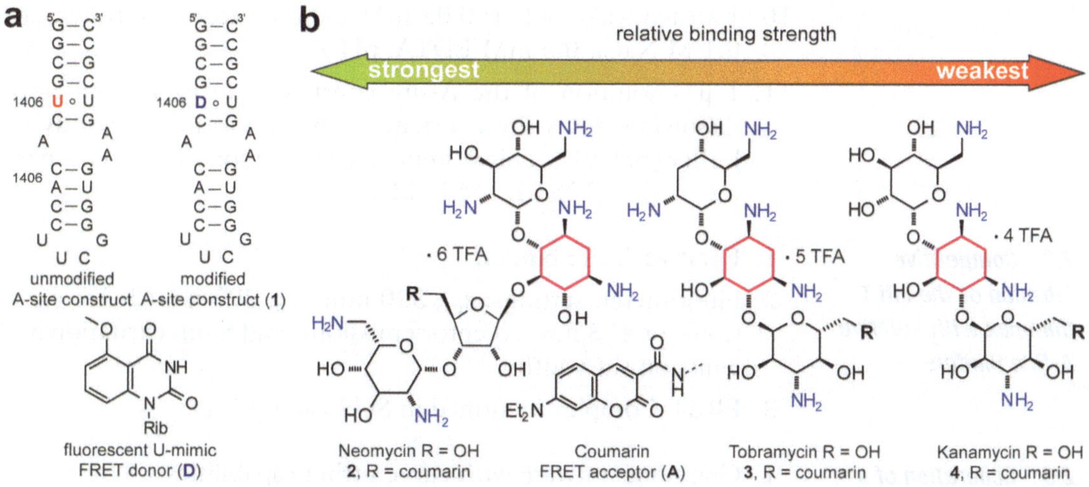


**Fig. 2** (**a**) Non-modified and the modified A-site construct containing the **U**1406 substitute (**1**), (**b**) structures for modified nucleoside (**D**), and coumarin-labeled aminoglycosides neomycin (**2**), tobramycin (**3**), and kanamycin (**4**)

radius of 27 Å. The protocol outlined here provides a step-by-step description of the assembly and application of this assay for assessing the affinity of ligands for the bacterial A-site.

## 2 Materials

### *2.1 Formation of A-Site–Antibiotic FRET Complex*

1. Coumarin-labeled neomycin (**2** · 6 TFA, Mw: 1437.09 g/mol) [28] (*see* **Note 1**).
2. Coumarin-labeled tobramycin (**3** · 5 TFA, Mw: 1193.18 g/mol) [28] (*see* **Note 1**).
3. Coumarin-labeled kanamycin (**4** · 4 TFA, Mw: 1183.33 g/mol) [29] (*see* **Note 1**).
4. 5, 2, 1, 0.5, 0.1, 0.05, and 0.02 mM stock solutions of coumarin-labeled neomycin (**2**) in water.
5. 10, 5, 2, 1, 0.5, 0.2, and 0.1 mM stock solutions of coumarin-labeled tobramycin (**3**) in water.
6. 20, 10, 5, 2, 1, 0.5, and 0.2 mM stock solution of coumarin-labeled kanamycin (**4**) in water.
7. Fluorimeter; settings: $\lambda_{ex}$, 320 nm; $\lambda_{em}$, 395 nm (donor emission), or 473 nm (acceptor emission), and 8 nm excitation and emission slit-widths (*see* **Notes 2** and **3**).
8. A small-volume quartz fluorescence cell (100 μL) with a path length of 1.0 cm.
9. A-site construct (**1**, Mw: 8720.5 g/mol) [28].

10. $1\times$ cacodylate buffer: 0.02 mM sodium cacodylate trihydrate, 0.1 M NaCl, 0.5 mM EDTA, pH 7.
11. 1 μM solution of the A-site construct (**1**) in $1\times$ buffer by dissolving the A-site construct (0.65 nmol) in $1\times$ cacodylate buffer (650 μL). This volume is sufficient for 5 titration experiments (5 × 125 μL = 625 μL).

#### 2.2 Competitive Titration of the FRET Complex with Putative A-Site Binders

1. Putative A-site binder.
2. Fluorimeter; settings: $\lambda_{ex}$, 320 nm; $\lambda_{em}$, 395 nm (donor emission), or 473 nm (acceptor emission), and 8 nm excitation and emission slit-widths.
3. FRET Complex identified in Subheading 3.1.

#### 2.3 Generation of a Titration Curve

1. Graphing software with curve fitting capabilities.

## 3 Method

#### 3.1 Formation of A-Site–Antibiotic FRET Complex

1. Anneal the RNA construct by heating the solution to 75 °C for 5 min, followed by cooling to room temperature over 2 h.
2. Place the solution on ice for 30 min prior to performing the titration experiments.
3. Bring the cuvette holder to and keep at 21 °C (room temperature).
4. Use an excitation wavelength of 320 nm for all emission spectra. Follow changes in emission at 395 nm (donor emission) and 473 nm (acceptor emission).
5. Fill the cuvette with $1\times$ cacodylate buffer (125 μL).
6. Record a background spectrum (buffer), and (automatically) subtract from each subsequently taken sample spectrum.
7. Rinse the cuvette, with plenty of deionized water followed by ethanol, and dry the cuvette with pressurized air.
8. Add the A-site construct (**1**) in $1\times$ cacodylate buffer (125 μL) to the cuvette.
9. Record the emission spectrum.
10. Choose a labeled aminoglycoside FRET acceptor based on the desired relative binding strength to the A-site (neomycin > tobramycin > kanamycin). If the binding strength of the competitor ligand is not known or cannot be easily estimated, formation of a FRET complex with kanamycin A, the weakest place-holder antibiotic, is recommended for a first trial.

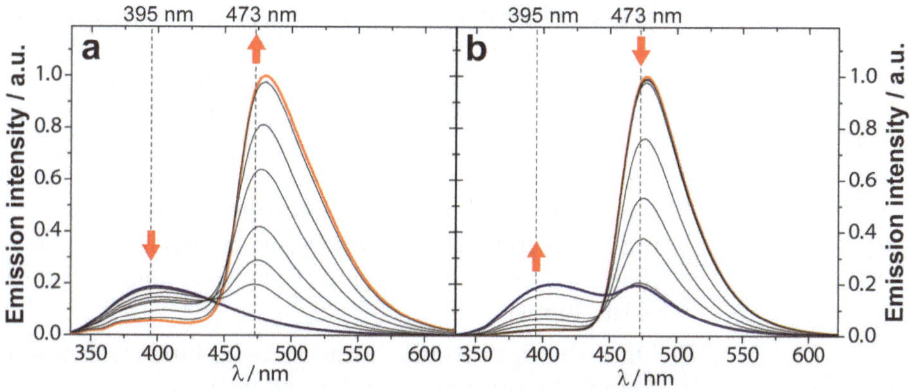


**Fig. 3** (**a**) Formation of the FRET complex by addition of coumarin-labeled aminoglycoside to the modified A-site construct (**1**) (blue line) until saturation of the FRET emission at 473 nm (orange line) with the loss of donor emission at 395 nm and (**b**) example of a displacement study revealing how the FRET emission ($\lambda_{max}$: 473 nm) decreases with concomitant restoration of the donor emission ($\lambda_{max}$: 395 nm)

11. Titrate from the lowest stock solution of coumarin-labeled aminoglycoside (**2**, **3**, or **4**) (1 μL), prepared as described in Subheading 2.1, to the cuvette.
12. Gently mix the solution in the cuvette by using a 30 μL pipette, fitted with a clean pipette tip, to suck up and push back the solution multiple times while preventing air bubble formation.
13. Record the emission spectrum after each addition.
14. Repeat **steps 9–11** with increasingly more concentrated aminoglycoside stock solutions until the FRET signal is saturated (minimum intensity at 395 nm and maximum emission intensity at 473 nm) in 10 addition steps (Fig. 3a).

### 3.2 Competitive Titration of the FRET Complex with Putative A-Site Binders

Since the affinity of A-site binders not previously studied is unknown, a trial titration is performed to determine the concentration of A-site binder stock solutions to be used.

1. Prepare a FRET complex of choice (with either the coumarin-labeled neomycin, tobramycin, or kanamycin as the FRET acceptor) in a clean and dry cuvette (*see* Subheading 3.1).
2. Titrate 0.01 μM stock solution made with the putative A-site binder (1 μL) into the cuvette.
3. Gently mix the solution by using a 30 μL pipette, fitted with a clean pipette tip, by sucking up and pushing back the solution multiple times while preventing air bubble formation.
4. Record the emission spectrum.
5. Determine the fluorescence response induced by A-site binder addition by monitoring the increase in the 395 nm emission and concomitant decrease in the 473 nm emission.

6. Repeat **steps 2–5** with increasingly more concentrated A-site binder stock solutions until the FRET signal (473 nm) is minimized and the restored donor emission (395 nm) is maximized in ten 1 μL additions (Fig. 3b). Importantly, the choice of A-site binder stock solution concentrations will depend on the binding strength of the new A-site binder relative to the binding strength of the labeled aminoglycoside chosen to form the FRET complex (neomycin > tobramycin > kanamycin) and therefore needs to be empirically determined for each new A-site binder.
7. Once the right combination of A-site binder stock solutions is found, the titration is performed in triplicate.

### 3.3 Generation of a Titration Curve

1. Calculate the fractional fluorescence saturation ($F_s$) at 395 nm and 473 nm for each titration point according to Eq. 1 where $F_i$ is the fluorescence intensity at each titration data point. $F_0$ and $F_\infty$ are the fluorescence intensities in the absence of A-site binder or at saturation, respectively:

$$F_s = \frac{F_i}{F_0 - F_\infty} \tag{1}$$

2. Average the $F_s$ values for the triplicate measurement and calculate their standard error of mean (use, e.g., MS Excel or OriginPro 8).
3. Calculate the A-site binder concentration in the cuvette for each titration point, taking into account the cumulative change of the total sample volume in the cuvette with each titration of A-site binder stock solution.
4. Plot the averaged $F_s$ value, for the increasing fluorescence intensity at 395 nm and decreasing fluorescence intensity at 473 nm, as a function of the log of the A-site binder concentration (Fig. 4).

### 3.4 Curve Fitting and $IC_{50}$ Calculation

1. The plotted data points are fitted to a dose-response curve (using OriginPro 8 software), using Eq. 2, where $F_0$ and $F_\infty$ are the fluorescence intensity in the absence of A-site binder or at saturation, respectively, and *n* is the Hill coefficient or degree of cooperativity associated with the binding (Fig. 4):

$$F_s = F_0 + \left( \frac{F_\infty [\mathrm{AG}]^n}{[\mathrm{EC}_{50}]^n} + [\mathrm{AG}]^n \right) \tag{2}$$

2. The IC50 value is the concentration A-site binder at half the fractional saturation ($F_s = 0.5$). One can graphically determine the value (Fig. 4) or use the fit to calculate it (recommended).

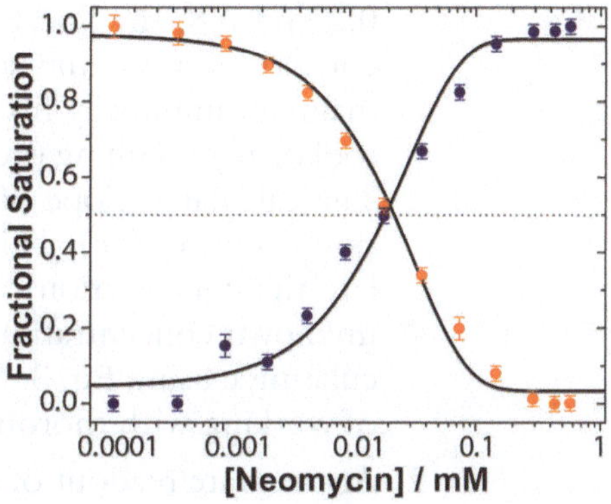


**Fig. 4** A representative plot of the fractional saturation (data points) and their standard error of the mean (error bars) of the decreasing FRET signal (orange circles) and increasing donor signal (blue circles) vs. titrated A-site binder concentration (e.g., neomycin). The solid lines represent the fit of the data points to a dose-response curve

## 4 Notes

1. Concentrations (c) of stock solutions of fluorescently labeled components (A-site constructs or labeled aminoglycosides) are calculated using Beer–Lambert law, Eq. 3, where A (sometimes E is used for extinction) is the measured absorption of a given solvent at a given wavelength, l is the path length of the cuvette (1 for a 1 cm cuvette), and ε is the molar absorption coefficient (also called extinction coefficient) of the fluorescent label at the same wavelength in the same solvent:

$$\varepsilon = \frac{A}{c \cdot l} \tag{3}$$

Hence, when new fluorescent probes are used, it is essential to correctly determine their molar absorptivity ($\varepsilon$) to minimize errors in the calculated concentration of diluted stock solutions. This can be done following the subsequent procedure. The concentrations of the stock solutions can be determined with absorption spectroscopy based on the molar absorptivity of the coumarin label ($\varepsilon_{\text{water, 420 nm}} = 20{,}000\ \text{M}^{-1}\ \text{cm}^{-1}$). Prepare a stock solution of an accurately determined concentration of the fluorescent probe in the solvent that will be used for the titration experiment (ideally by accurately weighing dried material), in this case water or buffer. Make minimally five dilutions of different concentrations from this stock solution that result in absorption ($A$) values that span the range between

0.2 and 0.8, the linear regime of Beer–Lambert law. Plot the concertation vs. the absorption (remove clear outliers but maintain minimally five data points); use the linear regression tool of any graphing program to calculate the best straight line. The calculated slope of this line represents the molar absorption coefficient ($\varepsilon$). This value for ε can now be used to calculate the concertation of the fluorescent probe in solutions of unknown concentrations by measuring its absorption and calculating c using Eq. 3. *See* also a note below on the importance of working with thoroughly purified compounds.

2. An accurate readout of any photophysical experiment is reliant on many factors including proper stock solution preparation and its serial dilutions, temperature control of the cuvette during the measurements, instrument setup, and purity of solvent and reagents used. For the competitive binding studies described here, it is noteworthy that common impurities in commercial aminoglycosides, e.g., the presence of neomycin C in commercial preparations of neomycin B and (potentially fluorescent) by-products in other antibiotics obtained by fermentation, will introduce, often unnoticeable, undesirable errors in the readout. Analysis of all reagents used in the experiment, and their appropriate purification if needed, is thus critically important.

3. Photophysical studies with fluorescently tagged DNA or RNA oligonucleotides benefit from selecting an excitation wavelength removed from the absorption wavelengths of native nucleosides (*ca.* 250–270 nm). The latter are practically non-emissive, but their concentration relative to the fluorescent probe could be very high. This may result in undesirable background emission, lower excitation intensity, and even photochemistry in duplexed oligonucleotides where the nucleobases are in close proximity. For FRET studies, it is important to choose an excitation wavelength for the FRET donor that does not result in excitation of the FRET acceptor. The undesirable consequence of exciting the latter is a background acceptor emission that compromises the accurate determination of the FRET process. In this experiment, it was empirically determined that excitation at 320 nm minimized native nucleosides absorption while preventing direct coumarin excitation. For newly synthesized fluorophores, where their photochemical features are not always known for every sequence context, it is frequently desirable to excite at lower energies (i.e., red-shifted to their absorption maximum), provided the FRET condition listed above is met.

## Acknowledgment

We thank the National Institutes of Health (grant number GM 069773) for generous support and Dr. Yun Xie for her insight and assistance.

### References


1. McCoy LS, Xie Y, Tor Y (2011) Antibiotics that target protein synthesis. WIREs RNA 2:209–232
2. Knowles DJC, Foloppe N, Matassova NB, Murchie AIH (2002) The bacterial ribosome, a promising focus for structure-based drug design. Curr Opin Pharmacol 2:501–506
3. Hermann T (2005) Drugs targeting the ribosome. Curr Opin Struct Biol 15:355–366
4. Auerbach T, Bashan A, Harms J, Schluenzen F, Zarivach R, Bartels H, Agmon I, Kessler M, Pioletti M, Franceschi F, Yonath A (2002) Antibiotics targeting ribosomes: crystallographic studies. Curr Drug Targets Infect Disord 2:169–186
5. Carter AP, Clemons WM, Brodersen DE, Morgan-Warren RJ, Wimberly BT, Ramakrishnan V (2000) Functional insights from the structure of the 30S ribosomal subunit and its interactions with antibiotics. Nature 407:340–348
6. Gale EF, Cundliffe E, Renolds PE, Richmond MH, Waring MJ (1981) The molecular basis of antibiotic action. Wiley, London
7. Moazed D, Noller HF (1987) Interaction of antibiotics with functional sites in 16S ribosomal-RNA. Nature 327:389–394
8. Brodersen DE, Clemons WM Jr, Carter AP, Morgan-Warren RJ, Wimberly BT, Ramakrishnan V (2000) The structural basis for the action of the antibiotics tetracycline, pactamycin, and hygromycin B on the 30S ribosomal subunit. Cell 103:1143–1154
9. Harms JM, Bartels H, Schlünzen F, Yonath A (2003) Antibiotics acting on the translational machinery. J Cell Sci 116:1391–1393
10. Wirmer J, Westhof E, Minoru F (2006) Molecular contacts between antibiotics and the 30S ribosomal particle. Methods Enzymol 415:180–202
11. Schlünzen F, Zarivach R, Harms J, Bashan A, Tocilj A, Albrecht R, Yonath A, Franceschi F (2001) Structural basis for the interaction of antibiotics with the peptidyl transferase centre in eubacteria. Nature 413:814–821
12. Vicens Q, Westhof E (2003) RNA as a drug target: the case of aminoglycosides. Chembiochem 4:1018–1023
13. Francois B, Russell RJM, Murray JB, Aboul-ela F, Masquida B t, Vicens Q, Westhof E (2005) Crystal structures of complexes between aminoglycosides and decoding A site oligonucleotides: role of the number of rings and positive charges in the specific binding leading to miscoding. Nucleic Acids Res 33:5677–5690
14. Purohit P, Stern S (1994) Interactions of a small RNA with antibiotic and RNA ligands of the 30S subunit. Nature 370:659–662
15. Fourmy D, Recht MI, Blanchard SC, Puglisi JD (1996) Structure of the A site of *escherichia coli* 16S ribosomal RNA complexed with an aminoglycoside antibiotic. Science 274:1367–1371
16. Yoshizawa S, Fourmy D, Puglisi JD (1998) Structural origins of gentamicin antibiotic action. EMBO J 17:6437–6448
17. Vicens Q, Westhof E (2001) Crystal structure of paromomycin docked into the eubacterial ribosomal decoding A site. Structure 9:647–658
18. Vicens Q, Westhof E (2002) Crystal structure of a complex between the aminoglycoside tobramycin and an oligonucleotide containing the ribosomal decoding A-site. Chem Biol 9:747–755
19. Kaul M, Barbieri CM, Pilch DS (2006) Aminoglycoside-induced reduction in nucleotide mobility at the ribosomal RNA A-site as a potentially key determinant of antibacterial activity. J Am Chem Soc 128:1261–1271
20. Hofstadler SA, Griffey RH (2001) Analysis of noncovalent complexes of DNA and RNA by mass spectrometry. Chem Rev 101:377–390
21. Haddad J, Kotra LP, Llano-Sotelo B, Kim C, Azucena EF, Liu M, Vakulenko SB, Chow CS, Mobashery S (2002) Design of novel antibiotics that bind to the ribosomal acyltransfer site. J Am Chem Soc 124:3229–3237

22. Kaul M, Barbieri CM, Pilch DS (2004) Fluorescence-based approach for detecting and characterizing antibiotic-induced conformational changes in ribosomal RNA: comparing aminoglycoside binding to prokaryotic and eukaryotic ribosomal RNA sequences. J Am Chem Soc 126:3447–3453
23. Parsons J, Hermann T (2007) Conformational flexibility of ribosomal decoding-site RNA monitored by fluorescent pteridine base analogues. Tetrahedron 63:3548–3552
24. Chao P-W, Chow CS (2007) Monitoring aminoglycoside-induced conformational changes in 16S rRNA through acrylamide quenching. Bioorg Med Chem 15:3825–3831
25. Wang Y, Hamasaki K, Rando RR (1997) Specificity of aminoglycoside binding to RNA constructs derived from the 16S rRNA decoding region and the HIV-RRE activator region. Biochemistry 36:768–779
26. Hamasaki K, Rando RR (1997) Specific binding of aminoglycosides to a human rRNA construct based on a DNA polymorphism which causes aminoglycoside-induced deafness. Biochemistry 36:12323–12328
27. Hamasaki K, Ueno A (2001) Aminoglycoside antibiotics, neamine and its derivatives as potent inhibitors for the RNA-protein interactions derived from HIV-1 activators. Bioorg Med Chem Lett 11:591–594
28. Xie Y, Dix AV, Tor Y (2009) FRET enabled real time detection of RNA-small molecule binding. J Am Chem Soc 131:17605–17614
29. Xie Y, Dix AV, Tor Y (2010) Antibiotic selectivity for prokaryotic vs. eukaryotic decoding sites. Chem Commun 46:5542–5544

# Chapter 17

# The Use of Serinol Nucleic Acids as Ultrasensitive Molecular Beacons

**Keiji Murayama, Hiromu Kashida, and Hiroyuki Asanuma**

## Abstract

Molecular beacons composed of the artificial serinol nucleic acid (SNA) have demonstrated utility as novel fluorescence probes for visualization of RNA in fixed cells using both conventional fluorescence in situ hybridization (FISH) and wash-free FISH protocols. The SNA molecular beacons have higher affinity for target RNA and greater sensitivity than molecular beacons composed of DNA. Here we describe facile synthesis of the SNA using a conventional DNA synthesizer and protocols for purification by PAGE and HPLC as well as methods for use of the SNA molecular beacon in FISH.

**Key words** DNA, Artificial nucleic acid, SNA, Molecular beacon, Fluorescent probe, FISH, RNA imaging

## 1 Introduction

Noncoding and coding RNAs play important roles in cells [1–3], and various fluorescent probes that can visualize RNAs in cells in a sequence-specific manner have been developed. Since the probes must be resistant to nucleases when employed to detect RNA in cells, a number of different artificial nucleic acids have been used to prepare fluorescent probes [4–8]. Our group has developed an artificial nucleic acid based on an acyclic scaffold: serinol nucleic acid (SNA) (Fig. 1) [9, 10]. SNA forms extremely stable antiparallel homo-duplexes and also cross-hybridizes with DNA and RNA. Since the acyclic scaffold is entirely different from the natural ribose scaffold, SNA has extremely high resistance to nuclease degradation. These properties mean that SNA has potential for use as probes in selective RNA visualization.

A molecular beacon is an oligonucleotide or modified oligonucleotide or artificial oligomer with a fluorophore at one terminus and a quencher at the other and an internal region complementary to a nucleic acid target of interest such as an mRNA. The MB can adopt a hairpin structure, and formation of a hairpin results in

Nathaniel Shank (ed.), *Non-Natural Nucleic Acids: Methods and Protocols*, Methods in Molecular Biology, vol. 1973, https://doi.org/10.1007/978-1-4939-9216-4_17, © Springer Science+Business Media, LLC, part of Springer Nature 2019

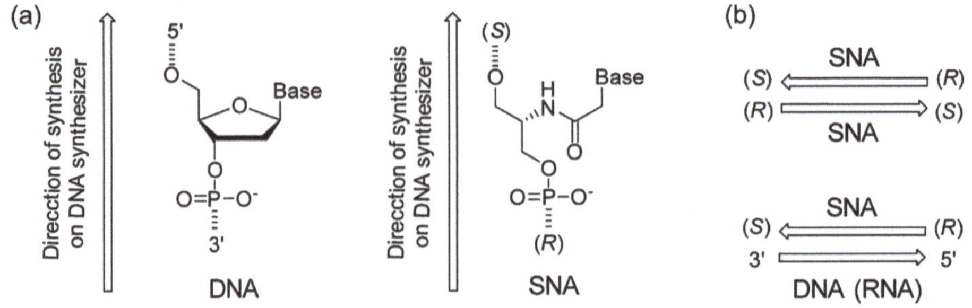


**Fig. 1** (**a**) Chemical structure of DNA and SNA. (**b**) Direction of synthesis and duplex manner are shown

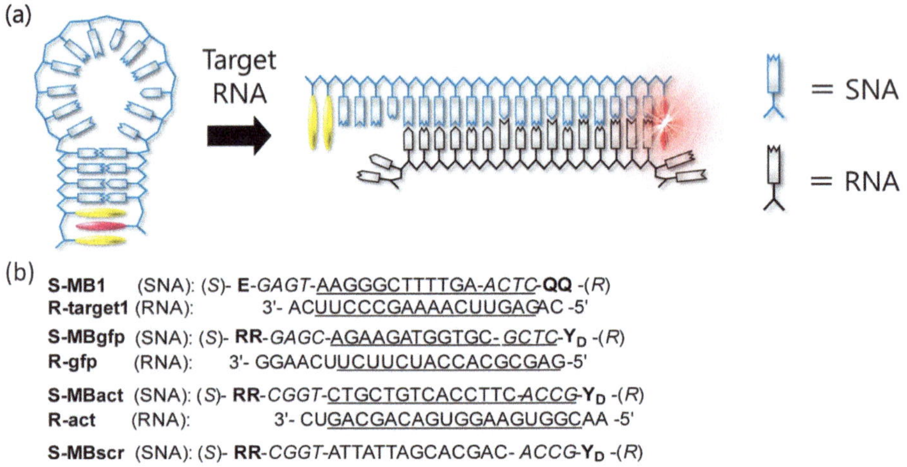


**Fig. 2** (**a**) Schematic illustration of RNA detection by SNA-MB. (**b**) Sequences of SNA-MBs and target RNAs used in this study. Underlining indicates region of complementarity between MB and target RNA. Italicized letters indicate stem region. **R-gfp** is the sequence of nucleotides 286–307 of the coding region of *eGFP* mRNA. **R-act** is the sequence of nucleotides 153–174 of the 3′-UTR of human *β-actin*. S-MB-scr is a control MB with sequence that is not complementary to any human RNAs

quenching of fluorescence emission (Fig. 2a) [11]. Upon hybridization of the MB with the target nucleic acid, the fluorophore can emit light due to separation from the quencher. In the MB design used here (Fig. 2b), all the residues are composed of SNA, and the fluorophore and quencher are also tethered to serinol scaffolds.

**S-MB1** tethers perylene and anthraquinone and has complementarity to **R-target1**, which is a model sequence. **S-MBgfp** and **S-MBact** tether Cy3 and nitromethyl red and have regions complementary to **R-gfp** and **R-act**, which contain sequences of mRNAs encoding eGFP and β-actin, respectively. **S-MBscr** has no complementarity to any human RNAs and was used as a negative control. The stem region of an SNA-MB should form four or five base pairs and should contain more than two GC pairs to ensure effective quenching in the hairpin state. Loop region of the SNA-MB is designed to be complementary to the RNA of interest.

It might be necessary for one of the stem-forming regions to be complementary to the target in order to ensure opening of the hairpin structure in the presence of target RNA under physiological conditions. To provide specificity and sufficient affinity, the length of the recognition site should be 16 to 20 residues in length.

The high stability of the SNA/SNA homo-duplex reduces background emission in the closed state as it facilitates stacking interactions between the chromophores. The high affinity of SNA for RNA results in sensitive detection of RNA. SNA phosphoramidite monomers are easily synthesized, and long SNA oligomers can be obtained through solid-phase synthesis on a standard DNA synthesizer with low cost compared with other artificial nucleic acids. Additionally, the nuclease resistance of the SNA is of great advantage in live-cell applications. Thus, the SNA-MB has many advantages compared with other probe chemistries [12].

## 2 Materials

All reagents should be prepared and stored at room temperature unless indicated otherwise. All waste disposal regulations should be diligently followed.

### 2.1 Synthesis of SNA Oligomers

1. Amidite monomers of SNA (Fig. 3): synthesized following procedures described in the literature [9, 12] (*see* **Note 1**).
2. Vials for DNA synthesizer.
3. Dry acetonitrile (ACN), super dehydrated (water content under 10 ppm).
4. Dry dichloromethane (DCM).
5. Column for 1 μmol synthesis of DNA.

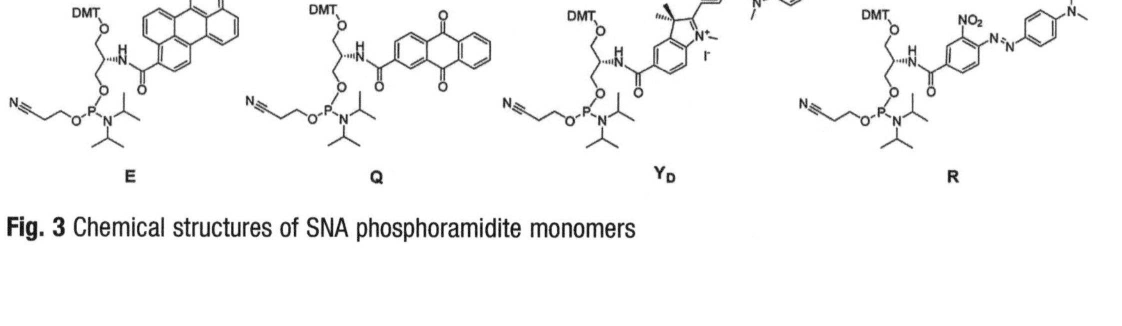


**Fig. 3** Chemical structures of SNA phosphoramidite monomers

6. DNA synthesizer.
7. *N*-Methyl-succinimido[3,4-b]-7-oxabicyclo[2.2.1]heptane-6-(4,4′-dimethoxytrityloxy)-5-succinoyl long chain alkylamino-controlled pore glass (CPG 500 Å) or synthesized CPG (*see* **Note 2**).
8. Deblocking mix: 2.5% dichloroacetic acid in DCM.
9. Oxidizing solution: 0.02 M iodine in tetrahydrofuran/water/pyridine (*see* **Note 3**).
10. Activator: 0.25 M 5-benzylthio-1*H*-tetrazole in ACN.
11. Cap Mix A: tetrahydrofuran/2,6-lutidine/acetic anhydride (*see* **Note 4**).
12. Cap Mix B: 10% 1-methylimidazole in tetrahydrofuran.
13. 5 mL vials with screw caps.
14. 28% aqueous ammonia.
15. Water bath (maintained at 55 °C).
16. Poly-Pak II cartridges.
17. Deionized water (18 MΩ-cm).
18. 2 M triethylamine acetate (TEAA), pH 7.0.
19. 1.3% aqueous ammonia solution.
20. 1:1 dry ACN/deionized water solution.
21. 2.0 mL tube.
22. Liquid nitrogen.
23. Centrifugal evaporator.

### 2.2 Gel Purification of SNA with PAGE

1. Gel cassette (buffer container, glass plates, comb, spacer, and clips).
2. 10% APS: 10% ammonium peroxydisulfate solution in deionized water. Stored at 4 °C.
3. *N*,*N*,*N*′,*N*′-Tetramethyl ethylenediamine (TEMED). Stored at 4 °C.
4. 15% Polyacrylamide gel solution: 14.4% acrylamide, 0.6% bis-acrylamide, 8 M urea, 20% 5× TBE buffer in deionized water.
5. TBE running buffer: 90 mM tris, 90 mM boric acid, 2 mM EDTA.
6. Electrophoresis system.
7. 6× loading buffer: 0.75% bromophenol blue, 50 mM EDTA, in 2:1 formamide/glycerol.
8. Gel cutter.
9. Thermostat.

10. Deionized water (18 MΩ-cm) autoclaved to inactivate nucleases.
11. Centrifugal evaporator.
12. Liquid nitrogen.

### 2.3 HPLC Purification and MALDI-TOF Mass Analysis of SNA

1. HPLC system equipped with reversed-phase C18 column.
2. Buffer A: 50 mM ammonium formate (*see* **Note 5**).
3. Buffer B: 25 mM ammonium formate in 1:1 deionized water/dry ACN solution (*see* **Note 6**).
4. Centrifugal evaporator.
5. MALDI-TOF mass spectrometer.
6. 200 μL tube.
7. Matrix A: 0.5 M diammonium hydrogen citrate.
8. Matrix B: saturated 3-hydroxypicolinic acid in 1:1 water/ACN solution.

### 2.4 Determination of the Concentration of SNA Oligomers

1. NanoDrop UV-vis spectrophotometer.

### 2.5 Measurement of Fluorescent Spectra

1. 2× measurement buffer: 200 mM NaCl, 20 mM phosphate (pH 7.0) (*see* **Note 7**).
2. Quartz cell: 3 mm × 3 mm.
3. Fluorescence spectrometer equipped with temperature controller.

### 2.6 Cell Culture and FISH of eGFP

1. 12-well cell culture plates.
2. Glass plates to fit the bottoms of wells in 12-well plates.
3. HeLa cells: cells should be dissociated by conventional passage procedure using 0.25% trypsin solution. The number of cells is counted by trypan blue method (*see* **Note 8**).
4. Dulbecco's modified Eagle's medium (DMEM): 10% fetal bovine serum, 80 μg/mL penicillin, 90 μg/mL streptomycin.
5. Water bath filled with deionized water and maintained at 37 °C.
6. $CO_2$ incubator.
7. OptiMEM Reduced Serum Media.
8. 1.5 mL tube.
9. Lipofectamine 2000.
10. pSicoR-eGFP: plasmid encoding eGFP.
11. Phosphate-buffered saline (PBS): 144 mg/L $KH_2PO_4$, 9.00 g/L NaCl, 795 mg/L $Na_2HPO_4$-$7H_2O$ (pH 7.4).

12. PFA-PBS: 4% paraformaldehyde in PBS.
13. PTGN: 0.2% Triton-X100, 10 mM glycine, 0.1% $NaN_3$ in PBS.
14. PGN: 10 mM glycine, 0.01% $NaN_3$ in PBS.
15. Paper towels.
16. Mowiol: 10% Mowiol 4-88, 100 mM Tris–HCl (pH 8.5) in 1:1 water/glycerol; store at −20 °C.
17. Glass plate for the observation.
18. Confocal laser microscope.

#### 2.7 Cell Culture and Wash-Free FISH of β-Actin

1. Glass-based dish: 35 mm dish with 12 mm glass.
2. HeLa cells.
3. Dulbecco's modified Eagle's medium (DMEM): 10% fetal bovine serum, 80 μg/mL penicillin, 90 μg/mL streptomycin.
4. Water bath filled with deionized water and maintained at 37 °C.
5. $CO_2$ incubator.
6. Phosphate-buffered saline (PBS): 144 mg/L $KH_2PO_4$, 9.00 g/L NaCl, 795 mg/L $Na_2HPO_4$-$7H_2O$ (pH 7.4).
7. PFA-PBS: 4% paraformaldehyde in PBS.
8. PTGN: 0.2% Triton-X100, 10 mM glycine, 0.1% $NaN_3$ in PBS.
9. PGN: 10 mM glycine, 0.01% $NaN_3$ in PBS.
10. Confocal laser microscope.

#### 2.8 Synthesis of CPG-Tethered Terminal SNA Monomer [Optional]

1. Compound 1: synthesized following procedures described in the literature [9, 12].
2. Dry dichloromethane (DCM).
3. Triethylamine.
4. Succinic anhydride.
5. 50 mL round-bottom flask.
6. Magnetic stirrer.
7. Methanol.
8. Separatory funnel.
9. 4% aqueous citric acid solution.
10. Magnesium sulfate.
11. Filtration system with aspirator.
12. Rotary evaporator.
13. Vacuum pump.
14. Native Amino Lcaa CPG (500 Å).
15. 20 mL reaction container with micro filter.

16. 1-Hydroxybenzotriazole (HOBt).
17. Dry ACN.
18. Dry pyridine.
19. *N,N'*-Diisopropylcarbodiimide: stored at 4 °C.
20. Shaker.
21. Tetrahydrofuran (THF).
22. 2,6-Lutidine.
23. Acetic anhydride.
24. Cap Mix B: 10% 1-methylimidazole in tetrahydrofuran.

## 3 Methods

### 3.1 Synthesis of SNA Oligomers

1. Dissolve the phosphoramidite monomers (Fig. 3) in DNA synthesizer vials. Compounds **A**, **C**, **T**, **R**, and **Q** should be dissolved in dry ACN to 0.08 M. Compound **G** should be prepared at 1.0 M in dry ACN. Compound **E** should be dissolved in dry dichloromethane to 0.08 M. Compound **Y** should be dissolved in dry ACN to a concentration of 2.0 M (*see* **Note 9**).
2. Equip the DNA synthesizer with vials containing solutions of each of these phosphoramidite monomers, ACN, DCM, deblocking mix, oxidizing solution, activator, Cap Mix A, and Cap Mix B.
3. Place a column containing 1.0 μmol of UnySupport CPG or modified CPG on the DNA synthesizer, and input the sequence (*see* **Note 10**). Start the synthesis of the SNA at a scale of 1.0 μmol (*see* **Note 11**).
4. After completion of the SNA synthesis, remove the DMT group at the (*S*)-terminus using the DMT-off mode on the synthesizer.
5. Dry the CPG under vacuum for 3 min, and transfer the CPG support to a 5 mL screw-cap vial.
6. Add 1 mL 28% aqueous ammonia. Incubate at 55 °C in a water bath for 8–12 h to deprotect the nucleobases and cleave the SNA from the support (*see* **Note 12**).
7. After cooling to room temperature, add 3 mL deionized water and 0.5 mL 2 M TEAA to the deprotected SNA in the ammonia solution (*see* **Note 13**).
8. Flush the Poly-Pak II cartridge with 4 mL ACN followed by 4 mL 2 M TEAA using a disposable syringe (*see* **Note 14**).

9. Load the SNA solution onto the cartridge. Collect the eluted fraction and reload it twice. Flush the cartridge with 8 mL 1.3% aqueous ammonia solution (*see* **Note 15**).
10. Flush the cartridge with 8 mL deionized water.
11. Elute the SNA by flushing the cartridge with 1:1 ACN/water solution. Discard the first several colorless drops, and collect the colored drops (approximately 3 mL) in two or three 2 mL tubes (*see* **Note 16**).
12. Freeze the collected solutions in tubes by immersing in liquid nitrogen. Remove the solvent from the frozen samples by centrifugal evaporation overnight.

### 3.2 PAGE Purification of SNA

1. Assemble the gel cassette (*see* **Note 17**).
2. Add 240 μL 10% APS and 24 μL TEMED to 40 mL 15% polyacrylamide gel solution in a gel injection container, and mix well. Immediately pour the solution into the gel cassette and set the comb (*see* **Note 18**).
3. Wait for at least 2 h to ensure complete polymerization of acrylamide.
4. Pull out the comb, set up the gel cassette, and pre-run using TBE running buffer at a constant voltage (700 V) for 15 min (*see* **Note 19**).
5. Dissolve the dried SNA in 400 μL deionized water. Transfer 200 μL into another tube, add 40 μL 6× loading buffer, and mix well. Store the remaining 200 μL SNA at −20 °C (*see* **Note 20**).
6. Wash the well of gel. Load the 240 μL sample solution slowly onto the top of the gel (*see* **Note 21**).
7. Electrophorese using 1× TBE running buffer at a constant voltage (750 V) for 2 h (*see* **Note 22**).
8. Remove the glass plate from the gel, and cut out the most intensely colored band using a gel cutter. Transfer to a 2 mL tube. The less intensely colored bands probably contain shorter SNA oligomers that are byproducts of SNA synthesis (*see* **Note 23**).
9. Add deionized water (at least three times the gel volume) to the tubes containing the gel slice, and incubate for 2 h at 50 °C in thermostat to extract the SNA. Transfer the extracted solution to another 2 mL tubes. Repeat this extraction two more times (*see* **Note 24**).
10. Freeze the three collected samples by immersion in liquid nitrogen, and remove the solvent from the frozen samples by centrifugal evaporation overnight.
11. Procedures 6–10 are repeated with stored remaining 200 μL SNA.

### 3.3 HPLC Purification of SNA

1. Set up the HPLC with reversed-phase column. Set the wavelength of the UV detector at 260 nm. Stabilize the HPLC column by flushing with appropriate buffer solution at a flow rate of 0.5 mL $min^{-1}$ for more than 15 min. The column should be heated to 50 °C (*see* **Note 25**).
2. Dissolve the dried SNA in deionized water, and collect into one tube. Total volume brings to 300 μL (*see* **Note 26**).
3. Inject 1.5 μL sample solution into the HPLC column, and analyze using a linear gradient (e.g., from 47% buffer B to 62% buffer B in 30 min). Alter the gradient until the retention time of the SNA is between 15 and 30 min (*see* **Note 27**).
4. Inject remaining sample solution into the HPLC, and collect the fractions containing the SNA into 2 mL tubes (*see* **Note 28**).
5. Freeze the collected samples by immersion in liquid nitrogen, and remove the solvent from the samples by centrifugal evaporation overnight.
6. Add 300 μL deionized water to the dried sample, and dissolve to make the stock solution of the SNA.
7. Analyze the purity of the stock solution by HPLC using the same column and gradient conditions as selected above. Only one peak attributed to the full-length SNA should be observed (*see* **Note 29**).
8. Analyze the sample by MALDI-TOF mass spectroscopy (MS) in positive mode. Mix 1.1 μL sample and 1.1 μL Matrix A in a 200 μL tube. Add 1.1 μL Matrix B, mix, and immediately place on measurement plate. The results of MS analysis of SNA oligomers used in this study are shown in Table 1.

### 3.4 Determination of the Concentration of SNA Oligomers

1. Measure the absorbance of the stock solution of SNA at 260 nm using a NanoDrop UV-visible spectrometer (*see* **Note 30**).

**Table 1**
**Results of MALDI-TOF MS analyses of SNA**

| | Calcd. | Obsd. |
|---|---|---|
| S-MB1 | 7676 | 7678 |
| S-MBgfp | 7940 | 7922 |
| S-MBact | 8401 | 8398 |
| S-MBscr | 8482 | 8489 |

2. Based on the absorbance, calculate the concentration of the SNA according to the following formula:

$$c = A/d\left(\varepsilon_{\text{nat}} + n_{\text{E}}\varepsilon_{\text{E}} + n_{\text{Q}}\varepsilon_{\text{Q}} + n_{\text{Y}}\varepsilon_{\text{Y}} + n_{\text{R}}\varepsilon_{\text{R}}\right)$$

where $c$ is the concentration of SNA (in M); $A$ is the absorbance at 260 nm; $d$ is the path length of the cell (cm); $\varepsilon_{\text{nat}}$ is the molar extinction coefficient of the nucleobases of SNA ($M^{-1}$ $cm^{-1}$), calculated based on the nearest neighbor model for DNA [13, 14] (*see* **Note 31**); $n_{\text{E}}$, $n_{\text{Q}}$, $n_{\text{Y}}$, and $n_{\text{R}}$ are the numbers of perylene, anthraquinone, Cy3, and nitromethyl red dyes in the SNA; and $\varepsilon_{\text{E}}$, $\varepsilon_{\text{Q}}$, $\varepsilon_{\text{Y}}$, and $\varepsilon_{\text{R}}$ are the molecular absorbances at 260 nm of perylene (18,000 $M^{-1}$ $cm^{-1}$), anthraquinone (38,420 $M^{-1}$ $cm^{-1}$), Cy3 (28,000 $M^{-1}$ $cm^{-1}$), and nitromethyl red (19,980 $M^{-1}$ $cm^{-1}$), respectively (*see* **Note 32**).

### 3.5 Measurement of Fluorescent Spectra in the Presence and Absence of Target RNA

1. Mix 100 μL of 2× measurement buffer and oligomer stock solution; bring to 200 μL with deionized water (sample solution).
2. Add 200 μL of sample solution containing SNA oligomer and target RNA in measurement buffer (pH 7.0) to the quartz cell (*see* **Note 33**).
3. Heat the cell at 80 °C for 3 min, and record the fluorescent spectrum (excitation wavelength: 426 nm or 546 nm) at 80 °C. Choose an appropriate wavelength for excitation (426 nm for E, 546 nm for Y).
4. Cool to 70 °C, wait for 3 min, and then measure fluorescence spectrum at 70 °C.
5. Repeat the procedure at 60, 50, 40, 30, and 20 °C. The spectra of SNAs used in this study are shown in Fig. 4.
6. Calculate the signal-to-background (S/B) ratio as an index of sensitivity. The S/B ration is the ratio of fluorescent intensity at the peak in the presence relative to the absence of target at a chosen temperature. The S/B ratios for the SNA-MBs used in this study are listed in Table 2.

### 3.6 Cell Culture and FISH of eGFP

1. Transfer suspension of 4 × $10^4$ HeLa cells to a cover glass placed on the bottom of the well of a 12-well plate (*see* **Note 34**). Bring volume per well to 500 μL with DMEM (*see* **Note 35**).
2. Incubate plate at 37 °C in 5% $CO_2$ in humidified air for 24 h.
3. Remove DMEM and gently wash cells with 250 μL OptiMEM. Remove OptiMEM, and add 500 μL OptiMEM to the well (*see* **Note 36**).

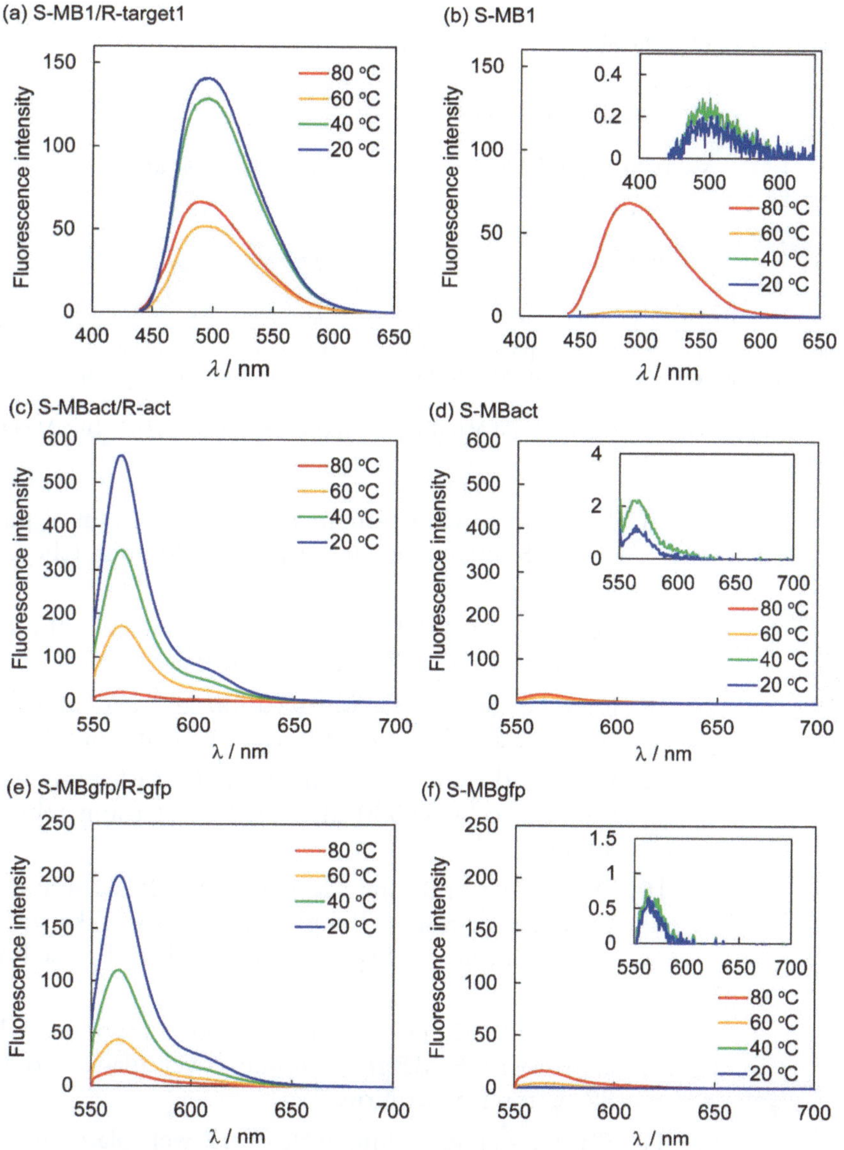


**Fig. 4** Fluorescence spectra of (**a**) **S-MB1/R-target1**, (**b**) **S-MB1**, (**c**) **S-MBact/R-act**, (**d**) **S-MBact**, (**e**) **S-MBgfp/R-gfp**, and (**f**) **S-MBgfp**. Conditions for (**a**) and (**b**): ex. = 426 nm, 1.0 μM MB, 2.0 μM target RNA, 100 mM NaCl, 10 mM phosphate buffer (pH 7.0). Conditions for (**c–f**): ex. = 546 nm, 0.5 μM MB, 1.0 μM target RNA, 100 mM NaCl, 10 mM phosphate buffer (pH 7.0)

4. 2.0 μL Lipofectamine is added into 100 μL OptiMEM in a 1.5 mL tube and mixes by pipetting (solution A).
5. 1.0 μg of pSicoR-eGFP encoding eGFP is added into 100 μL OptiMEM in another 1.5 mL tube and mixes by pipetting (solution B).

**Table 2**
**Fluorescent intensities of SNA-MBs**

| | Emission intensities | | |
|---|---|---|---|
| | With target RNA | Without target RNA | S/B[a] |
| S-MB1[b] | 140 | 0.15 | 930 |
| S-MBgfp[c] | 200 | 0.55 | 360 |
| S-MBact[c] | 561 | 1.3 | 430 |

[a]S/B is the ratio of fluorescent intensity in the presence of target to that in the absence of target
[b]Conditions: ex. = 426 nm; em. = 500 nm, 1.0 μM MB, 2.0 μM target RNA, 100 mM NaCl, 10 mM phosphate buffer (pH 7.0), 20 °C
[c]Conditions: ex. = 546 nm; em. = 564 nm, 0.5 μM MB, 1.0 μM target RNA, 100 mM NaCl, 10 mM phosphate buffer (pH 7.0), 20 °C

6. After 5 min incubation, mix solutions A and B by pipetting in a new 1.5 mL tube, and wait for 20 min (the transfection solution).
7. Remove OptiMEM from the well, and add the transfection solution to the well. Incubate the plate at 37 °C for 4 h.
8. Remove the transfection solution and add 500 μL DMEM. Incubate at 37 °C for 24 h in 5% $CO_2$ in humidified air.
9. Remove the solution from the well; wash twice with 500 μL PBS. Fix cells in 500 μL PFA-PBS at room temperature for 30 min.
10. Discard PFA-PBS and add 500 μL PTGN to permeabilize cells. Incubate for 5 min. Remove the PTGN. Take the cover glass on which cells are fixed from the well, and place on wet paper towel.
11. Place 100 μL PGN containing 200 pmol of MB (for this experiment **S-MBgfp**) on the cover glass, and incubate at 37 °C for 1 h (*see* **Note 37**).
12. Place cover glass into well of 12-well plate and rinse twice with PGN.
13. Place 10 μL of Mowiol on glass plate for the observation. Cover glass taken from the well is embedded on this glass plate, and incubate overnight in the dark.
14. Visualize stained HeLa cells using confocal laser microscopy. Images are taken with a 100× oil emission objective lens. The 488 nm laser is used to excite eGFP with emission collected using a 500–530 nm band-pass filter. The 543 nm laser is used to excite the Cy3 with emission collected using a 555–655 nm band path filter (*see* **Note 38**). Representative images of fixed HeLa cells that express eGFP stained with **S-MBgfp** are shown in Fig. 5.

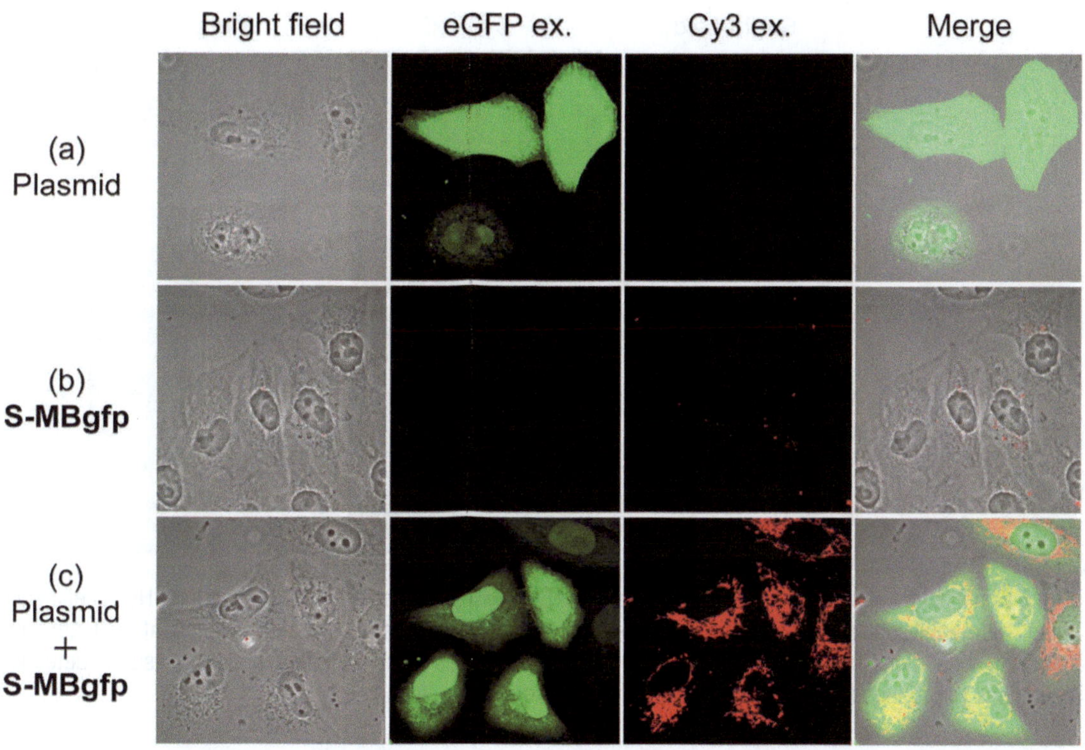


**Fig. 5** Visualization of mRNA encoding eGFP in fixed HeLa cells using SNA-MB. (**a**) Representative images of cells transfected with plasmid encoding eGFP after fixation. (**b**) Representative images of control cells incubated with **S-MBgfp** after fixation. (**c**) Representative images of cells transfected with eGFP expression plasmid, fixed, and incubated with **S-MBgfp**. eGFP was excited at 488 nm, and emission was collected at 500–530 nm. Cy3 was excited at 543 nm, and emission was monitored at 555–655 nm. Each fluorescent image was obtained under the same detection conditions (reproduced from ref. 12 with permission from Wiley-VCH)

### *3.7 Cell Culture and Wash-Free FISH of β-Actin*

1. Transfer $1.2 \times 10^5$ HeLa cells to the center of a glass-based dish. Dilute with DMEM to 2000 μL (*see* **Note 34**).
2. Incubate at 37 °C with 5% $CO_2$ in humidified air for 24 h.
3. Remove the solution from the dish; rinse twice with 2000 μL PBS. Fix cells with 2000 μL PFA-PBS at room temperature for 20 min.
4. Discard PFA-PBS and add 1000 μL PTGN to dish and incubate for 5 min. After washing with 1000 μL PTGN, 200 μL PGN containing 200 pmol of MB is placed on the well of dish, and the sample is incubated at 37 °C for 1 h. For the experiment shown, cells were incubated with **S-MBact** or **S-MBscr**.
5. Prior to imaging, 1000 μL PBS should be added to prevent drying. Note that the sample is used for observation without further washing (*see* **Note 39**).

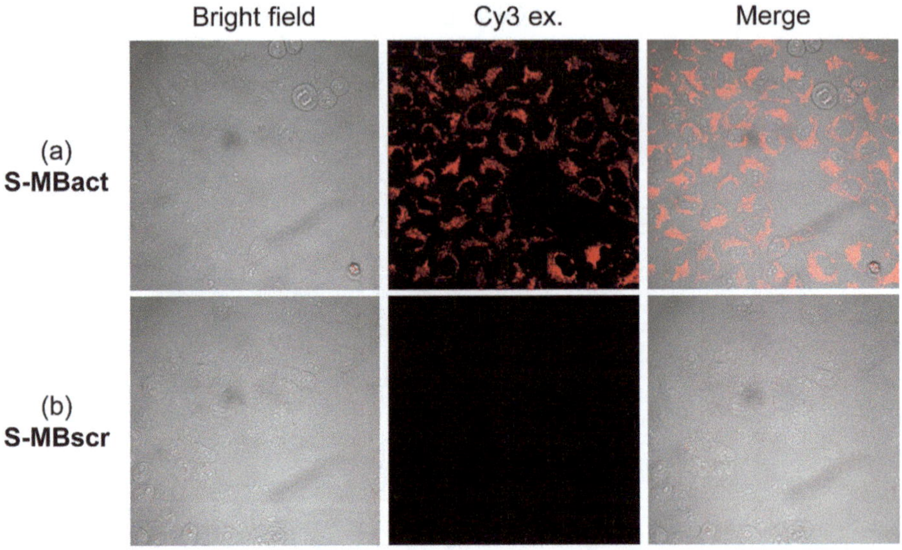


**Fig. 6** Visualization of an endogenous mRNA using an SNA-MB in a wash-free FISH procedure. HeLa cells were fixed, permeabilized, and treated with (**a**) **S-MBact** and (**b**) **S-MBscr**. Cy3 was excited at 543 nm, and emission was monitored at 555–655 nm. Fluorescent images were obtained under the same detection conditions (reproduced from ref. 12 with permission from Wiley-VCH)

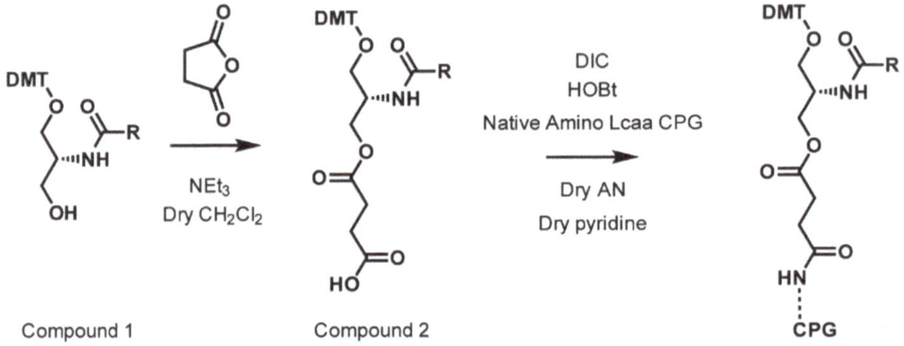


**Fig. 7** Synthetic scheme of CPG tethered to SNA monomer

6. The stained HeLa cells are visualized by using confocal laser microscopy. Images are taken with a 60× oil emission objective lens. The 543 nm laser is used to excite the Cy3 with emission collected using a 555–655 nm band path filter (*see* **Note 38**). Representative images of HeLa cells stained with **S-MBact** or **S-MBscr** are shown in Fig. 6.

#### *3.8 Synthesis of CPG Tethered to SNA Monomer (See Fig. 7) [Optional]*

1. Add dry dichloromethane (4.0 mL) and triethylamine (0.5 mL) to the compound 1 (0.15 mmol) and succinic anhydride (40 mg) in a 50 mL round-bottom flask under nitrogen atmosphere.

2. After vigorous stirring at room temperature for 5 h, check the progress of reaction by TLC with 10:1 or 5:1 (vol/vol) chloroform/methanol (*see* **Note 40**).
3. After the reaction is complete, transfer mixture to a separatory funnel, rinse with dichloromethane, and wash the organic phase once with 4% aqueous citric acid solution.
4. Dry the organic phase over magnesium sulfate for 10 min. Filter through filter paper with 1 μm pore. Concentrate the filtrate using a rotary evaporator, and dry completely under vacuum. Compound 2 is obtained as solid.
5. Place 500 mg Native Amino Lcaa CPG (500 Å) in a reaction container equipped with a filter for peptide synthesis. Add 5.0 mg HOBt, 2.0 mL dry ACN, 0.1 mL dry pyridine, and 25 μL *N*,*N*′-diisopropylcarbodiimide, and shake for 30 min.
6. Add 0.06 mmol dried compound 2 to the reaction container and shake for 12 h.
7. Remove the solvent by filtration through the filter on the reaction container. Wash the CPG with methanol and dichloromethane and dry under vacuum.
8. Mix 4 mL THF, 0.5 mL lutidine, and 0.5 mL acetic anhydride with 5 mL Cap Mix B and dried CPG in the reaction container, and shake for 1 h.
9. Remove solvent by filtration through the filter on the reaction container. Wash CPG with methanol and dichloromethane, and dry under vacuum.
10. Loading is determined by using the following procedures described in the literature [15].

## 4 Notes

1. Dried amidite monomers covered with a layer of argon gas can be stored at −20 °C for at least 1 year.
2. Store at −20 °C. If CPG tethering an SNA monomer is used, follow procedure 3.8.
3. The ratio of tetrahydrofuran/water/pyridine is undefined, purchased from Glen Research.
4. The ratio of tetrahydrofuran/2,6-lutidine/acetic anhydride is undefined, purchased from Glen Research.
5. Use deionized water and degas by an aspirator with vigorous stirring.
6. Degassing should be performed before addition of ACN. Degassing after addition of ACN causes evaporation of the solution, resulting in change of the buffer ratio.

7. Deionized water containing 200 mM NaCl and 20 mM disodium phosphate is mixed with 200 mM NaCl and 20 mM sodium dihydrogen phosphate solution in an appropriate ratio to obtain a solution of pH 7.0.
8. Mixture of 10 μL cell suspension and 10 μL trypan blue is measured by automated cell counter.
9. Even a small amount of water inhibits a coupling reaction. Use dry ACN and keep under argon atmosphere. This concentration of phosphoramidite monomer is optimized for a DNA synthesizer that requires 0.05 M for conventional DNA monomers. The concentration necessary will depend on the type of DNA synthesizer. After dissolving in solvent, monomers should be used immediately to avoid degradation.
10. The use of UnySupport CPG tends to reduce yield of desired full-length SNA. If higher yield is required, use modified CPG as described in Subheading 3.8. When CPG tethering an SNA monomer is used, CPG conjugated to the (*R*)-terminus. 5′ and 3′ termini of DNA correspond to (*S*) and (*R*) termini of SNA, respectively. The synthesis direction of SNA on a DNA synthesizer is (*R*) to (*S*).
11. Several modifications should be used to the standard DNA synthesis steps: at all SNA coupling steps except for those with the **Y** monomer, the reaction time should be elongated to 300 s twice (0.2 mL × 2). For the **Y** monomer, coupling time should be elongated to 300 s three times (0.2 mL × 3). If the phosphoramidite monomer is coupled efficiently, the waste solvent at the detritylation step will be colored orange, and CPG should be the color of the conjugated dye.
12. Wear glasses and gloves when dealing with ammonia solution. Vials should be tightly sealed with resin tape to inhibit the release of ammonia gas. Do not use a higher temperature for this step as high temperatures cause degradation of amide bonds.
13. This procedure should be performed in a fume hood to prevent inhalation and diffusion of ammonia gas.
14. During Poly-Pak purification, the flow rate of the solvents through the cartridge should be one to two drops per second.
15. If a deep-colored solution is eluted in this process, collect elution, and dilute with 0.2 M TEAA and load onto cartridge again. After loading, move on to the next step without further washing.
16. Color of drops depends on fluorescent dye in the sequence. **Y** monomer is red and **E** monomer is yellow.
17. Thickness of combs and gel should be 1 mm for good separation.

18. Gloves should be worn when handling acrylamide solutions and gels as it is a neurotoxin.
19. Glass plates holding the gel become hot during electrophoresis. For the cooling and dispersion of heat, a metal plate can be fitted over the glass plate.
20. If all of SNA is loaded at once, separation ability becomes low.
21. Wells of the gel should be washed with buffer using a syringe before loading. Turn the power of electrophoresis system off when washing the wells and when loading the SNA solution to prevent an electric shock.
22. Electrophoresis time will depend on length and sequence of SNA.
23. Mobility is affected by not only length of sequence but also type of dyes. As a result, full-length SNA cannot be identified based on mobility. If several bands of approximately equal intensity are observed, collect all bands separately, and analyze by HPLC.
24. Repeat extraction process again if color remains in the gel. All three extractions can be put in the same tube. If extraction volume is too much, use different tubes.
25. The column should be heated to 50 °C using a column heater to denature SNA structure. This will enhance resolution.
26. If fragments of gel remain in the SNA sample, remove using a spin column before the injection.
27. Gradient should be 10–20% for 30 min to obtain good separation. The retention time of the SNA depends on numbers, positions, type of incorporated dyes, and sequence of SNA.
28. If the signal (absorbance) is too strong, the detector wavelength can be set to 280 nm, and the sample solution can be subdivided into aliquots for injection.
29. If a shoulder is observed when collected SNA is analyzed, separation is not sufficient. Repeat the HPLC purification. Note that structures formed by the SNA, which depend on the sequence, may result in several peaks even when SNA is sufficiently purified. If a sample from a fraction corresponding to a single peak gives multiple peaks when analyzed by HPLC and if MALDI-TOF MS is indicative of a single compound, a structural effect is indicated.
30. When the absorbance exceeds 1.0, dilute the stock solution adequately.
31. Calculation of extinction coefficients of the nucleobases can also be performed by using a web tool on IDT Biophysics website (http://biophysics.idtdna.com/UVSpectrum.html).

32. Stock solutions of SNA can be stored at −20 °C for at least 1 year.
33. Seal the top of cell using resin tape or a cap to avoid an evaporation of water during measurement. In the case of measurement in the absence of target, RNA should not be included in sample solution. Concentration of oligos should be around 0.1–2.0 μM for performing the fluorescence experiments.
34. An aliquot of HeLa cells from a culture is used directly. The number of cells should be determined by analysis of a sample stained with trypan blue.
35. All reagents and solutions for cell culture should be warmed at 37 °C before using.
36. If wells dry, cells will die. After removal of solvent, immediately add the next solution. Pipetting operations should be performed slowly to avoid dislodging adhered cells.
37. Check the condition of surface of cover glass after 20 min. If it is too dry, add a small amount of PGN.
38. Observation conditions, including HV, gain, and offset, should be identical for all measurements.
39. Excess dilution or prolonged incubation time may cause dissociation of probe from the target RNA. After dilution with 1000 μL PBS, samples should be observed as soon as possible.
40. Product will appear below starting material on TLC plate. If starting material remains, add 0.4 mL triethylamine and 20 mg succinic anhydride, and stir for 1 h.

## Acknowledgments

This work was supported by a Grant-in-Aid for Young Scientists (B) [no. 17K14514] (to K.M.) and by PRESTO, Japan Science and Technology Agency [no. JPMJPR14F7] (to H.K.). Supports by Japan Science and Technology Agency (JST) under Adaptable and Seamless Technology Transfer Program through Target-Driven R&D (A-STEP); a Grant-in-Aid for Challenging Exploratory Research [no. 16 K12522], Technology, Japan; and the Asahi Glass Foundation are also acknowledged.

## References

1. Ambros V (2004) The functions of animal microRNAs. Nature 431:350–355
2. Bartel DP (2004) MicroRNAs: genomics, biogenesis, mechanism, and function. Cell 116:281–297
3. Pitchiaya S, Heinicke LA, Custer TC, Walter NG (2014) Single molecule fluorescence approaches shed light on intracellular RNAs. Chem Rev 114:3224–3265

4. Tyagi S, Alsmadi O (2004) Imaging Native β-Actin mRNA in Motile Fibroblasts. Biophys J 87:4153–4162
5. Yang CJ, Wang L, Wu Y, Kim Y, Medley CD, Lin H, Tan W (2007) Synthesis and investigation of deoxyribonucleic acid/locked nucleic acid chimeric molecular beacons. Nucleic Acids Res 35:4030–4041
6. Kummer S, Knoll A, Socher E, Bethge L, Herrmann A, Seitz O (2011) Fluorescence imaging of influenza H1N1 mRNA in living infected cells using single-chromophore FIT-PNA. Angew Chem Int Ed 50:1931–1934. Angew Chem 123:1972–1975
7. Kubota T, Ikeda S, Yanagisawa H, Yuki M, Okamoto A (2011) Cy5-conjugated hybridization-sensitive fluorescent oligonucleotides for ratiometric analysis of nuclear poly(A)+ RNA. Bioconjug Chem 22:1625–1630
8. Catrina IE, Marras SAE, Bratu DP (2012) Tiny molecular beacons: LNA/2′-O-methyl RNA chimeric probes for imaging dynamic mRNA processes in living cells. ACS Chem Biol 7:1586–1595
9. Kashida H, Murayama K, Toda T, Asanuma H (2011) Control of the chirality and helicity of oligomers of serinol nucleic acid (SNA) by sequence design. Angew Chem Int Ed 50:1285–1288. Angew Chem 123:1321–1324
10. Murayama K, Tanaka Y, Toda T, Kashida H, Asanuma H (2013) Highly stable duplex formation by artificial nucleic acids aTNA and SNA with acyclic scaffolds. Chem Eur J 19:14151–14158
11. Tyagi S, Kramer FR (1996) Molecular beacons: probes that fluoresce upon hybridization. Nat Biotechnol 14:303–308
12. Murayama K, Kamiya Y, Kashida H, Asanuma H (2015) Ultra-sensitive molecular beacon designed with totally serinol nucleic acid (SNA) for monitoring mRNA in cell. Chembiochem 16:1298–1301
13. Cantor CR, Warshaw MM, Shapiro H (1970) Oligonucleotide interactions. III. Circular dichroism studies of the conformation of deoxyoligonucleotides. Biopolymers 9:1059–1077
14. Tataurov AV, You Y, Owczarzy R (2008) Predicting ultraviolet spectrum of single stranded and double stranded deoxyribonucleic acids. Biophys Chem 133:66–70
15. Pon RT (2001) Attachment of nucleosides to solid-phase supports. Curr Protoc Nucleic Acid Chem 00:3.2.1–3.2.23

# Chapter 18

# Oligonucleotide Primers with $G^{8AE}$-Clamp Modifications for RT-qPCR Detection of the Low-Copy dsRNA

**Timofei S. Zatsepin, Anna M. Varizhuk, Vladimir G. Dedkov, German A. Shipulin, and Andrey V. Aralov**

## Abstract

We developed a new technique suitable for improved detection of low-copy dsRNA using modified oligonucleotides as primers in RT-qPCR. Insertion of $G^{8AE}$-clamp residues into primers significantly improves thermal stability of duplexes with RNA without decrease of hybridization selectivity. The applicability of modified primers is demonstrated for detection of low-copy Kemerovo virus dsRNA.

**Key words** G-clamp, Modified primer, RT-qPCR, Oligonucleotide, dsRNA

## 1 Introduction

Specific amplification of DNA or reverse transcription of single-stranded RNA is a routine procedure now and can be performed with numerous commercial or in-house kits. However reverse transcription of highly structured RNA or dsRNA is a challenge. RNA-RNA duplexes are more stable than DNA-RNA ones [1] so application of common DNA primers provides poor results. There are several sophisticated approaches to overcome this problem: (a) use of modified oligonucleotide primers, (b) use of thermostable reverse transcriptases at elevated temperatures, and (c) targeted partial hydrolysis of RNA by RNAse H or tailoring to improve reverse transcription [2]. Among them the use of modified primers that improve duplex stability without affecting specificity and enzyme functioning is preferable especially in case of low-copy targets like rare mRNA [3] or for highly structured RNAs [4]. Primers with LNA nucleotides [5, 6] and 5′-polyamine conjugates (ZNA) [7] are available from commercial suppliers. However only a limited number of LNA residues per primer are acceptable by polymerases, while LNA residues close to 3′-end totally block the primer elongation [8]. Recently primers modified with classical

Nathaniel Shank (ed.), *Non-Natural Nucleic Acids: Methods and Protocols*, Methods in Molecular Biology, vol. 1973, https://doi.org/10.1007/978-1-4939-9216-4_18, © Springer Science+Business Media, LLC, part of Springer Nature 2019

G-clamp that significantly increases duplex stability were applied for detecting and genotyping rotavirus strains [9].

Here we present a protocol for improved detection of Kemerovo virus dsRNA by qPCR using modified primers for enhanced dsRNA amplification. We demonstrated that introduction of phenoxazine-based nucleotide analogue ($G^{8AE}$-clamp) into primers improves detection of low-copy dsRNA by RT-qPCR using previously described assay [10].

## 2 Materials

### 2.1 Synthesis of $G^{8AE}$-Clamp Phosphoramidite

The solvents were high-performance liquid chromatography (HPLC) grade and were used without further purification unless otherwise noted.

1. 4-(2-Bromoethoxy)phenol **1** (*see* **Note 1**).
2. 3′,5′-*O*-Acetyl-5-bromo-2′-deoxyuridine **4** (*see* **Note 2**).
3. DIEA (*N*,*N*-diisopropylethylamine).
4. 2-Cyanoethyl *N*,*N*-diisopropylchlorophosphoramidite.
5. TEA (triethylamine).
6. DMTr-Cl (4,4′-dimethoxytrityl chloride).
7. 1,2,4-Triazole.
8. Ethyl trifluoroacetate.
9. $POCl_3$ (phosphoryl chloride).
10. 10% Pd/C.
11. Calcium hydride ($CaH_2$).
12. Potassium hydroxide (KOH).
13. Saturated aqueous $NH_3$ solution.
14. 56% aqueous $HNO_3$ solution.
15. Saturated aqueous $NaHCO_3$ solution.
16. Brine.
17. Anhydrous $Na_2SO_4$.
18. $NaN_3$.
19. DCM (dichloromethane) (*see* **Note 3**).
20. DMF (dimethylformamide) (*see* **Note 4**).
21. Anhydrous pyridine (*see* **Note 5**).
22. EtOAc (ethyl acetate).
23. EtOH (ethanol).
24. MeOH (methanol).
25. Toluene.

26. Hexane.
27. Acetone.
28. Benzene.
29. Acetonitrile.
30. Hydrogen balloon.
31. Argon balloon.
32. Round-bottom flasks.
33. Rotary evaporator.
34. 254 nm UV lamp.
35. High vacuum oil pump.
36. TLC plates (aluminum sheets coated with silica gel 60$F_{254}$) (*see* **Note 6**).
37. Silica gel for column chromatography (Merck Kieselgel 60 0.040–0.063 mm).

### 2.2 Oligonucleotide Synthesis, Purification, and Characterization

#### 2.2.1 Reagents and Equipment for Oligonucleotide Synthesis

1. Protected 2′-deoxynucleosidephosphoramidites.
2. Deprotection solution (3% trichloroacetic acid in DCM).
3. Oxidizing solution (0.02 M iodine in THF/pyridine/water (80/10/10, v/v/v)).
4. Capping solutions (Cap A, 10% acetic anhydride, 16% *N*-methylimidazole in THF; Cap B, 10% pyridine in THF).
5. Anhydrous acetonitrile (*see* **Note 7**).
6. Saturated aqueous $NH_3$ solution.
7. 40% aqueous $MeNH_2$ solution.
8. Solution of activator (0.25 M 5-ethylthiotetrazole in acetonitrile).
9. 2′-Deoxynucleoside solid supports.
10. DNA synthesizer (e.g., ABI 3400).
11. SpeedVac vacuum concentrator.

#### 2.2.2 Reagents for PAGE Oligonucleotide Purification

1. Acrylamide.
2. bis-Acrylamide.
3. Urea.
4. Ammonium persulfate.
5. 80% formamide.
6. PAGE buffer A (50 mM Tris–HCl, 50 mM boric acid, 1 mM EDTA, pH 8.3).
7. PAGE buffer B (5 mM Tris–HCl, 5 mM boric acid, pH 8.3).

*2.2.3 Buffers for LC-ESI-MS Characterization*

1. ESI buffer A: add 0.7 ml of HPLC-grade diisopropylamine to 500 ml of MS-grade water, mix thoroughly, add 0.8 ml of MS-grade 1,1,1,3,3,3-hexafluoroisopropanol, and mix thoroughly.
2. ESI buffer B: add 0.7 ml of HPLC-grade diisopropylamine to 100 ml of MS-grade water, mix thoroughly, add 0.8 ml of MS-grade 1,1,1,3,3,3-hexafluoroisopropanol, mix thoroughly, and slowly add 400 ml of MS-grade acetonitrile.
3. Degas the solution and transfer to bottle attached to LC-ESI-MS instrument.

*2.2.4 Buffers for HPLC Purification and Analysis*

1. Stock 200 mM ammonium acetate solution: Add 15.42 g of ammonium acetate to 400 ml of HPLC-grade water and mix thoroughly until all salts are dissolved. Filter solution through a 0.2 μm HPLC-certified Nylon filter. Add HPLC-grade water to 950 ml and check pH. Adjust pH to 7.0. Add water to final volume of 1 l, degas solution, and transfer to bottle and store at +4 °C not more than 1 month.
2. HPLC buffer A: Add 350 ml of HPLC-grade water to 125 ml of 200 mM ammonium acetate solution, then add 25 ml HPLC gradient grade acetonitrile, and mix thoroughly. Degas solution and transfer to bottle attached to HPLC instrument.
3. HPLC buffer B: Add 25 ml of HPLC-grade water to 75 ml of 200 mM ammonium acetate solution, then add 400 ml HPLC gradient grade acetonitrile, and mix thoroughly.
4. Degas the solution and transfer to bottle attached to HPLC instrument.

### 2.3 RT-qPCR

1. Primers for PCR: 5′-tccgccaccctggaatgagac-3′ (rtKem4f) and 5′-tcaggatcggtcaaggccattc-3′ (rtKem4r).
2. TaqMan probe 5′-R6G-agccgtcttctgtccacgcagacgc-BHQ1 (Kemprb4).
3. dNTPs mixture.
4. Restriction enzyme BamHI.
5. 10× buffer BamHI.
6. DEPC-treated water.
7. BamHI reaction mixture: 2 μl 10× buffer BamHI, 1 μg template DNA, 1 μl BamHI, and DEPC-treated water (up to 20 μl).
8. Set of NTPs 100 mM each (ATP, GTP, CTP, UTP).
9. T7 RNA polymerase supplied with 5× transcription buffer.
10. DNase I supplied with 10× reaction buffer with $MgCl_2$.

11. RNAlater Stabilization Solution.
12. RT-PCR-mix-2 FEP/FRT, RT-G-mix-2 additional reagent.
13. TM-reverse transcriptase (MMLv).
14. Taq polymerase.
15. ddPCR Supermix For Probes kit.
16. pGEM-T plasmids with the insert of VP1 gene fragment either in forward or reverse direction (*see* **Note 8**).
17. RT-ddPCR reaction mix (per 1 reaction): mix 10 μl of 2× -one-step RT-ddPCR supermix, 0.8 μl of 25 mM manganese acetate solution, 10 pmol of rtKem4f primer, 10 pmol of rtKem4r primer, 10 pmol of Kemprb4 specific probe, and DEPC-treated water up to 15 μl.
18. Phenol.
19. Chloroform.
20. Ethanol.
21. Isoamyl alcohol.
22. System for RT Droplet Digital PCR.
23. Thermal cycler for RT-qPCR.

## 3 Methods

### 3.1 Synthesis of $G^{8AE}$-Clamp Phosphoramidite (See Fig. 1)

#### 3.1.1 4-(2-Bromoethoxy)-2-Nitrophenol (2)

1. In a 1 l round-bottom flask, dissolve 4-(2-bromoethoxy)phenol **1** (10.85 g, 50 mmol) in benzene (500 ml).
2. Cool the solution to 0 °C with an ice bath, and add 56% aqueous $HNO_3$ solution (3 ml, 64 mmol) in one portion with vigorous stirring.
3. Stir the mixture for 5 min.
4. Wash the mixture with water (100 ml) and brine (50 ml).
5. Dry the organic layer over $Na_2SO_4$.
6. Evaporate under reduced pressure to afford brown oil.
7. Purify the oil by flash column chromatography on silica gel using 30% $CH_2Cl_2$ in hexane ($R_f = 0.40$, 30% $CH_2Cl_2$ in hexane) as the eluent.
8. Recrystallize the yellow solid obtained from the mixture of EtOAc (10 ml) and hexane (20 ml) to afford **2** as orange crystals in 42% yield (5.50 g, 21 mmol, 42%).
9. $^1H$ NMR (600 MHz, $CDCl_3$): δ 10.33 (s, 1H), 7.53 (d, $J = 3.1$ Hz, 1H), 7.28 (dd, $J = 3.1$ Hz, $J = 9.3$ Hz, 1H), 7.12 (d, $J = 9.3$ Hz, 1H), 4.31 (t, $J = 6.0$ Hz, 2H), 3.67 (t, $J = 6.0$ Hz, 2H). $^{13}C$ NMR (151 MHz, $CDCl_3$): δ 150.9,

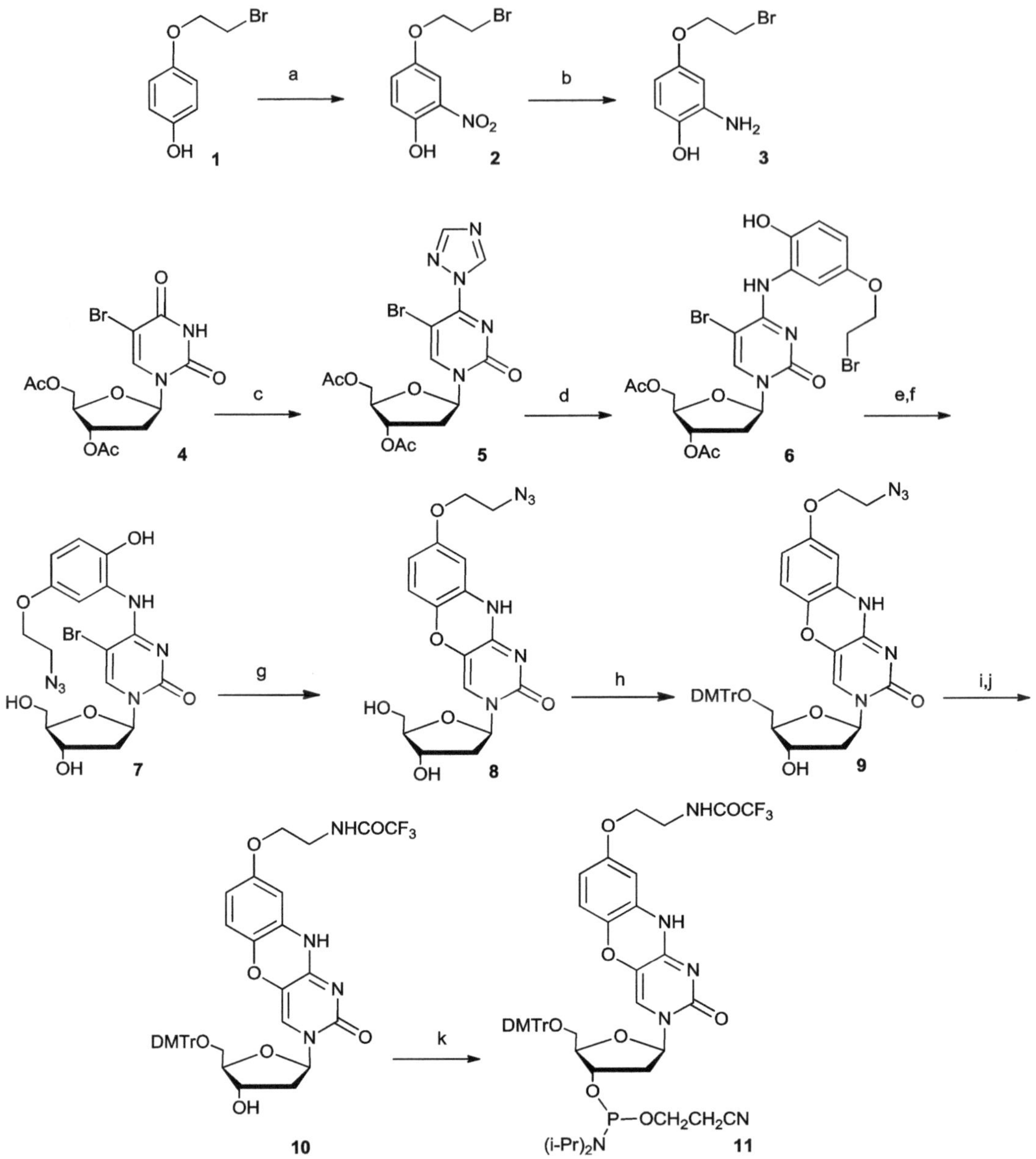


**Fig. 1** Synthesis of G[8AE]-clamp phosphoramidite. *Reagents and conditions*: (*a*) $HNO_3$, benzene, 42%; (*b*) $H_{2(g)}$, 10% Pd/C, MeOH, quantitative; (*c*) 1,2,4-triazole, $POCl_3$, TEA, $CH_3CN$, 89%; (*d*) **3**, DIEA, $CH_2Cl_2$, 72%; (*e*) $NaN_3$, DMF; (*f*) $NH_3$, MeOH, $H_2O$, 76% for two steps; (*g*) TEA, $C_2H_5OH$, 51% (*see* **Note 9**); (*h*) DMTr-Cl, pyridine, 83%; (*i*) $H_{2(g)}$, 10% Pd/C, TEA, EtOAc; (*j*) ethyl trifluoroacetate, pyridine, 70% for two steps; (*k*) $NCCH_2CH_2OP(Cl)N$-*i*$Pr_2$, DIEA, $CH_2Cl_2$, 79%

150.3, 132.8, 127.5, 121.0, 107.2, 68.7, 28.5. HRMS (ESI) *m/z*: calcd for $C_8H_7BrNO_4^-$ $[M-H]^-$: 259.9564; found 259.9561.

#### 3.1.2 2-Amino-4-(2-Bromoethoxy)Phenol (3)

1. In a 250 ml round-bottom flask, dissolve **2** (3.54 g, 13.5 mmol) in MeOH (100 ml).
2. Add Pd/C (10%, 150 mg) to the stirring solution in one portion.
3. Evacuate the flask twice and place it under hydrogen atmosphere (balloon pressure).
4. Stir the suspension at room temperature until completion by TLC ($R_f$ = 0.40, 5% MeOH in $CH_2Cl_2$).
5. Filter off Pd/C.
6. Evaporate under reduced pressure to afford **3** as a brownish solid in quantitative yield (3.13 g).
7. This compound was used directly in the preparation of 3′,5′-*O*-acetyl-4-*N*-(5-(2-bromoethoxy)-2-hydroxyphenyl)-5-bromo-2′-deoxycytidine **6**.

#### 3.1.3 3′,5′-O-Acetyl-4-N-(1,2,4-Triazol-1-yl)-5-Bromo-2′-Deoxycytidine (5)

1. In a 1 l round-bottom flask, suspend 1,2,4-triazole (6.7 g, 99 mmol) in dry MeCN (80 ml).
2. Cool the suspension to 0 °C with an ice bath.
3. Add $POCl_3$ (2.0 ml, 22 mmol) and cold triethylamine (22.7 ml, 163 mmol) under argon atmosphere.
4. Stir the mixture for 30 min at 0 °C.
5. Add the solution of 3′,5′-*O*-acetyl-5-bromo-2′-deoxyuridine **4** (4.30 g, 11.0 mmol) in dry MeCN (40 ml) dropwise at 0 °C.
6. Stir the mixture for 4 h at room temperature.
7. Evaporate under reduced pressure to afford an orange solid.
8. Dissolve the solid in $CH_2Cl_2$ (400 ml).
9. Wash the organic layer with saturated $NaHCO_3$ (300 ml).
10. Wash the organic layer with water (100 ml) and brine (50 ml).
11. Dry the organics over $Na_2SO_4$.
12. Evaporate under reduced pressure to afford an orange solid.
13. Purify the solid by flash column chromatography on silica gel using a gradient of 0–1% MeOH in $CH_2Cl_2$ ($R_f$ = 0.80, 5% MeOH in $CH_2Cl_2$) to yield **5** as yellow foam (4.32 g, 9.78 mmol, 89%).
14. $^1H$ NMR (600 MHz, $CDCl_3$): δ 9.11 (s, 1H), 8.48 (s, 1H), 8.13 (s, 1H), 6.20 (t, $J$ = 6.5 Hz, 1H), 5.23–5.19 (m, 1H), 4.43–4.35 (m, 3H), 2.90–2.84 (m, 1H), 2.25–2.17 (m, 1H), 2.11 (s, 3H), 2.08 (s, 3H). $^{13}C$ NMR (151 MHz, $CDCl_3$): δ

170.1, 169.9, 155.7, 153.3, 151.9, 148.6, 145.1, 88.2, 86.7, 83.6, 73.4, 63.1, 39.1, 20.7, 20.6. HRMS (ESI) $m/z$: calcd for $C_{15}H_{15}BrN_5O_6^-$ $[M-H]^-$: 440.0211; found 440.0218.

#### 3.1.4 3′,5′-O-Acetyl-4-N-(5-(2-Bromoethoxy)-2-Hydroxyphenyl)-5-Bromo-2′-Deoxycytidine (6)

1. In a 250 ml round-bottom flask, suspend **3** (3.14 g, 13.5 mmol) in dry $CH_2Cl_2$ (100 ml).
2. Add **5** (4.0 g, 9.0 mmol) and DIEA (2.45 ml, 14 mmol) under argon atmosphere.
3. Stir the solution for 4 h at room temperature.
4. Concentrate under reduced pressure to afford brown oil.
5. Purify the oil by flash column chromatography on silica gel using a gradient of 0–1% MeOH in $CH_2Cl_2$ ($R_f = 0.80$, 5% MeOH in $CH_2Cl_2$) to yield **6** as beige foam (3.92 g, 6.48 mmol, 72% relative to **5**).
6. $^1H$ NMR (600 MHz, $CDCl_3$): δ 8.89 (s, 1H), 7.95 (s, 1H), 7.90 (s, 1H), 7.48 (d, $J = 2.9$ Hz, 1H), 6.93 (d, $J = 8.7$ Hz, 1H), 6.60 (dd, $J = 2.9$ Hz, $J = 8.7$ Hz, 1H), 6.26 (dd, $J = 7.3$ Hz, $J = 6.0$ Hz, 1H), 5.23–5.18 (m, 1H), 4.37 (d, $J = 3.2$ Hz, 2H), 4.33–4.30 (m, 1H), 4.15 (t, $J = 6.1$ Hz, 2H), 3.51 (t, $J = 6.1$ Hz, 2H), 2.73–2.70 (m, 1H), 2.17–2.11 (m, 4H), 2.09 (s, 3H). $^{13}C$ NMR (151 MHz, $CDCl_3$): δ 170.5, 170.3, 157.1, 153.8, 151.8, 142.9, 140.4, 126.3, 118.8, 113.5, 109.4, 89.1, 87.1, 83.0, 74.0, 68.9, 63.8, 39.0, 29.7, 21.0, 21.0. HRMS (ESI) $m/z$: calcd for $C_{21}H_{22}Br_2N_3O_8^-$ $[M-H]^-$: 601.9779; found 601.9768.

#### 3.1.5 4-N-(5-(2-Azidoethoxy)-2-Hydroxyphenyl)-5-Bromo-2′-Deoxycytidine (7)

1. In a 100 ml round-bottom flask, dissolve **6** (3.64 g, 6.0 mmol) in dry DMF (25 ml).
2. Add $NaN_3$ (0.52 g, 8 mmol).
3. Stir the mixture overnight at room temperature.
4. Partition the mixture between EtOAc (100 ml) and water (100 ml).
5. Separate the organic layer.
6. Wash it with water (100 ml) and brine (50 ml).
7. Dry the organic layer over $Na_2SO_4$.
8. Evaporate under reduced pressure to afford brown oil.
9. Dissolve the oil in MeOH (25 ml).
10. Add saturated aqueous $NH_3$ solution (5 ml).
11. Stir the solution for 6 h at room temperature.
12. Concentrate under reduced pressure to afford brown oil.
13. Purify the oil by flash column chromatography on silica gel using a gradient of 5–12% MeOH in $CH_2Cl_2$ ($R_f = 0.25$, 10%

MeOH in $CH_2Cl_2$) to yield **7** as brown solid (2.21 g, 4.57 mmol, 76%).

14. $^1H$ NMR (600 MHz, DMSO-$d_6$): δ 9.88 (s, 1H), 8.47 (s, 1H), 8.30 (s, 1H), 8.02 (d, $J$ = 3.1 Hz, 1H), 6.84 (d, $J$ = 8.8 Hz, 1H), 6.63 (dd, $J$ = 3.1 Hz, $J$ = 8.8 Hz, 1H), 6.10 (t, $J$ = 6.3 Hz, 1H), 5.22 (d, $J$ = 4.4 Hz, 1H), 5.16 (t, $J$ = 5.0 Hz, 1H), 4.26–4.22 (m, 1H), 4.07 (t, $J$ = 5.0 Hz, 2H), 3.84–3.82 (m, 1H), 3.69–3.63 (m, 3H), 3.61–3.57 (m, 1H), 2.25–2.20 (m, 1H), 2.11–2.06 (m, 1H). $^{13}C$ NMR (151 MHz, DMSO-$d_6$): δ 156.6, 152.9, 150.5, 141.9, 141.6, 126.5, 114.8, 109.5, 108.8, 87.4, 87.4, 85.7, 69.5, 67.0, 60.5, 49.5, 40.7. HRMS (ESI) *m/z*: calcd for $C_{17}H_{18}BrN_6O_6^-$ $[M-H]^-$: 481.0477; found 481.0482.

#### 3.1.6 3-(β-D-2-Deoxyribofuranosyl)-8-(2-Azidoethoxy)-1,3-Diaza-2-Oxophenoxazine (8)

1. Dissolve **7** (2.00 g, 4.13 mmol) in absolute EtOH (80 ml).
2. Add freshly distilled TEA (40 ml).
3. Reflux the solution for 72 h under argon atmosphere.
4. Concentrate under reduced pressure to afford brown foam.
5. Purify the foam by flash column chromatography on silica gel using a gradient of 7–12% MeOH in $CH_2Cl_2$ ($R_f$ = 0.20, 10% MeOH in $CH_2Cl_2$) to yield **8** as brown solid (0.85 g, 2.10 mmol, 51%).
6. $^1H$ NMR (600 MHz, DMSO-$d_6$): δ 10.58 (br s, 1H), 7.54 (s, 1H), 6.74 (d, $J$ = 8.6 Hz, 1H), 6.44 (dd, $J$ = 2.7 Hz, $J$ = 8.6 Hz, 1H), 6.41 (d, $J$ = 2.7 Hz, 1H), 6.12 (t, $J$ = 6.8 Hz, 1H), 5.19 (d, $J$ = 4.1 Hz, 1H), 5.07 (t, $J$ = 5.1 Hz, 1H), 4.24–4.19 (m, 1H), 4.07 (t, $J$ = 4.8 Hz, 2H), 3.79–3.75 (m, 1H), 3.64–3.52 (m, 2H), 3.60 (t, $J$ = 4.8 Hz, 2H), 2.12–1.96 (m, 2H). $^{13}C$ NMR (151 MHz, DMSO-$d_6$): δ 159.0, 153.9, 153.1, 152.4, 149.6, 140.1, 136.5, 127.0, 115.3, 108.4, 87.2, 84.8, 70.3, 67.1, 61.1, 49.4, 40.0.HRMS (ESI) *m/z*: calcd for $C_{17}H_{17}N_6O_6^-$ $[M-H]^-$: 401.1215; found 401.1216.

#### 3.1.7 3-(5′-O-Dimethoxytrityl-β-D-2-Deoxyribofuranosyl)-8-(2-Azidoethoxy)-1,3-Diaza-2-Oxophenoxazine (9)

1. Dissolve **8** (0.81 g, 2.00 mmol) in anhydrous pyridine (15 ml) (*see* **Note 10**).
2. Concentrate under reduced pressure to afford brown oil.
3. Dissolve the oil in anhydrous pyridine (20 ml).
4. Add 4,4′-dimethoxytrityl chloride (0.78 g, 2.2 mmol).
5. Stir the mixture for 3 h at room temperature.
6. Dilute with $CH_2Cl_2$ (50 ml).
7. Wash the organics with saturated $NaHCO_3$ (50 ml).
8. Separate the organic layer.
9. Wash it with water (50 ml) and brine (25 ml).

10. Dry the organic layer over $Na_2SO_4$.
11. Concentrate under reduced pressure to afford brown oil.
12. Dissolve the oil in toluene (25 ml) (*see* **Note 11**).
13. Evaporate under reduced pressure to afford brown foam.
14. Purify the foam by flash column chromatography on silica gel using a gradient of 25–100% EtOAc in $CH_2Cl_2$ with 0.1% TEA ($R_f$ = 0.60, 10% MeOH in $CH_2Cl_2$) to yield **9** as yellowish foam (1.16 g, 1.65 mmol, 83%).
15. $^1$H NMR (600 MHz, DMSO-$d_6$):δ 10.63 (br s, 1H), 7.43–6.87 (m, 14H), 6.52 (d, $J$ = 8.8 Hz, 1H), 6.45 (dd, $J$=8.8 Hz, $J$=2.9 Hz, 1H), 6.42 (d, $J$=2.9 Hz, 1H), 6.11 (t, $J$ = 6.6 Hz, 1H), 5.28 (d, $J$ = 4.5 Hz, 1H), 4.29–4.25 (m, 1H), 4.07 (t, $J$ = 4.8 Hz, 2H), 3.91–3.87 (m, 1H), 3.72 (s, 3H), 3.71 (s, 3H), 3.60 (t, $J$= 5.3 Hz, $J$= 4.8 Hz, 2H), 3.26 (dd, $J$ = 10.4 Hz, $J$ = 5.3 Hz, 1H), 3.12 (dd, $J$ = 10.4 Hz, $J$ = 2.9 Hz, 1H), 2.19–2.10 (m, 2H). $^{13}$C NMR (151 MHz, DMSO-$d_6$): δ 157.9, 153.8, 144.5, 136.4, 135.4, 135.2, 131.4, 129.6, 129.5, 128.5, 127.7, 127.5, 126.9, 126.5, 115.0, 113.1, 113.0, 108.3, 85.7, 85.3, 84.7, 70.1, 67.1, 63.5, 54.8, 49.3, 39.8. HRMS (ESI) $m/z$: calcd for $C_{38}H_{35}N_6O_8^-$ $[M-H]^-$: 703.2522; found 703.2546.

*3.1.8 3-(5′-O-Dimethoxytrityl-β-D-2-Deoxyribofuranosyl)-8-(2-Trifluoroacetamido-Ethoxy)-1,3-Diaza-2-Oxophenoxazine (**10**)*

1. Dissolve **9** (1.06 g, 1.50 mmol) in EtOAc (40 ml).
2. Add TEA (1.5 ml).
3. Add Pd/C (10%, 50 mg) to the stirring solution in one portion.
4. Evacuate the flask twice and place it under hydrogen atmosphere (balloon pressure).
5. Stir the suspension at room temperature until completion by TLC.
6. Filter off Pd/C.
7. Concentrate the filtrate under reduced pressure to afford yellowish foam.
8. Dissolve the foam in anhydrous pyridine (15 ml).
9. Cool the solution to 0 °C.
10. Add ethyl trifluoroacetate (0.36 ml, 3 mmol) dropwise.
11. Stir the solution overnight at room temperature.
12. Evaporate under reduced pressure to afford a yellowish oil.
13. Dissolve the oil in toluene (20 ml) (*see* **Note 11**).
14. Evaporate under reduced pressure to afford yellowish foam.
15. Purify the foam by flash column chromatography on silica gel using a gradient of 50–100% EtOAc in $CH_2Cl_2$ with 0.1% TEA

($R_f = 0.50$, 10% MeOH in $CH_2Cl_2$) to yield **10** as yellowish foam (0.84 g, 1.05 mmol, 70%).

16. $^{1}H$ NMR (400 MHz, DMSO-$d_6$): δ 10.61 (br s, 1H), 9.62 (t, $J = 5.3$ Hz, 1H), 7.43–6.86 (m, 14H), 6.50 (d, $J = 8.8$ Hz, 1H), 6.42 (dd, $J = 2.7$ Hz, $J = 8.8$ Hz, 1H), 6.39 (d, $J = 2.7$ Hz, 1H), 6.11 (t, $J = 6.7$ Hz, 1H), 5.29 (d, $J = 4.3$ Hz, 1H), 4.30–4.24 (m, 1H), 3.99 (t, $J = 5.4$ Hz, 2H), 3.91–3.87 (m, 1H), 3.71 (s, 3H), 3.71 (s, 3H), 3.53 (dt, $J = 5.3$ Hz, $J = 5.4$ Hz, 2H), 3.26 (dd, $J = 10.4$ Hz, $J = 5.3$ Hz, 1H), 3.11 (dd, $J = 10.4$ Hz, $J = 2.5$ Hz, 1H), 2.18–2.12 (m, 2H). $^{13}C$ NMR (101 MHz, DMSO-$d_6$): δ 157.9, 156.2, 153.5, 152.9, 145.9, 144.5, 142.2, 136.9, 135.5, 135.2, 129.6, 129.5, 128.2, 127.7, 127.6, 126.5, 123.0, 122.3, 116.8, 115.5, 114.9, 113.1, 113.0, 107.7, 106.9, 85.7, 85.5, 84.9, 70.2, 66.7, 63.5, 54.8, 40.0, 38.5. $^{19}F$ NMR (376 MHz, DMSO-$d_6$): δ −74.4. HRMS (ESI) *m/z*: calcd for $C_{40}H_{36}F_3N_4O_9^-$ $[M-H]^-$: 773.2440; found 773.2433.

#### 3.1.9 3-(3′-O-β-Cyanoethyl-*N*,N-Diisopropylphosphoramidyl-5′-*O*-Dimethoxytrityl-β-D-2-Deoxyribofuranosyl)-8-(2-Trifluoroacetamidoethoxy)-1,3-Diaza-2-Oxophenoxazine (11)

1. Dissolve **10** (0.77 g, 1.00 mmol) in anhydrous $CH_2Cl_2$ (10 ml).
2. Add DIEA (0.70 ml, 4.00 mmol).
3. Add 2-cyanoethyl diisopropylchlorophosphoramidite (0.33 ml, 1.50 mmol) to the stirring solution at 0 °C under argon atmosphere in one portion.
4. Stir the reaction mixture for 30 min.
5. Quench the reaction by the addition of MeOH (0.5 ml).
6. Dilute the solution with EtOAc (25 ml).
7. Wash the organics with saturated $NaHCO_3$ (25 ml).
8. Wash the organic layer with water (25 ml) and brine (25 ml).
9. Dry it over $Na_2SO_4$.
10. Concentrate under reduced pressure to afford pale yellow oil.
11. Purify the foam by flash column chromatography on silica gel using a mixture of $CH_2Cl_2$:EtOAc:TEA (45:45:10, v/v/v) ($R_f = 0.50$, 80% EtOAc in acetone) to yield **11** as a pale yellow foam (0.77 g; 0.79 mmol; 79%).
12. $^{1}H$ NMR (400 MHz, DMSO-$d_6$): δ 10.64 (br s, 1H), 9.61 (t, $J = 5.1$ Hz, 1H), 7.44–6.84 (m, 14H), 6.54, 6.51 (d, $J = 8.8$ Hz, 1H), 6.47–6.37 (m, 2H), 6.14; 6.11 (t, $J = 6.5$ Hz, 1H), 4.56–4.46 (m, 1H), 4.09–3.95 (m, 1H), 3.99 (t, $J = 5.1$ Hz, 2H), 3.80–3.45 (m, 12H), 3.36–3.14 (m, 2H), 2.76, 2.66 (t, $J = 5.9$ Hz, 2H), 2.38–2.25 (m, 2H), 1.20–0.98 (m, 12H). $^{13}C$ NMR (101 MHz, DMSO-$d_6$): δ 158.0, 156.1, 153.5, 152.9, 145.8, 144.4, 144.4, 142.2,

136.8, 135.3, 135.1, 129.6, 129.5, 128.2, 127.7, 127.5, 127.5, 126.5, 123.1, 122.3, 118.7, 118.5, 116.7, 115.4, 114.8, 113.0, 113.0, 107.7, 106.9, 85.9, 85.8, 84.8, 84.4, 84.1, 72.9, 72.8, 72.5, 72.4, 67.8, 66.7, 65.3, 63.1, 62.8, 58.2, 58.1, 54.8, 54.7, 44.4, 44.3, 42.5, 42.4, 38.5, 24.2, 24.1, 24.0, 24.0, 22.4, 19.7, 19.6, 19.6. $^{31}$P NMR (162 MHz, DMSO-$d_6$): δ 150.51, 147.62.$^{19}$F NMR (376 MHz, DMSO-$d_6$): δ −74.4. HRMS (ESI) $m/z$: calcd for $C_{49}H_{53}F_3N_6O_{10}P^-$ $[M-H]^-$: 973.3518; found 973.3496.

### 3.2 Synthesis of Oligonucleotide Primers Bearing $G^{8AE}$-Clamp

The procedure includes solid-phase oligonucleotide synthesis, purification, and characterization of modified primers by LC-ESI-MS.

#### 3.2.1 Solid-Phase Oligonucleotide Synthesis

1. Primers should be assembled in DNA synthesizer at 1 μmol scale (e.g., ABI 3400) with the phosphoramidite method according to the manufacturer's recommendations (*see* **Note 12**).
2. After completion of the synthesis, dry column in vacuo, and transfer CPG with protected oligonucleotide into the screw-capped tube (1.5 ml).
3. Add 500 μl of the mixture of saturated aqueous $NH_3$ solution and 40% aqueous $MeNH_2$ solution (v/v 1:1), and perform deprotection for 3 h at room temperature under gentle shaking.
4. Cool the solution for 20 min at −20 °C, filter off solids, wash them thoroughly by water (2 × 200 μl), and evaporate the combined washings to dryness.

#### 3.2.2 Purification of Modified Primers by Denaturing Polyacrylamide Gel Electrophoresis (PAGE)

At first oligonucleotides should be purified by denaturing PAGE (*see* Fig. 2a).

1. The denaturing gel electrophoresis of oligonucleotides is performed in 20% PAGE containing 7 M urea in PAGE buffer A.
2. Prepare large vertical gel chamber (gel size, 200 × 200 × 1.5 mm) and pre-run the gel for 3 h at 200 V to remove salts.
3. Dissolve ~5 $OD_{260}$ of the primer in 50 μl of 80% formamide, denature the sample by heating at 90 °C for 3 min, and then rapidly cool on ice.
4. Flush the wells of the gel with water and load the solution of the primer in 80% formamide to the gel.
5. Run PAGE at constant voltage (500 V) (*see* **Note 13**).
6. Oligonucleotides are recovered from the gel by electroelution with Elutrap (Whatman) for 3 h in PAGE buffer B.

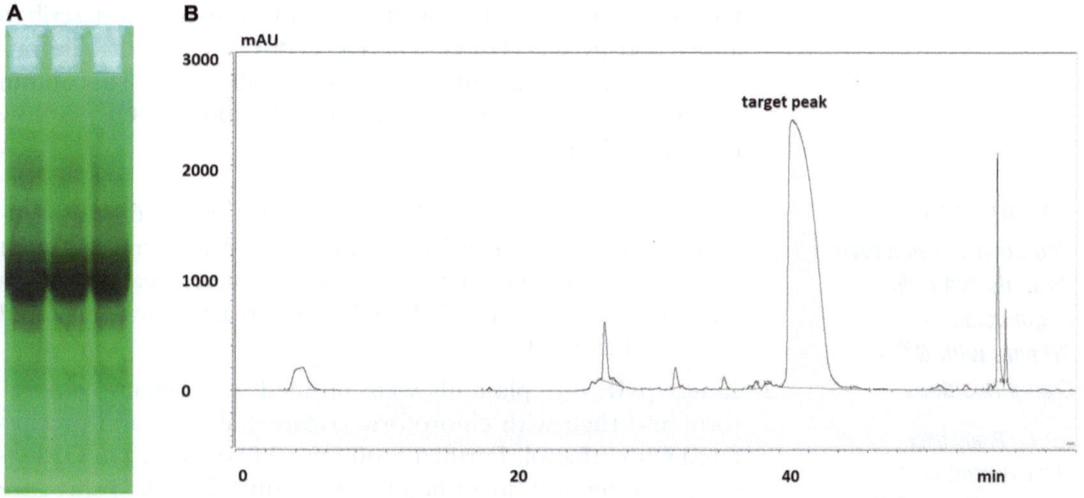


**Fig. 2** Example of purification of a modified primer rtKem4f. (**a**) Denaturative PAGE; (**b**) RP-HPLC

*3.2.3 LC-ESI-MS QC*

Check the molecular mass of the product by running LC-ESI-MS (*see* **Note 14**).

1. The HPLC is equipped with the 2.1 × 50 mm C18 column (5 μm) (*see* **Note 15**); ESI buffer A, 10 mM diisopropylamine, 15 mM 1,1,1,3,3,3-hexafluoroisopropanol in water; ESI buffer B, 10 mM diisopropylamine, 15 mM 1,1,1,3,3,3-hexafluoroisopropanol, 80% acetonitrile.
2. Salts are washed out with ESI buffer A (4 CV) followed by a step of 100% B (2 CV) with a flow rate of 0.3 ml/min; temperature 45 °C.

*3.2.4 Reverse-Phase HPLC Purification and Analysis of Primers*

1. The HPLC purification of primers is carried out on a 4.6 × 250 mm C18 column (5 μm) (*see* **Note 16**); a gradient of buffer B: 0–15% (1 CV), 15–50% (10 CV); a flow rate of 1 ml/min; temperature 45 °C. An example of the HPLC trace for primer is presented in Fig. 2b.
2. Pool the appropriate fractions, evaporate to ~100 μl, and add 1 ml of ethanol (molecular biology grade) to precipitate a primer and to remove most of ammonium acetate.
3. A mixture is cooled in a freezer at −20 °C for 1 h, and the primer is isolated by centrifugation for 5 min at 10,000 × *g* followed by final drying of ethanol traces.
4. Dissolve the product in water, and quantify the oligonucleotide by measuring the UV absorbance at 260 nm. Store the solution frozen.

5. Check the purity of the product by LC-ESI-MS as described above and by RP-HPLC. The RP-HPLC analysis of oligonucleotides is carried out on a 4.6 × 250 mm C18 column (5 μm); a linear gradient of buffer B: 0–100% (8 CV); a flow rate of 1 ml/min; temperature 45 °C.

### 3.3 RT-qPCR Detection of Kemerovo Virus dsRNA Using Oligonucleotide Primers with $G^{8AE}$-Clamp Residues

#### 3.3.1 Preparation of the Recombinant Kemerovo Virus dsRNA Control

1. Linearization of template DNA. pGEM-T plasmids that contain either the direct or inversed insert of VP1 gene fragment under T7 RNA polymerase promoter were treated with a restriction enzyme BamHI in BamHI reaction mixture and incubated at 37 °C for 2 h.
2. Linear pGEM-T plasmids were purified with phenol/chloroform and then with chloroform/isoamyl alcohol and precipitated with ethanol. Purified products were dissolved in DEPC-treated water and quantified by measuring the UV absorbance at 260 nm. Store the solution frozen.
3. IVT (in vitro transcription) reaction mixture contains 1 μg purified linear pGEM-T plasmid (either with direct or inversed insertion of VP1 gene fragment), 4 μl ATP/GTP/CTP/UTP mix (25 mM each), 10 μl 5× transcription buffer, 40 units T7 RNA polymerase, and DEPC-treated water up to 20 μl. Incubate at 37 °C for 2 h.
4. Add 2 units of DNase I directly to IVT reaction mixture. Incubate at 37 °C for 15 min. Inactivation of DNase I and purification of sense or antisense RNAs were carried out by phenol/chloroform extraction, followed by chloroform/isoamyl alcohol extraction and ethanol precipitation. Purified RNAs were dissolved in DEPC-treated water and quantified by measuring the UV absorbance at 260 nm. Store the solution frozen.
5. Sense or antisense RNAs were mixed in equal amounts. dsRNA was diluted with RNAlater Stabilization Solution to 100 μl. Store the solution at 2–8 °C.
6. Quantification of Kemerovo dsRNAs was carried out by RT Droplet Digital PCR using QX100 system. RT-ddPCR reaction mix was dispensed in equal aliquots of 15 μl in each reaction tube, and 5 μl of RNA template (0.1–1.0 ng) was added in each reaction tube.
7. The reaction mixture (each 20 μl) was loaded into DG8™ Cartridge. Cartridge with samples was loaded into QX100 droplet generator to create an emulsion according to manufacturer manual. After droplet generation emulsified samples were transferred into 96-well PCR plate.
8. PCR plate with samples was sealed with PX1 PCR Plate Sealer. RT-ddPCR was performed on C1000 Touch™ Thermal Cycler according to manufacturer recommendations. Droplets

were counted with QX100 Droplet Reader; data were analyzed using QuantaSoft™ v. 1.3.2.0. The final concentration of Kemerovo dsRNAs was determined in copies/μl.

9. Calibrator preparation. Recombinant dsRNA Kemerovo-specific control of known concentration was diluted to the final concentrations of $1 \times 10^7$, $1 \times 10^6$, $1 \times 10^4$, and $1 \times 10^3$ copies/ml by DEPC-treated water and used then as a standard to measure the RNA template concentration in the Kemerovo-specific RT-qPCR analysis.

### *3.3.2 RT-qPCR Analysis of Kemerovo Virus dsRNA*

RT-qPCR analysis was performed using Rotor-Q (Qiagen, Germany) according to manufacturer recommendations.

1. Preparation of qRT-PCR mixture. We mixed 10 μl solution of analyzed RNA, 5 pmol of rtKem4f primer, 5 pmol of rtKem4r primer, 3 pmol of Kemprb4-specific probe, 2.5 μl of dNTPs (1.76 mM), 5 μl RT-PCR-mix-2 FEP/FRT, 0.25 μl of RT-G-mix-2, 0.25 μl of TM-reverse transcriptase (MMLv), 0.5 μl of Taq polymerase, and DEPC-treated water up to 25 μl.
2. Thermal cycling parameters. 55 °C for 15 min, followed by 95 °C for 15 min; and then 45 cycles of 95 °C for 10 s, 60 °C for 20 s, and 72 °C for 10 s with fluorescence detection at 60 °C.
3. Calculation and interpretation of the results: The fluorescence data from the yellow channel was analyzed using Rotor-Q software. After normalization the threshold value in linear mode (0.1) was chosen as the middle of the linear increase of fluorescence in logarithmic mode. Amplification results were considered as positive if the level of fluorescence crossed the threshold.
4. Verification and development of a calibration curve. Using samples of dsRNA with concentrations validated by ddPCR, we performed a set of experiments (*see* table below) resulted in calibration curve ($C_t = -4.92 \times \text{Log } C + 52.16$). Measurements for each concentration were done for two independent preparations using two different instruments in triplicate to get an average value for $C_t$.

| Concentration of RNA, copies/ml ($1 \pm 0.01$) | $C_t$, quantification cycle |
|---|---|
| $1 \times 10^7$ | $17.6 \pm 0.2$ |
| $1 \times 10^6$ | $22.4 \pm 0.2$ |
| $1 \times 10^5$ | $28.3 \pm 0.2$ |
| $1 \times 10^4$ | $32.2 \pm 0.2$ |
| $1 \times 10^3$ | $37.3 \pm 0.2$ |

5. Quantification of the RNA template. Quantification of the RNA template was carried out in the region of linear dependence of $C_t$ (quantification cycle) on the concentration using the equation obtained for calibrators that allowed us to determine concentration of the unknown sample with error less than 5%.

## 4 Notes

1. The starting material could be obtained according to the known procedure [11].
2. The compound could be synthesized according to the literature [12].
3. DCM is always used freshly distilled over $CaH_2$.
4. DMF is freshly distilled under reduced pressure.
5. Pyridine was dried by the distillation over KOH.
6. Spots are visualized under UV light (254 nm).
7. Acetonitrile should be of DNA synthesis grade.
8. They were provided by Dr. G. Karganova, Chumakov Institute of Poliomyelitis and Viral Encephalitides, Russia.
9. We compared different cyclization conditions ($KF/C_2H_5OH$ [13], $CsF/Cs_2CO_3/C_2H_5OH$ [14], and $TEA/C_2H_5OH$ [15]) to prepare phenoxazine **8** and found that the latter mixture gave the highest yield.
10. This procedure is used to remove all traces of water.
11. It is generally used to remove traces of residual pyridine.
12. $G^{8AE}$-clamp phosphoramidite should be used as 0.05 M solution in acetonitrile (DNA synthesis grade), and the coupling time should be increased to 10 min.
13. Electrophoresis should be stopped when the marker dyes have migrated an appropriate distance. For 25–30-mer primers a running time of 3 h is sufficient. Slab PAGE should be visualized under UV lamp using fluorescent TLC plate (TLC Silica gel 60G F254 or analog) as a substrate.
14. LC-ESI-MS analysis for oligonucleotides can be performed using Agilent 1260-Bruker Maxis Impact system as described earlier with minor modifications [16] or any other suitable method.
15. Jupiter C18 column (2.1 × 50 mm, 5 μm, Phenomenex) could be used.
16. Jupiter C18 column (4.6 × 250 mm, 5 μm, Phenomenex) could be used.

## Acknowledgments

This work was supported by the Russian Science Foundation [project no. 18-74-00051].

### References


1. Okamoto A (2011) ECHO probes: a concept of fluorescence control for practical nucleic acid sensing. Chem Soc Rev 40:5815–5828
2. Van Pelt-Vercuil E, van Belkum A, Hays JP (2008) Principles and technical aspects of PCR amplification. Springer, Dordrecht
3. Malgoyre A, Banzet S, Mouret C, Bigard AX, Peinnequin A (2007) Quantification of low-expressed mRNA using 5′ LNA-containing real-time PCR primers. Biochem Biophys Res Commun 354:246–252
4. Fratczak A, Kierzek R, Kierzek E (2009) LNA-modified primers drastically improve hybridization to target RNA and reverse transcription. Biochemistry 48:514–516
5. Ikenaga M, Sakai M (2014) Application of locked nucleic acid (LNA) oligonucleotide-PCR clamping technique to selectively PCR amplify the SSU rRNA genes of bacteria in investigating the plant-associated community structures. Microbes Environ 29:286–295
6. Chen R, Gao XB, Yu XL, Song CX, Qiu Y (2016) Novel multiplex PCR assay using locked nucleic acid (LNA)-based universal primers for the simultaneous detection of five swine viruses. J Virol Methods 228:60–66
7. Moreau V, Voirin E, Paris C, Kotera M, Nothisen M, Rémy JS, Behr JP, Erbacher P, Lenne-Samuel N (2009) Zip nucleic acids: new high affinity oligonucleotides as potent primers for PCR and reverse transcription. Nucleic Acids Res 37:e130
8. Latorra D, Arar K, Hurley JM (2003) Design considerations and effects of LNA in PCR primers. Mol Cell Probes 17:253–259
9. Gautam R, Esona MD, Mijatovic-Rustempasic S, Tam KI, Gentsch JR, Bowen MD (2014) Real-time RT-PCR assays to differentiate wild-type group A rotavirus strains from Rotarix(®) and RotaTeq(®) vaccine strains in stool samples. Hum Vaccin Immunother 10:767–777
10. Dedkov VG, Markelov ML, Gridneva KA, Bekova MV, Gmyl AP, Kozlovskaya LI, Karganova GG, Romanova LI, Pogodina VV, Yakimenko VV, Shipulin GA (2014) Prevalence of Kemerovo virus in ixodid ticks from the Russian Federation. Ticks Tick Borne Dis 5:651–655
11. Bell-Horwath TR, Vadukoot AK, Thowfeik FS, Li G, Wunderlich M, Mulloy JC, Merino E (2013) Novel ROS-activated agents utilize a tethered amine to selectively target acute myeloid leukemia. J Bioorg Med Chem Lett 23:2951–2954
12. Rodgers BJ, Elsharif NA, Vashisht N, Mingus MM, Mulvahill MA, Stengel G, Kuchta RD, Purse BW (2014) Functionalized tricyclic cytosine analogues provide nucleoside fluorophores with improved photophysical properties and a range of solvent sensitivities. Chem Eur J 20:2010–2015
13. Holmes SC, Arzumanov AA, Gait MJ (2003) Steric inhibition of human immunodeficiency virus type-1 Tat-dependent trans-activation in vitro and in cells by oligonucleotides containing 2′-O-methyl G-clamp ribonucleoside analogues. Nucleic Acids Res 31:2759–2768
14. Rajeev KG, Maier MA, Lesnik EA, Manoharan M (2002) High-affinity peptide nucleic acid oligomers containing tricyclic cytosine analogues. Org Lett 4:4395–4398
15. Koga Y, Fuchi Y, Nakagawa O, Sasaki S (2011) Optimization of fluorescence property of the 8-oxodGclamp derivative for better selectivity for 8-oxo-2′-deoxyguanosine. Tetrahedron 67:6746–6752
16. Apffel A, Chakel JA, Fischer S, Lichtenwalter K, Hancock WS (1997) Analysis of oligonucleotides by HPLC-electrospray ionization mass spectrometry. Anal Chem 69:1320–1325

# Chapter 19

# Determining Steady-State Kinetics of DNA Polymerase Nucleotide Incorporation


**Hailey L. Gahlon and Shana J. Sturla**



## Abstract

Polymerase enzymes catalyze the replication of DNA by incorporating deoxynucleoside monophosphates (dNMPs) into a primer strand in a 5′ to 3′ direction. Monitoring kinetic aspects of this catalytic process provides mechanistic information regarding polymerase-mediated DNA synthesis and the influences of nucleobase structure. For example, a range of polymerases have different capacities to synthesize DNA depending on the structure of the inserted dNMP (natural or synthetic) and also depending on the templating DNA base (modified vs. unmodified). Under steady-state conditions, relative rates depend on the deoxynucleoside triphosphate (dNTP) residence times in the ternary (polymerase-DNA-dNTP) complex. This chapter describes a method to measure steady-state incorporation efficiencies by which polymerase enzymes insert dNMPs into primer-template (P/T) oligonucleotides. The method described involves the use of a primer oligonucleotide 5′ radiolabeled with [$\gamma$-$^{32}$P]ATP. Significant established applications of this experiment include studies regarding mechanisms of nucleotide misincorporation as a basis of chemically induced DNA mutation. Further, it can provide information important in various contexts ranging from biophysical to medical-based studies.



**Key words** DNA polymerase, Steady-state enzyme kinetics, Primer extension assay, Nucleoside triphosphates


## 1 Introduction

DNA polymerases are remarkable catalysts responsible for replicating the genome. During DNA polymerization, replicative polymerases are both flexible and selective allowing them to work at fast rates (e.g., 600–1000 bp/s in *Escherichia coli*) [1] and low error frequencies (e.g., 1 mistake every $10^6$ bases) [2]. Chemically modified bases, either engineered or toxicologically relevant DNA lesions, may stall replicative polymerases and act as substrates for specialized polymerases [3–6]. For all functional polymerases, during each catalytic cycle, the enzyme must discriminate between four different templating DNA bases (G, A, T, or C) while, at the same time, inserting the correct pairing partner from a pool of four different nucleoside triphosphates. Furthermore, a synthetic


Nathaniel Shank (ed.), *Non-Natural Nucleic Acids: Methods and Protocols*, Methods in Molecular Biology, vol. 1973, https://doi.org/10.1007/978-1-4939-9216-4_19, © Springer Science+Business Media, LLC, part of Springer Nature 2019

dNTP may be an additional option for the enzyme, for example, either as a chemical probe in mechanistic studies or as a polymerase-interacting drug [7, 8]. Understanding the complexities of this intricate biochemical process is important since any dysfunctions can strongly impact human health and disease [9]. Measuring steady-state polymerase kinetics is useful for quantitatively characterizing the activity of polymerases and to elucidate mechanisms related to DNA synthesis as a function of changes in base or enzyme structure.

Steady-state kinetic parameters, and selectivity factors for particular bases, can be obtained by performing polymerase-mediated extension of oligonucleotide primers in the presence of dNTP substrates and monitoring reaction progress using a gel-based assay. Quantitative determination of $K_m$ and $V_{max}$ for DNA polymerase steady-state kinetics was first described in a seminal paper by Goodman and co-workers where polymerase-mediated nucleoside insertion fidelity of *Drosophila* DNA polymerase α and M13 primer/template DNA was evaluated [10]. To date, many investigators have adopted similar methods to determine polymerase reaction kinetics leading to a better understanding of mechanisms of biological processes controlled by these enzymes [11].

In the basic experiment described in this chapter, primer oligonucleotides are 5′ radiolabeled with [γ-$^{32}$P]ATP and subject to polymerase-mediated DNA synthesis. Oligonucleotides with 5-′-fluorescent dyes can also be utilized for DNA polymerase kinetic analysis and are reported elsewhere [12, 13]. Following the enzymatic reaction, primer oligonucleotides are resolved on denaturing polyacrylamide gels, such that primers and extension ($n + 1$) products are separated and quantified. Plotting of reaction velocity (product formation/time) vs. dNTP concentration yields a hyperbola from which $K_{m,dNTP}$ and $V_{max}$ values can be derived. Polymerase incorporation efficiencies ($k_{cat}/K_{m,dNTP}$) can be determined, and mutational propensities can be found by comparing the $(k_{cat}/K_{m,dNTP})_{wrong}/(k_{cat}/K_{m,dNTP})_{right}$. Here, "wrong" represents incorrect dNTP insertion, and "right" represents the correct formation of the canonical DNA base pair or, in the case of damaged DNA, the non-mutagenic nucleoside insertion (i.e., C opposite the DNA adduct $O^6$-MeG) [6, 14]. In some reports, a detailed and practical explanation of how to perform these experiments is often omitted. Therefore, with the goal of promoting further research involving new applications of polymerase-mediated primer extension, it is the focus of this chapter to outline the techniques and methods necessary to obtain polymerase steady-state kinetic parameters from [γ-$^{32}$P]ATP radiolabeled primer oligonucleotides.

## 2 Materials

Oligonucleotides and dNTP solutions should be prepared using DNase/RNase-free water and stored at −20 °C. Purchase ultrapure or molecular biology grade reagents, and follow storage conditions provided by the manufacturer. Enzymes should be stored in frozen aliquots to reduce freeze-thaw exposure. For long-term storage, enzyme samples should be kept at −80 °C. This assay requires the use of radioactive $^{32}P$; safe handling and proper waste management guidelines for this material should be diligently followed.

### 2.1 Primer Radiolabeling and Duplex Annealing Materials

1. T4 polynucleotide kinase (T4PNK) (*see* **Note 1**).
2. 5× reaction buffer (5×).
3. DNase/RNase-free water.
4. [$\gamma$-$^{32}P$]ATP.
5. Primer and template DNA (*see* **Note 2**).
6. Micro spin columns (*see* **Note 1**).
7. Geiger counter.

### 2.2 Primer Extension Assay Materials

1. DNA polymerase enzyme.
2. dNTPs.
3. Stop-dye solution (*see* **Note 3**).
4. Assay buffer (*see* **Note 4**).

### 2.3 Gel Electrophoresis Materials

1. 7 M urea (*see* **Note 5**).
2. 40% acryl-bisacrylamide, 19:1 (*see* **Note 6**).
3. 10× Tris-Borate-EDTA (TBE) buffer (*see* **Note** 7).
4. 10% ammonium persulfate (APS) (*see* **Note 8**).
5. *N,N,N,N′*-tetramethyl-ethylenediamine (TEMED).
6. Sigmacote (*see* **Note 9**).
7. DNA sequencing gel apparatus.
8. Square-tooth comb.
9. Casting clamp.
10. Paper clamps.
11. High-voltage power supply with power cords.
12. Gel-loading tips (*see* **Note 10**).

### 2.4 Quantitative Gel Analysis Materials

1. Phosphor screen and storage cassette (*see* **Note 11**).
2. Plastic cling wrap (*see* **Note 12**).
3. Phosphorimager.

4. Quantitative analysis software.
5. Screen/image eraser.

## 3 Methods

### 3.1 Polymerase-Mediated Primer Extension

1. Polymerase kinetic parameters are determined by monitoring ($n + 1$) primer extension products over a range of dNTP concentrations and a range of times. The general experimental approach is to prepare two separate solutions (1) containing the dNTPs and (2) containing the enzyme-P/T. The primer should be radiolabeled beforehand with $^{32}P$; follow the radiolabeling instructions as outlined (*see* **Note 1**). Each sample, dNTP and enzyme-P/T, should be diluted in the appropriate assay buffer, which depends on the respective enzyme under investigation. Prepare the dNTP and enzyme-P/T samples at a concentration that is double the desired final concentration. Incubate the enzyme-P/T at the reaction temperature for at least 1 min before initiating the reaction. Initiate the reaction by adding a portion of dNTP solution of equal volume as the enzyme-P/T solution, i.e., such that the final concentration of each is half the initial volume. Quench samples by transferring aliquots (e.g., 4 μL) of the reaction mixture, at specific time points, into Eppendorf tubes containing an approximately ninefold volume excess (e.g., 36 μL) of the stop-dye solution relative to the aliquot.
2. A variety of features should be considered that will have an impact on the progress of the polymerase-mediated reaction. The type of polymerase (i.e., high fidelity vs. low fidelity), structure of the templating nucleobase (i.e., modified vs. unmodified), dNTP (synthetic vs. natural), temperature, and metal counterion ($Mg^{2+}$ or $Mn^{2+}$) are some of the major factors that can impact DNA synthesis rates.
3. An important element of determining kinetics is to first optimize the reaction conditions for each system under interrogation. The enzyme, P/T, and dNTP concentrations need to be adjusted to keep ($n + 1$) primer extension product levels to less than 20% of the starting material to maintain initial velocity conditions.
4. When initially performing a kinetic assay, it can be helpful to first perform a titration experiment of the polymerase. This initial study is especially important if the activity of the enzyme is not well established (e.g., a mutant polymerase). Thus, as a starting point, test primer extension with polymerase concentrations 1- to 50-fold less than the [P/T] at a fixed [dNTP] and a fixed time. This test will gauge the activity of the enzyme and

help generate a starting point for reaction conditions of the assay. These values offer a rule of thumb, but polymerase concentrations may need to be adjusted depending on the efficiency of the enzyme.

5. Under the steady-state approximation, the enzyme-substrate complex should remain constant during the reaction. Therefore, the concentration of reactants should be in excess of the enzyme and not be limiting. [P/T] and [dNTP] should be in at least tenfold excess relative to the polymerase concentration.
6. Reaction time is a crucial factor in performing reliable kinetic analysis. For example, depending on the activity of the enzyme or the type of DNA (modified vs. unmodified), the time points can vary significantly, from seconds to minutes. As a starting point, test a range of time points from seconds (15 s) to minutes (5 min or more) using a relatively high (i.e., saturating) concentration of dNTP.
7. Once an estimate of the $K_m$ for a given polymerase reaction is found, subsequent experiments should be performed that contain [dNTP] concentrations that flank the $K_m$ value (Fig. 1). For example, if the $K_m$ is 3 μM, run various [dNTP] concentrations that surround this value such as 2, 3, 4, and 5 μM, thus ensuring that the data produces a reliable $K_m$ value. A good rule of thumb is to try a range of concentrations that covers a couple of orders of magnitude (100-fold). Additionally, higher [dNTP] concentrations that exceed the $K_m$ should be tested in order to determine the maximum velocity of the reaction.
8. Approximately six different dNTP concentrations (each with approximately 4–6 time points) should be tested to obtain a good representation of initial velocity conditions from which

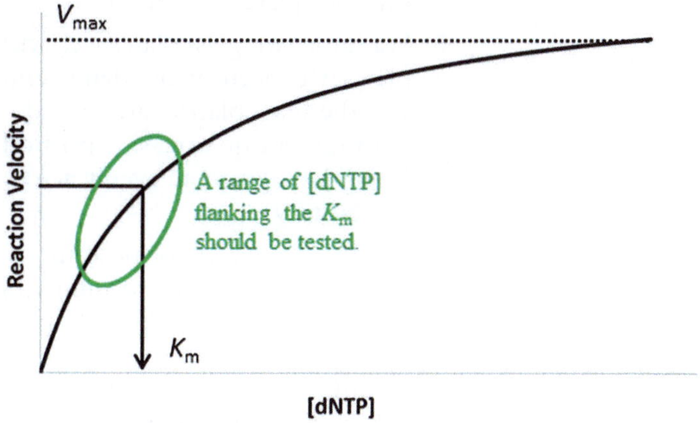


**Fig. 1** Velocity curve generated when plotting reaction velocity vs. [dNTP] in the form of a rectangular hyperbola. The area in green represents concentrations flanking the $K_m$ value

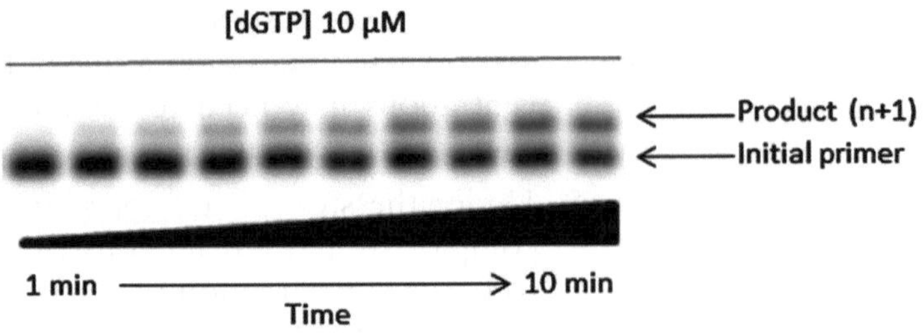


**Fig. 2** Gel image of a representative time-course experiment with DNA polymerase IV (Dpo4) from *Sulfolobus solfataricus*, dGTP, and a [γ-$^{32}$P]ATP radiolabeled primer. The primer/template oligonucleotides consisted of a 28mer template (5′-ACTCGTCTCCCTATAGTGAGTCGTATTA-3′) and a 24mer primer (5-′-TAATACGACTCACTATAGGGAGAC-3′). Ten time points, increasing from left to right, were each taken at 1 min intervals starting at 1 min and ending at 10 min. Final buffer conditions include 50 mM Tris–HCl (pH 8.0 at 25 °C), 2.5 mM $MgCl_2$, 50 mM NaCl, 5 mM DTT, 100 μg/mL BSA, and 5% glycerol

$K_m$, $V_{max}$, and $k_{cat}$ values can be calculated. It is important to measure product formation within the linear portion of the initial velocity curve in order to obtain an accurate rate estimate for any given [dNTP] (Fig. 2). Kinetic experiments should be performed in triplicate to obtain standard deviation values.

### 3.2 Polyacrylamide Gel Electrophoresis

1. Prepare 10× TBE (1 L) (*see* **Note** 7).
2. Prepare a denaturing, 7 M urea, 15% polyacrylamide gel for a large sequencing gel (gel dimensions (W × H) 31 × 38.5 cm) by mixing the following components together in a large-mouth Erlenmeyer flask: 42 g urea, 37.5 mL 19:1 acryl-bisacrylamide 40%, 10 mL TBE, and 22.5 mL distilled water. Stir until the urea completely dissolves. Holding the flask in warm water or on a hot plate will help the urea dissolve faster.
3. Wash the two glass plates well with warm water. Rinse the glass plates with acetone or ethanol and wipe dry with a paper towel. Lay the glass plates flat, and add Sigmacote or another silane derivative (a quarter size portion) to one of the glass plates. Wipe the Sigmacote evenly across the plate and allow it to dry (*see* **Note** 9).
4. Ready glass plates for pouring by placing the spacers between the glass plates and forming the casting clamp around the plates. It is often helpful to place large paper clamps on the glass plates before putting on the casting clamp. Also, the casting clamp should be placed with the props on the same side as the large glass plate, thus ensuring slight elevation to facilitate polymerization.

5. Initiate gel polymerization by adding 880 μL of 10% APS and 88 μL TEMED to the gel solution, and stir. Without delay, transfer the gel mix into a beaker, and draw into a disposable syringe, making sure the syringe has sufficient volume. APS is a radical initiator that starts the free radical chain reaction for gel polymerization; therefore, the reaction begins immediately and the gel must be poured at once.
6. Pouring the gel can be done one of two ways: (1) by holding the glass plates at a 45° angle and administering the gel solution at the top of the glass plates or (2) by laying the glass plates on a table and pushing, via a syringe, the gel solution through a preformed hole at the bottom of the casting clamp.
7. Once the gel is poured, insert the comb gently to prevent the formation of bubbles. Silanizing the comb (i.e., with Sigmacote, *see* **Note 9**) before inserting into the gel allows for easier removal post-polymerization. After inserting the comb, apply pressure between the glass plates and comb by placing large paper clamps between the glass plates. This step helps ensure nice lane formation during gel polymerization. Allow the gel to completely polymerize.
8. Once the gel is polymerized, remove the casting clamp, and place the gel into the gel apparatus and tighten the holding clamps. Add 1× TBE into the reservoirs and insert the power cords. Pre-run the gel for 20–25 min to allow the gel to equilibrate. 1500 V is a typical voltage for equilibrating large sequencing gels; however, this will need to be adapted for each specific gel system.
9. Turn off the power supply and subsequently remove the comb from the gel. Pull the comb straight up slowly and evenly to ensure good formation of lanes. The wells are often filled with debris and urea, so it is best to blow out the wells with 1× TBE using a plastic syringe and needle.
10. Load the samples onto the gel. It is typical to load approx. 0.01–0.06 pmoles of DNA per lane (this may need adjustment depending on the activity of the [$\gamma$-$^{32}$P]ATP), at a loading volume of 4 μL. Use gel-loading tips for easy loading and to ensure for each sample equal loading volumes, which is important for accurate subsequent quantification (*see* **Note 10**).
11. Insert the power cords and set the power supply to the appropriate voltage. Visually monitor the gel over time. Once the bromophenol blue dye reaches the bottom ¾ of the gel, it is safe to stop the power supply. It is typical to run large sequencing 15% gels at 1500 V for approximately 2–4 h.

### 3.3 Gel Exposure, Phosphorimage Scanning, and Densitometry Quantification

1. Disassemble the gel from the gel apparatus, and separate the glass plates slowly without disrupting the gel. Place plastic cling wrap over the gel followed by placing the Phosphor screen on top of the plastic wrap. The plastic wrap protects the screen from the gel. Next place a glass plate on top of the screen to apply pressure between the screen and gel; this helps transfer the image. Gels can also be dried before exposing onto the Phosphor screen.
2. The $^{32}P$ activity, and the amount that has been loaded onto the gel, will influence the optimal exposure time of the image onto the Phosphor screen.
3. After exposure, scan the Phosphor screens using a Phosphorimager.
4. Band quantification can be performed with a variety of densitometry-based software programs. Follow the instructions of the manufacturer to perform density analysis. Generally begin by circumscribing a band of interest with a box; copy the box, and resume circumscribing all bands, being careful to use the same size box for all bands being quantified.
5. Once the bands of interest have been circumscribed with a box, generate density values for each band. To manually calculate background signal, position a few boxes on areas that are absent of DNA. An average of the density value for background boxes can be subtracted from the density values of interest.
6. Export density values to a spreadsheet (such as an Excel document). In the spreadsheet, determine the intensity of the $(n + 1)$ extension product band for each time point with Eq. 1. Here $(B)$ is an average of three background values, $(P_{\text{intital}})$ is initial primer intensity, $(P_{\text{extended}})$ is $(n + 1)$ extension intensity, and $(x)$ is corrected ratio of total product band intensity (Eq. 1).

$$x = \frac{(P_{extended} - B)}{(P_{initial} - B) + (P_{extended} - B)} \tag{1}$$

### 3.4 Graphical and Kinetic Analysis

1. Using corrected total product intensities $(x)$ calculated from Subheading 3.3 (Eq. 1), generate a plot of product formation $(x)$ vs. time for each [dNTP] under investigation. Of interest is the linear portion of the velocity curve for the determination of $V_{\text{max}}$ and ultimately the specificity constant $(k_{\text{cat}}/K_{\text{m}})$ for the reaction under investigation.
2. Perform linear regression analysis for the plots of $x$ (Eq. 1) vs. time, for each [dNTP]. Calculate the slope of each line, which represents initial velocity and will be referred to as $k_{\text{obs}}$ from here on but may also be referred to as $\upsilon_0$ in some texts.

3. To determine $K_m$ and $V_{max}$ values, it is best to use nonlinear regression analysis. Double reciprocal plots, such as the Lineweaver-Burk (LB) plot, while often good to visualize data, are generally outdated by modern software programs and are more sensitive to experimental error in the data. Both types of analysis, namely, a LB plot and nonlinear regression, are outlined below.

### 3.4.1 Nonlinear Regression Analysis

1. Performing nonlinear regression analysis requires special software such as Graphpad Prism, KaleidaGraph, or SigmaPlot 12.
2. Perform nonlinear regression analysis by first plotting $k_{obs}$ vs. [dNTP]. The resulting data plot takes the form of a hyperbola (Fig. 3). Calculate the $K_m$ and $V_{max}$ values for nonlinear regression analysis per software instructions.
3. The $k_{cat}$ (turnover number) for a polymerase represents the number of dNTP molecules that are converted to product per unit time and can be determined on the basis of $V_{max}$ and total enzyme concentration $[E_T]$ by Eq. 2.

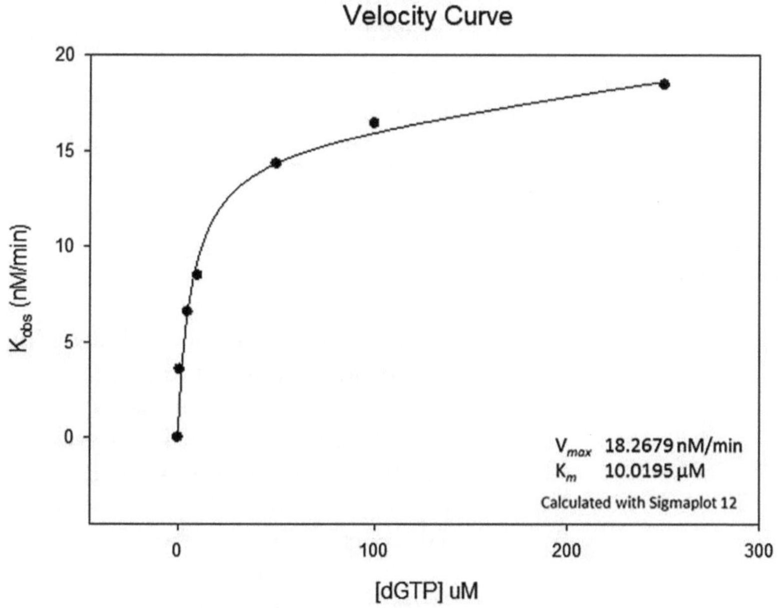


**Fig. 3** Representative velocity curve using SigmaPlot 12 software of a time-course study using DNA polymerase IV (Dpo4) from *Sulfolobus solfataricus*, a range of [dGTP], and primer/template oligonucleotides consisting of a 28mer template (5′-ACTCGTCTCCCTATAGTGAGTCGTATTA-3′) and a 24mer primer (5-′-TAATACGACTCACTATAGGGAGAC-3′). Final buffer conditions include 50 mM Tris–HCl (pH 8.0 at 25 °C), 2.5 mM $MgCl_2$, 50 mM NaCl, 5 mM DTT, 100 µg/mL BSA, and 5% glycerol

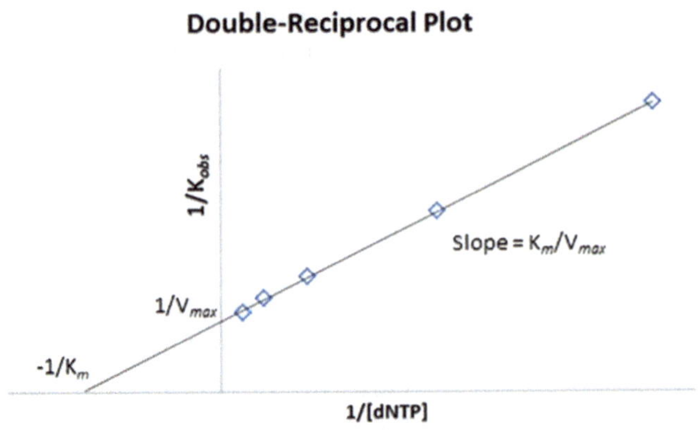


**Fig. 4** A double-reciprocal plot (Lineweaver-Burk) where the inverse of $k_{obs}$ and [dNTP] yields a linear graph depicting the $K_m$ and $V_{max}$ values as shown

$$k_{cat} = \frac{V_{max}}{[E_T]} \tag{2}$$

*3.4.2 Linear Regression Analysis*

1. Double reciprocal plots, such as the LB plot, are valuable for the visual representation of enzyme kinetic data. Before modern mathematical software was developed, these plots were utilized to determine kinetic parameters ($K_m$ and $V_{max}$) of enzyme-mediated reactions.
2. Plotting $1/k_{obs}$ vs. 1/[dNTP] will yield a linear plot from which the $K_m$, $x$-intercept, and the $V_{max}$, $y$-intercept, can be visualized (Fig. 4).

## 4 Notes

1. Detailed procedures for radiolabeling the primer strand can be found online under the manual and protocols of the T4PNK enzyme. This reaction adds the γ-phosphate group from [$\gamma$-$^{32}$P]ATP to the 5′ OH group of the primer oligonucleotide. For the labeling reaction, we generally incubate at 37 °C for 1 h followed by heat inactivation of the kinase at 95 °C for 10–15 min. During the heat inactivation step, close the cap on the microfuge tube to prevent loss of material from heat evaporation. Next, the sample is passed through a micro spin chromatography following the manufacturer's instructions. These columns remove unincorporated nucleotides from the labeling reaction. Finally, the template oligonucleotide is added along with DNase/RNase-free water to bring the primer/template solution to the desired final concentration. We often label 50–100 pmoles of primer strand per reaction and prepare a stock sample with a final [P/T] of 1 μM. Annealing the P/T

solution at 95 °C for 5 min followed by slow cooling to room temperature results in duplex DNA ready for the primer extension assay.

2. Store the oligonucleotides at −20 °C in DNase/RNase-free water. Oligonucleotides should be of high purity, such that after radiolabeling no other bands appear on the gel besides the primer strand. In general, polyacrylamide gel electrophoresis (PAGE) purification results in much higher purity samples than HPLC. Primer extension assays using radioactive labeling with $^{32}P$ are very sensitive, and small amounts of impurities, mainly from the primer strand, will make quantification extremely difficult. It is important to check oligonucleotide purity by PAGE and/or HPLC before beginning the assay. A good reference for handling and purification of oligonucleotides is the Sambrook and Russell laboratory guide [15].
3. The stop-dye solution (95% formamide, 20 mM EDTA, and 0.05% xylene cyanol and bromophenol blue) contains EDTA, which is responsible for terminating polymerase catalysis by chelation of the counterion, generally $Mg^{2+}$ or $Mn^{2+}$. The presence of the xylene cyanol and bromophenol blue assists in gel loading, for sample visualization, and also monitoring the progression of the gel. Xylene cyanol and bromophenol blue are dyes that migrate at different points on a gel depending on both the percentage of the gel and whether the gel is denaturing or non-denaturing.
4. Primer extension reactions are performed in buffer that is dependent on the enzyme under investigation. Often, Tris–HCl, magnesium chloride ($MgCl_2$), NaCl, dithiothreitol (DTT), bovine serum albumin (BSA), and glycerol are present. It is suggested to prepare the assay buffer master mix as a 5× stock solution. After 1 month, a fresh assay buffer master mix should be prepared if DTT is present as it is prone to oxidation.
5. Urea is hygroscopic and can aggregate over time; therefore, it is best to store it in a dry environment. A spatula often works well to break up the lumps that form. Care should be taken to reduce inhalation of airborne urea particles.
6. Acrylamide is a toxic and carcinogenic substance. Gloves, protective eye glasses, and a lab coat should always be worn when handling solutions of acrylamide. In the polymerized form, the toxic risk of acrylamide is greatly reduced.
7. For a 10× TBE buffer stock solution, dissolve 121.1 g Tris base, 61.8 g boric acid, and 7.4 g EDTA, and bring to a final volume of 1 L mL with distilled water.
8. Pre-made samples of 10% ammonium persulfate can be stored at −20 °C and thawed before use, but we find a fresh sample prepared each time works best.

9. Sigmacote is a silicone-based solution in heptane (Sigma Chemical Company, St. Louis, MO, USA). It forms a thin film on glass and will aid in the removal of the polyacrylamide gel from the glass plates. Caution should be taken when handling Sigmacote. It is highly flammable and volatile and may cause lung irritation.
10. Good gel-loading tips are important for accurate volume administration onto the gel and the subsequent quantitative determination of band intensities.
11. The Phosphorimaging screens can be used with radioisotopic emitters such as $^{32}P$, $^{33}P$, $^{35}S$, and $^{14}C$. Phosphor screens should be erased immediately after scanning for best care of the screen.
12. Used for protecting the Phosphor screens. The plastic cling wrap is to be placed directly over the acrylamide gel and, subsequently, the Phosphor screen on top of the plastic cling wrap.

## Acknowledgments

This work was supported by the European Research Council (260341) and the Swiss National Science Foundation (156280). We would like to thank Professors F. Peter Guengerich and Robert Eoff for sharing with us their knowledge and technical expertise of the experiments described here.

### References

1. Bloom LB, Chen X, Fygenson DK, Turner J, O'Donnell M, Goodman MF (1997) Fidelity of *Escherichia coli* DNA polymerase III holoenzyme. The effects of beta, gamma complex processivity proteins and epsilon proofreading exonuclease on nucleotide misincorporation efficiencies. J Biol Chem 272 (44):27919–27930
2. Kunkel TA, Bebenek K (2000) DNA replication fidelity. Annu Rev Biochem 69 (1):497–529
3. Hirao I, Mitsui T, Kimoto M, Yokoyama S (2007) An efficient unnatural base pair for PCR amplification. J Am Chem Soc 129 (50):15549–15555
4. Eoff RL, Stafford JB, Szekely J, Rizzo CJ, Egli M, Guengerich FP, Marnett LJ (2009) Structural and functional analysis of sulfolobus solfataricus Y-family DNA polymerase Dpo4-catalyzed bypass of the malondialdehyde–deoxyguanosine adduct. Biochemistry 48 (30):7079–7088
5. Gahlon HL, Schweizer WB, Sturla SJ (2013) Tolerance of base pair size and shape in postlesion DNA synthesis. J Am Chem Soc 135 (17):6384–6387
6. Gahlon HL, Boby ML, Sturla SJ (2014) O6-alkylguanine postlesion DNA synthesis is correct with the right complement of hydrogen bonding. ACS Chem Biol 9(12):2807–2814
7. Dahlmann HA, Vaidyanathan VG, Sturla SJ (2009) Investigating the biochemical impact of DNA damage with structure-based probes: abasic sites, photodimers, alkylation adducts, and oxidative lesions. Biochemistry 48 (40):9347–9359
8. Wyss LA, Nilforoushan A, Eichenseher F, Suter U, Blatter N, Marx A, Sturla SJ (2015) Specific incorporation of an artificial nucleotide opposite a mutagenic DNA adduct by a DNA polymerase. J Am Chem Soc 137(1):30–33
9. Berdis AJ (2009) Mechanisms of DNA polymerases. Chem Rev 109(7):2862–2879

10. Boosalis MS, Petruska J, Goodman MF (1987) DNA polymerase insertion fidelity. Gel assay for site-specific kinetics. J Biol Chem 262 (30):14689–14696
11. O'Flaherty DK, Guengerich FP (2014) Steady-state kinetic analysis of DNA polymerase single-nucleotide incorporation products. Curr Protoc Nucleic Acid Chem 59:7.21. 1–7.21.13
12. Schermerhorn KM, Gardner AF (2015) Pre-steady-state kinetic analysis of a family D DNA polymerase from Thermococcus sp. 9 degrees N reveals mechanisms for archaeal genomic replication and maintenance. J Biol Chem 290(36):21800–21810
13. Lahiri I, Mukherjee P, Pata JD (2013) Kinetic characterization of exonuclease-deficient *Staphylococcus aureus* PolC, a C-family replicative DNA polymerase. PLoS One 8(5):e63489
14. Guengerich FP (2006) Interactions of carcinogen-bound DNA with individual DNA polymerases. Chem Rev 106(2):420–452
15. Sambrook J, Russell DW (2001) Molecular cloning a laboratory manual. Cold Spring Harbor Laboratory Press, Cold Spring Harbor, New York

# Index



Nathaniel Shank (ed.), *Non-Natural Nucleic Acids: Methods and Protocols*, Methods in Molecular Biology, vol. 1973, https://doi.org/10.1007/978-1-4939-9216-4, © Springer Science+Business Media, LLC, part of Springer Nature 2019

## T



## U



## V



## W



## X

MIX
Papier aus verantwortungsvollen Quellen
Paper from responsible sources
FSC® C105338

If you have any concerns about our products,
you can contact us on
**ProductSafety@springernature.com**

In case Publisher is established outside the EU,
the EU authorized representative is:
**Springer Nature Customer Service Center GmbH**
**Europaplatz 3, 69115 Heidelberg, Germany**

Printed by Libri Plureos GmbH
in Hamburg, Germany